Fundamentals of Wastewater-Based Epidemiology

Fundamentals of Wastewater-Based Epidemiology

Biomonitoring of Bacteria, Protozoa, COVID-19, and Other Viruses

Frank R. Spellman

CRC Press is an imprint of the
Taylor & Francis Group, an **informa** business

First edition published 2021
by CRC Press
6000 Broken Sound Parkway NW, Suite 300, Boca Raton, FL 33487-2742

and by CRC Press
2 Park Square, Milton Park, Abingdon, Oxon, OX14 4RN

CRC Press is an imprint of Taylor & Francis Group, LLC

ISBN: 978-0-367-75806-6 (hbk)
ISBN: 978-0-367-77166-9 (pbk)
ISBN: 978-1-003-17010-5 (ebk)

Typeset in Times
by Deanta Global Publishing Services, Chennai, India

Contents

Preface

Fundamentals of Wastewater-Based Epidemiology: Biomonitoring of Bacteria, Protozoa, COVID-19, and Other Viruses provides a roadmap to detecting wastewater-borne pathogenic contaminants such as viruses, bacteria, fungi, and others. Today it is common practice to evaluate wastewater to understand drug consumption, from antibiotics to illegal drug consumption and even to analyze dietary habits or trends. Evaluating contaminants in wastewater enables researchers, environmental scientists, and water quality experts to gain valuable information and data. Wastewater-based epidemiology is an emerging science that has proven to be both a cost- and time-effective biomonitoring tool. This guide is an excellent resource for public health officials, water and wastewater professionals, environmental scientists, chemical engineers, microbiologists, educators, students, administrators, legal professionals, and general readers. Written in the author's characteristic conversational style, using plain English with little attention to gobbledygook, this resource is suitable and relevant for a wide audience … a very wide audience.

Frank R. Spellman
Norfolk, Virginia

Author Bio

Frank R. Spellman is a retired assistant professor of Environmental Health at Old Dominion University, Norfolk, VA, and author of over 150 books, with more to publish. Spellman has been cited in more than 400 publications, serves as a professional expert witness, incident/accident investigator for the U.S. Department of Justice and a private law firm, and consults on Homeland Security vulnerability assessments (VAs) for critical infrastructure including water/wastewater facilities nationwide. Dr. Spellman lectures on sewage treatment, water treatment, and homeland security and health and safety topics throughout the country and teaches water/wastewater operator short courses at Virginia Tech (Blacksburg, VA). He holds a BA in Public Administration; BS in Business Management; MBA; Master of Science, MS, in Environmental Engineering; and PhD in Environmental Engineering.

Part 1

Setting the Stage

1 And What's in the Pipes Will Tell

"And I believe that truth is mighty and shines by its own light."

—Vincent Bugliosi

"It can be said that Wastewater-Based Epidemiologists are the modern-day Sherlock Holmes of disease detection based on investigation and analysis of the biological contents contained in raw sewage ... including emerging studies of COVID-19" (see Figure 1.1).

—Frank R. Spellman

INTRODUCTION*

He wandered the foggy, filthy, corpse-ridden streets of 1854 London, searching, making notes, always looking, seeking a murdering villain—and find the miscreant, he did. He acted; he removed the handle from a water pump. And, fortunately for untold thousands of lives, his was the correct action—the lifesaving action.

He was a detective—of sorts. No, not Sherlock Holmes but absolutely as clever, as skillful, as knowledgeable, as intuitive, and definitely as driven. His real name: Dr. John Snow. His middle name? Common Sense. Snow's master criminal? His target—a mindless, conscienceless, brutal killer: cholera.

Let's take a closer look at this medical super sleuth and, at his quarry, the deadly killer cholera—and at Doctor Snow's actions to contain the spread of cholera. More to the point, let's look at Dr. Snow's subsequent impact on the treatment (disinfection) of potable water and raw water used for other purposes.

Dr. John Snow

An unassuming—and creative—London obstetrician, Dr. John Snow (1813–1858) achieved prominence in the mid-nineteenth century for proving his theory (in his *On the Mode of Communication of Cholera*) that cholera is a contagious disease caused by a "poison" that reproduces in the human body and is found in the vomitus and stools of cholera patients. He theorized that the main (though not the only) means of transmission was water contaminated with this poison. His theory was not held in high regard at first, because a commonly held and popular countertheory stated that diseases are transmitted by inhalation of vapors. Many theories of cholera's cause

* From F.R. Spellman (2020) *The Science of Water*, 4th ed. Boca Raton, FL: CRC Press.

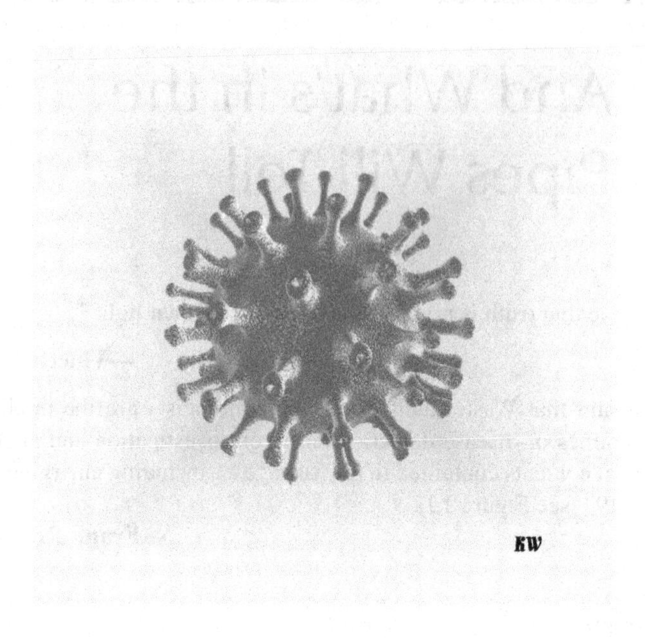

FIGURE 1.1 COVID-19. Illustration by Kat Welsh-Ware and F. Spellman.

were expounded. In the beginning, Snow's argument did not cause a great stir; it was only one of the many hopeful theories proposed during a time when cholera was causing great distress. Eventually, Snow was able to prove his theory. We describe how Snow accomplished this later, but for now, let's take a look at Snow's target: cholera.

Cholera

According to the U.S. Centers for Disease Control (CDC), cholera is an acute, diarrheal illness caused by infection of the intestine with the bacterium Vibrio cholera. The infection is often mild or without symptoms but sometimes can be severe. Approximately 1 in 20 infected persons have severe disease symptoms characterized by profuse watery diarrhea, vomiting, and leg cramps. In these persons, rapid loss of body fluids leads to dehydration and shock. Without treatment, death can occur within hours.

- **Note**: You don't need to be a rocket scientist to figure out just how deadly cholera was during the London cholera outbreak of 1854. Comparing the state of "medicine" at that time to ours is like comparing the speed potential of a horse and buggy to a state-of-the-art NASCAR race car today. Simply stated: cholera was the classic epidemic disease of the nineteenth century, as the Plague had been for the fourteenth. Its defeat reflected both common sense and progress in medical knowledge—and of the enduring changes in European and American social thought.

How does a person contract cholera? Good question. Again, we refer to the CDC for our answer. A person may contract cholera (even today) by drinking water or eating food contaminated with the cholera bacterium. In an epidemic, the source of the contamination is usually feces of an infected person. The disease can spread rapidly in areas with inadequate treatment of sewage and drinking water. Disaster areas often pose special risks. The aftermath of Hurricane Katrina in New Orleans, for example, caused concern for a potential cholera problem.

Cholera bacterium also lives in brackish rivers and coastal waters. Shellfish eaten raw have been a source of cholera, and a few people in the United States have contracted cholera after eating raw shellfish from the Gulf of Mexico. The disease is not likely to spread directly from one person to another; therefore, casual contact with an infected person is not a risk for transmission of the disease.

Flashback to 1854 London

The information provided in the preceding section was updated and provided by CDC in 1996. Basically, for our purposes, CDC confirms the fact that cholera is a waterborne disease. Today, we know quite a lot about cholera and its transmission, how to prevent infection, and how to treat it. But what did they know about cholera in the 1850s? Not much—however, one thing is certain: they knew cholera was a deadly killer. That was just about all they knew—until Dr. John Snow proved his theory. Recall that Snow theorized that cholera is a contagious disease caused by a poison that reproduces in the human body and is found in the vomitus and stools of cholera victims. He also believed that the main means of transmission was water contaminated with this poison.

Dr. Snow's theory was correct, of course, as we know it today. The question is, how did he prove his theory correct? The answer to this provides us with an account of one of the all-time legendary quests for answers in epidemiological research—and an interesting story.

Dr. Snow proved his theory in 1854, during yet another severe cholera epidemic in London. Though ignorant of the concept of bacteria carried in water, Snow traced an outbreak of cholera to a water pump located at an intersection of Cambridge and Broad Street (London).

How did he isolate this particular pump as the source? He accomplished this by mapping the location of deaths from cholera. His map indicated that the majority of the deaths occurred within 250 yards of that water pump. The water pump was used regularly by most of the area residents. Those who did not use the pump remained healthy. Suspecting the Broad Street pump as the plague's source, Snow had the water pump handle removed and ended the cholera epidemic.

Sounds like a rather simple solution, doesn't it? For us, it is simple, but recall in that era aspirin had not yet been formulated, to say nothing of other medical miracles we take for granted—antibiotics, for example. Dr. John Snow, by the methodical process of elimination and linkage (Sherlock Holmes would have been impressed—and he was), proved his point, his theory. Specifically, through painstaking documentation of cholera cases and correlation of the comparative incidence of cholera among subscribers to the city's two water companies, Snow showed that cholera occurred much more frequently in customers of the water company that drew its water from the lower Thames, where the river had become contaminated with London sewage. The other company obtained water from the upper Thames. Snow tracked and pinpointed the Broad Street pump's water source. You guessed it: the contaminated lower Thames, of course.

Dr. Snow, the obstetrician, became the first effective practitioner of scientific epidemiology. His creative use of logic, common sense, and scientific information enabled him to solve a major medical mystery—to discern the means by which cholera was transmitted.

Pump Handle Removal—to Water Treatment (Disinfection)

Dr. John Snow's major contribution to the medical profession, to society, and to humanity in general can be summarized rather succinctly: he determined and proved that the deadly disease cholera is a waterborne disease (Dr. John Snow's second medical accomplishment was being the first person to administer anesthesia during childbirth).

What does all of this have to do with water treatment (disinfection)? Actually, Dr. Snow's discovery—his stripping of a mystery to its barest bones—has quite a lot to do with water treatment. Combating any disease is rather difficult without a determination on how the disease is transmitted—how it travels from vector or carrier to receiver. Dr. Snow established this connection, and from his work, and the work of others, progress was made in understanding and combating many different waterborne diseases.

Today, sanitation problems in developed countries (those with the luxury of adequate financial and technical resources) deal more with the consequences that arise from inadequate commercial food preparation, and the results of bacteria becoming resistant to disinfection techniques and antibiotics. We simply flush our toilets to rid ourselves of unwanted wastes and turn on our taps to take in high-quality drinking water supplies, from which we've all but eliminated cholera and epidemic diarrheal diseases. This is generally the case in most developed countries today—but it certainly wasn't true in Dr. Snow's time.

The progress in water treatment from that notable day in 1854 (when Snow made the "connection" [actually the "disconnection" of handle from pump] between deadly cholera and its means of transmission) to the present reads like a chronology of discovery leading to our modern water treatment practices. This makes sense, of course, because with the passage of time, pivotal events and discoveries occur—events that have a profound effect on how we live today. Let's take a look at a few elements of the important chronological progression that evolved from the simple removal of a pump handle to the advanced water treatment (disinfection) methods we employ today to treat our water supplies.

After Snow's discovery (that cholera is a waterborne disease emanating primarily from human waste), events began to drive the water/wastewater treatment process. In 1859, four years after Snow's discovery, the British Parliament was suspended during the summer because the stench coming from the Thames was unbearable. According to one account, the river began to "seethe and ferment under a burning sun." As was the case in many cities at this time, storm sewers carried a combination of storm water, sewage, street debris, and other wastes to the nearest body of water. In the 1890s, Hamburg, Germany, suffered a cholera epidemic. Detailed studies by Koch tied the outbreak to the contaminated water supply. In response to the epidemic, Hamburg was among the first cities to use chlorine as part of a wastewater treatment regimen. About the same time, the town of Brewster, New York, became the first U.S. city to disinfect its treated wastewater. Chlorination of drinking water was used on a temporary basis in 1896, and its first known continuous use for water supply disinfection occurred in Lincoln, England, and Chicago in 1905. Jersey City, New Jersey, became one of the first routine users of chlorine in 1908.

Time marched on, and with it came an increased realization of the need to treat and disinfect both water supplies and wastewater. Between 1910 and 1915, technological improvements in gaseous and then solution feed of chlorine made the process more practical and efficient. Disinfection of water supplies and chlorination of treated wastewater for odor control increased over the next several decades. In the United States, disinfection, in one form or another, is now being used by more than 15,000 out of approximately 16,000 publicly owned treatment works (POTWs). The significance of this number becomes apparent when you consider that fewer than 25 of the 600+ POTWs in the United States in 1910 were using disinfectants.

The bottom line: Dr. John Snow was the first to prove that what's in the pipe(s) will tell.

FAST FORWARD TO THE PRESENT (2020)

Right up front in this presentation, it is important to provide definitions of a few pertinent terms used in this text. Let's begin with Doctor Snow's epic procedure to identify the source of cholera in London. This is the flagship example of what is (in many cases) practiced by today's epidemiologists. It must be said, however, that Dr. Snow's procedure of finding a contaminant—germs had not yet been identified, classified, or understood in his lifetime—in the drinking water conveyed to the Broad Street pump is different than the procedure and protocol used to conduct wastewater-based epidemiology biosurveillance related to assessing any and all pathological contents in wastewater. First of all, this book (explained in the simplest terms possible) discusses the use of modern epidemiological techniques to identify bacteria, fungi, COVID-19, and other viruses as a means of measurement. Within this text a "means of measurement" is defined as using wastewater to identify the predominance of pathogenic organisms specific to sources (populations) in an area, region, county, and/city.

Wastewater-based epidemiology (WBE), as used in this book, is another important term to define. WBE is sometimes labeled as, or discussed as, wastewater-based surveillance, sewage biological-information mining, or sewage chemical-information mining; it is the WBE surveillance used to mine for biological information in wastewater that is the focus herein. In light of this, WBE wastewater biological-information mining is the technique used for the consumption of, or exposure to, pathogens in a population. This is accomplished by measuring (the key word is "measuring") biomarkers (i.e., biological entities) in wastewater produced by people contributing to wastewater treatment plants.

Well, there are some who might believe that WBE is an emerging technology—that it must be new and developing. Actually, WBE has been used for years in the measurement of illicit drug use (and $100 bills) in populations in various locations. Where does the money come from? Well, when someone is a drug dealer with drugs and dollars in their possession, and then the law knocks on their door with a search warrant, it is amazing to many people what is suddenly flushed down the toilet, so to speak. Raw wastewater influent into a wastewater treatment plant is often and commonly used to measure illicit drug use and also measures the consumption of an assortment of personal care products, compounds, pharmaceuticals, nicotine, opioids, artificial sweeteners, alcohol, caffeine, and others. WBE has been modified or made to order to measure the pack of pathogens, such as SARS-CoV-2, in a particular area or region—note that this is a measure of a population whole. It should be pointed out that WBE is an interdisciplinary effort that pulls on input from specialists such as wastewater treatment plant professionals, chemists, and epidemiologists.

Signatures of Viruses

WBE of pathogenic organisms has the possibility to inform on the presence of disease outbreak when or where it is not assumed or expected. At the present time we are concerned with the virus COVID-19 pandemic. Wastewater-based epidemiology is being employed to test for the presence of SARS-CoV-2 in wastewater. The point is, wastewater can be tested for signatures of viruses including SARS-CoV-2 excreted via feces (Medema et al., 2020; Okoh et al., 2010; Gundy et al., 2008).

In August 2020, the author conducted an informal survey of wastewater treatment plants in the United States and found more than 100 sanitation districts and publicly owned treatment works that were conducting wastewater surveillance of SARS-CoV-2 as a potentially important source of information on the prevalence and chronological trends of COVID-19 in communities. For example, one of the globe's premier wastewater treatment operations, Hampton Roads Sanitation District (HRSD) in southeastern Virginia, is actively pursuing collection, study, research, and testing of wastewater influent in the major treatment plants.

With regard to HRSD and its realization that sewage can suggest a wider spread of COVID or no spread at all, the environmental scientists, microbiologists, technical services specialists, and chemists in the organization are actively taking samples for COVID-19 testing. The fact is wastewater has proven to be an early indicator of the next outbreak. It is interesting to note that the 2020 summer increase in COVID cases in the Hampton Roads region of southeastern Virginia was being detected not through nose swabs or other medical testing, but instead in the wastewater (sewer) pipes—"And what's in the pipes will tell."

The professionals saw an increase in genetic material from the virus surge in wastewater being produced by hundreds of thousands of Hampton Roads residents when samples were analyzed from the district's treatment plants. As the district's data was plotted over the health department's data, pretty close similarities were seen, although there was a lag time of one week from the time signals were apparent in the wastewater analyzed and the time clinical data started to change.

The process employed by HRSD in sampling, testing, and analyzing raw sewage samples is wastewater-based epidemiology. The advantage of using WBE is the detection of the virus's presence soon after people flush their toilets. It is a heck of a lot quicker than waiting for people to be tested and get their results back. While SARS-CoV-2—its official scientific name—is primarily thought of as a respiratory pathogenic virus, scientists have found that it can affect the digestive system, and the virus' genetic makeup can be detected in stool specimens even before symptoms manifest and are observable or felt. Tracking the coronavirus through wastewater can be an early indicator of where the next outbreak could happen. In other words, using WBE-provided information can allow decisions to be made before things get worse. And that is the beauty, the goal, the objective, and the purpose of employing WBE.

REFERENCES

Gundy, P.M., et al., 2008. Survival of Coronaviruses in water and wastewater. *Food and Environmental Virology*, 1(1): 10.

Medema, G., et al., 2020. Presence of SARS-Coronavirus-2 RNA in sewage and correlation with reported COVID-19 prevalence in the early stage of the epidemic in the Netherlands. *Environmental Science & Technology Letters*, 7(7): 511–516.

Okoh, A.I., et al., 2010. Inadequately treated wastewater as a source of human enteric viruses in the environment. *International Journal of Environmental Research and Public Health*, 7(6): 2620–2637.

2 Fundamental Epidemiology

Sewage suggests wider spread of virus.

—Associated Press, 2020

INTRODUCTION

In order to get close to obtaining a fundamental understanding of wastewater-based epidemiology (WBE) used in gathering data about COVID-19 and the studying of concentrations of pathogens in wastewater, a fundamental or at least a basic knowledge of epidemiology is necessary. In simple terms, WBE is an approach to gathering crucial public health data that other traditional approaches might miss. During the coronavirus pandemic, U.S. agencies such as the Centers for Disease Control and Prevention (CDC) have partnered with the U.S. Environmental Protection Agency (USEPA) to establish a formalized, federal-led approach to WBE (WEF, 2020). Although WBE used to detect and provide data about COVID-19 is currently a work in progress, it has been used for a couple of decades to track use of legal and illegal drugs and other chemicals in pesticides, personal care products, and several other products. With regard to WBE and its use in gathering data about COVID-19, keep in mind that it is being studied, practiced, and refined by several wastewater treatment operations throughout the United States; it is predicted that WBE will provide the WHY/HOW about COVID-19 and other pathogenic organisms in wastewater.

WHY/HOW

In characterizing epidemiologic events such as the outbreak of COVID-19 the *5 W*'s come into play: what, who, where, when, and why (sometimes stated and/or cited as why/how). Practicing epidemiologists employ the *5 W*'s extensively in their work. However, epidemiologists tend to use synonyms for the *5 W*'s: diagnosis or health event (what), person (who), place (where), time (when), and causes, risk factors, and modes of transmission (why/how) (CDC, 2012).

EPIDEMIOLOGY: WHAT IS IT?

Epidemiology is all about science, reasoning, and common sense. It was stated earlier that Dr. John Snow (the first disease detective) was like a Sherlock Holmes at the Broad Street pump in London during an epic cholera event. This characterization of Dr. Snow as a Sherlock Holmes is fitting (when you disregard the weak attributes of Sherlock Holmes, his drug abuse and low regard for women). Dr. Watson stated that Holmes' best traits were his 190 IQ, his ability to think outside the box, his unmatched reasoning capabilities, and his power of deduction. How does this compare to Dr. John Snow? Well, as far as we know, Snow did not use drugs and he loved women (Dr. Snow was an expert in the use of ether and chloroform to help women endure pain during childbirth). So, when it is said that Dr. John Snow, the father of modern epidemiology, was like a Sherlock Holmes, it is because he used innovative reasoning (thinking outside the box) and approaches in investigating disease events and/or medical emergencies.

So, the question is whether or not the modern epidemiologist is a Sherlock Holmes or a Dr. John Snow, or a combination of both? Well, that certainly should be, and mostly is, the goal of all practicing or want-to-be epidemiologists.

Having staged the groundwork for a definition of epidemiology, let's get to it. *Epidemiology* is a quantitative basic science built on a working knowledge of probability, statistics, and sound research methods. Moreover, it is a method of causal reasoning based on developing and testing hypotheses pertaining to occurrence and prevention of morbidity and mortality. Beyond being all about science epidemiology, it is also about promoting and protecting the public's health.

How is this accomplished?

Well, to the point, epidemiologists study the distribution and determinants of health-related states or events in specific populations, and they apply this study to the control of health problems (CDC, 2012). Epidemiologists have sound methods of scientific inquiry in their toolboxes. The tools in the epidemiologist's toolbox are used to determine the frequency and pattern of health events in a population. *Frequency* includes not only the number of such events in a population but also the rate or risk of disease in the population. Thus, a critical tool for use by epidemiologists is the ability to determine the number of events divided by the size of the population (rate of) because it allows legitimate comparisons across various populations.

DID YOU KNOW?

In epidemiology, "distribution" refers to who, when, and where.

DID YOU KNOW?

In epidemiology, "determinants" generally includes agents, causes, risk factors, and sources.

Okay, what is pattern of health events all about?

The occurrence of health-related events by time, place, and personal characteristics is what the pattern of health events is all about.

How so?

Let's consider *time* first. Time characteristics involve occurrence frequencies (i.e., annual, seasonal, daily, or hourly occurrence during an epidemic). The characteristics of *place* include geographic variation, urban–rural differences, and locations of schools or worksites. Demographic factors such as age, race, sex, marital status, and socioeconomic status, including behaviors and environmental exposures, are the *personal* characteristics of concern.

Like Sherlock Holmes and Dr. John Snow when they were on the prowl, so to speak, modern epidemiologists on the prowl are in a search for *determinants*, which are the factors and causes that influence an event. In epidemiology, specifically, the determinants are associated with the causes and other factors that influence the occurrence of disease and other health-related events. Epidemiologists are trained to assume that illness does not occur randomly in a population; instead, it takes place only when the right accumulation of risk factors or determinants exists in an individual. This brings us to the "why" and "how" of health-related events. To get to the why and how of events, epidemiologists use analytic epidemiology or epidemiologic studies. What this means is that they assess whether groups with different rates of disease differ in their demographic characteristic, genetic or immunologic make-up, behaviors, environmental exposures, or other potential risk factors. The hope from the findings is that they will lead to findings that provide sufficient evidence to direct prompt and effective public health control and prevention measures (CDC, 2012).

DID YOU KNOW?

Determinant is any factor, whether event, characteristic, or other definable entity, that brings about a change in a health condition or other defined characteristics (CDC, 2012).

It is interesting to note that epidemiology was originally focused exclusively on epidemics of communicable infectious diseases (Greenwood, 1935) but was subsequently expanded to address endemic communicable diseases and non-communicable infectious diseases. By the middle of the twentieth century, however, additional epidemiologic methods had been advanced and applied to chronic diseases, injuries, birth defects, maternal-child health, occupation health, and environmental health (CDC, 2012). Subsequently, epidemiologists began to look at behaviors related to health and well-being, such as diet, vehicle safety (seat belt use), amount of sleep, and amount of exercise. Currently, with the on-going expansion of molecular methods, epidemiologists can make important advances in finding and examining genetic markers of disease risk (i.e., looking for specific inherited variants in a person's genes). As a matter of fact, the term "health-related states" or events may be seen as anything that affects the well-being of a population. On the other hand, many epidemiologists still use the term "disease" as shorthand for the wide range of health-related states and events that are studied.

DID YOU KNOW?

When an epidemiologist performs an activity such as comparing food histories between persons with *Staphylococcus* food poisoning and those without, this is an action known as determinants. A *determinant* is any factor, whether event, characteristic, or other definable entity, that brings about change in a health condition, or in other defined characteristics (CDC, 2012).

It is important to point out (very important) the difference between health care providers and epidemiologists. While the health care provider is concerned about the health of an individual, the epidemiologist, on the other hand, is concerned about the collective health of the people in a community or population. Simply, the epidemiologist's patient is the community (population) and not the individual. Moreover, the health care provider focuses on treating and caring for the individual while the epidemiologist focuses on identifying the exposure or source that caused the illness. Additionally, the epidemiologist focuses on the number of other persons who may have been similarly affected, and the potential for further spread in the community. Finally, the epidemiologist provides or performs interventions to prevent additional causes or recurrences.

DID YOU KNOW?

When an epidemiologist performs an activity such as graphing the number of cases of congenital syphilis by year for the country, this is an action known as distribution. In epidemiology, *distribution* is the frequency and pattern of health-related characteristics and events in a population. In statistics, *distribution* is the observed or theoretical frequency of values of a variable (CDC, 2012).

Both a Science and an Art

Note that epidemiology is not just "the study of" health in a population; it also involves applying the knowledge gained by the studies to community-based practice. Similar to the practices of medicine, the practice of epidemiology and wastewater-based epidemiology is both a science and an art. To make the proper diagnosis and prescribe appropriate treatment for a patient, the health care professional combines medical (scientific) knowledge with experience, clinical judgment, and understanding of the patient. In the same way, the epidemiologist and the wastewater-based epidemiologist use the scientific methods of descriptive and analytic epidemiology or wastewater treatment operations as well as experience, judgment, common sense, and understanding of local conditions of the waste stream, conditions in "diagnosing" the health of a community, and/or the quality of the wastewater stream and proposing appropriate, practical, and acceptable public health interventions or level of disease, sources of disease, and location based on wastewater-based epidemiological findings.

DID YOU KNOW?

When an epidemiologist performs an activity such as recommending that persons who have come in close contact with a diseased person or a carrier of disease be administered specific medical treatment it is known as *application*. In epidemiological studies, application is the incorporation of findings to address public health issues (CDC, 2012).

KEY TERMS AND DEFINITIONS

Note: The famous French philosopher Voltaire, who was well known for saying what he thought and thinking what he said, made the statement: *If you wish to converse with me, define your terms.* This is good advice. It is especially important to define terms used in any technical discussion, such as the material presented in this book. It is common practice in technical books to include a glossary of terms and definitions at the end of the book. It is this author's common practice to list and define key terms early in the presentation because an early review of the terms and definitions enables the reader to imprint many of the terms and rough definitions as he or she proceeds through the text. Moreover, when terms are not listed and defined herein, they will be defined when presented in the text. Keep in mind that the definitions given are valid as they are used in this publication, but different definitions may be used in other contexts.

Terms and Definitions (CDC, 2012)

Agent. A factor, such as a microorganism, chemical substance, or form of radiation, whose presence, excessive presence, or (in deficiency diseases) relative absence is essential for the occurrence of a disease.

Analytic Study. A comparative study intended to identify and quantify associations, test hypotheses, and identify causes. Two common types are cohort study and case-control study.

Attributable Proportion. A measure of the public health impact of a causative factor; proportion of a disease in a group that is exposed to a particular factor which can be attributed to their exposure to that factor.

Bar Chart. A visual display of the size of the different categories of a variable. Each category or value of the variable is represented by a bar.

Bias. Deviation of results or inferences from the truth, or processes leading to such systematic deviation. Ay trend in the collection, analysis, interpretation, publication, or review of data that can lead to conclusions that are systematically different from the truth.

Biologic Transmission. The indirect vector-borne transmission of an infectious agent in which the agent undergoes biologic changes within the vector before being transmitted to a new host.

Box Plot. A visual display that summarizes data using a "box and whiskers" format to show the minimum and maximum values (ends of the whiskers), interquartile range (length of the box), and median (line through the box).

Carrier. A person or animal without apparent disease that harbors a specific infectious agent and is capable of transmitting the agent to others. The carrier state may occur in an individual with an infection that is not apparent throughout its course (known as asymptomatic carrier), or during the incubation period, convalescence, and post convalescence of an individual with a clinically recognizable disease. The carrier state may be of short or long duration (transient carrier or chronic carrier).

Case. In epidemiology, a countable instance in the population or study group of a particular disease, health disorder, or condition under investigation, sometimes, an individual with the particular disease.

Case-Control Study. A type of observational analytic study. Enrollment into the study is based on presence ("case") or absence ("control") of disease. Characteristics such as previous exposure are then compared between cases and controls.

Cause of Disease. A factor (characteristic, behavior, event, etc.) that directly influences the occurrence of disease. A reduction of the factor in the population should lead to a reduction in the occurrence of disease.

Chain of Infection. A process that begins when an agent leaves its reservoir or host through a portal or exit, and is conveyed by some mode of transmission, then enters through an appropriate portal of entry to infect a susceptible host.

Class Interval. A span of values of continuous variables which are grouped into a single category for a frequency distribution of that variable.

Cluster. An aggregation of cases of a disease or other health-related condition, particularly cancer and birth defects, which are closely grouped in time and place. The number of cases may or may not exceed the expected number; frequently the expected number is not known.

Cohort. A well-defined group of people who have had a common experience or exposure, who are then followed up for the incidence of new diseases or events, as in a cohort or prospective study. A group of people born during a particular period or year is called a birth cohort.

Cohort Study. A type of observational analytic study. Enrollment into the study is based on exposure characteristics or membership in a group. Disease, death, or other health-related outcomes are then ascertained and compared.

Confidence Interval. A range of values of a variable of interest, e.g., a rate, constructed so that this range has a specified probability of including the true value of the variable. The specified probability is called the confidence level, and the end points of the confidence interval are called the confidence limits.

Confidence Limit. The minimum or maximum value of a confidence interval.

Contact. Exposure to a source of an infection, or a person so exposed.

Determinant. Any factor, whether event, characteristic, or other definable entity, that brings about change in a health condition, or in other defined characteristics.

Distribution. In epidemiology, the frequency and pattern of health-related characteristics and events in a population. In statistics, the observed or theoretical frequency of values of a variable.

Dot Plot. A visual display of the actual data points of a noncontinuous variable.

Endemic Disease. The constant presence of a disease or infectious agent within a given geographic area or population group; it may also refer to the usual prevalence of a given disease within such area or group.

Epidemic. The occurrence of more cases of disease than expected in a given area or among a specific group of people over a particular period of time.

Evaluation. A process that attempts to determine as systematically and objectively as possible the relevance, effectiveness, and impact of activities in the light of their objectives.

Exposed (Group). A group whose members have been exposed to a supposed cause of disease or health state of interest or possess a characteristic that is a determinant of the health outcome of interest.

Frequency Distribution. A complete summary of the frequencies of the values or categories of a variable, often displayed in a two-column table: the left column lists the individual values or categories, while the right column indicates the number of observations in each category.

Health. A state of complete physical, mental, and social well-being and not merely the absence of disease or infirmity.

High-Risk Group. A group in the community with an elevated risk of disease.

Host. A person or other living organism that can be infected by an infectious agent under natural conditions.

Host Factor. An intrinsic factor (age, race, sex, behaviors, etc.) which influences an individual's exposure, susceptibility, or response to a causative agent.

Hypothesis. A supposition, arrived at from observation or reflection, that leads to refutable predictions. Any conjecture cast in a form that will allow it to be tested and refuted.

Hypothesis, Null. The first step in testing for statistical significance in which it is assumed that the exposure is not related to disease.

Immunity, Active. Resistance developed in response to stimulus by an antigen (infecting agent or vaccine) and usually characterized by the presence of antibody produced by the host.

Immunity, Herd. The resistance of a group to invasion and spread of an infectious agent, based on the resistance to infection of a high proportion of individual members of the group. The resistance is a product of the number susceptible and the probability that those who are susceptible will come into contact with an infected person.

Immunity, Passive. Immunity conferred by an antibody produced in another host and acquired naturally by an infant from its mother or artificially by administration of an antibody-containing preparation (antiserum or immune globulin).

Incidence Rate. A measure of the frequency with which an event, such as a new case of illness, occurs in a population over a period of time. The denominator is the population at risk; the numerator is the number of new cases occurring during a given time period.

Indirect Transmission. The transmission of an agent carried from a reservoir to a susceptible host by suspended air particles or by animate (vector) or inanimate (vehicle) intermediaries.

Infectivity. The proportion of persons exposed to a causative agent who become infected by an infectious disease.

Latency Period. A period of subclinical or in apparent pathologic changes following exposure, ending with the onset of symptoms of chronic disease.

Mean, Arithmetic. The measure of central location commonly called the average. It is calculated by adding together all the individual values in a group of measurements and dividing by the number of values in the group.

Mean, Geometric. The mean or average of a set of data measured on a logarithmic scale.

Measure of Association. A quantified relationship between exposure and disease; it includes relative risk, rate ratio, and odds ratio.

Measure of Central Location. A central value that best represents a distribution of data. Measures of central location include the mean, median, and mode. It is also called the measure of central tendency.

Measure of Dispersion. A measure of the spread of a distribution out from its central value. Measures of dispersion used in epidemiology include the interquartile range, variance, and standard deviation.

Median. The measure of central location which divides a set of data into two equal parts.

Mode. A measure of central location, the most frequently occurring value in a set of observations.

Morbidity. Any departure, subjective or objective, from a state of physiological or psychological well-being.

Mortality Rate. A measure of the frequency of occurrence of death in a defined population during a specified interval of time.

Nominal Scale. Classification into unordered qualitative categories—e.g., race, religion, and country of birth as measurements of individual attributes are purely nominal scales, as there is no inherent order to their categories.

Normal Curve. A bell-shaped curve that results when a normal distribution is graphed.

Normal Distribution. The symmetrical clustering of values around a central location. The properties of a normal distribution include the following: (1) it is a continuous, symmetrical distribution; both tails extend to infinity; (2) the arithmetic mean, mode, and median are identical; and (3) its shape is completely determined by the mean and standard deviation.

Outbreak. Synonymous with epidemic. Sometimes the preferred word, as it may escape sensationalism associated with the word "epidemic."

Pandemic. An epidemic occurring over a very wide area (several countries or continents) and usually affecting a large proportion of the population.

Pathogenicity. The proportion of persons infected, after exposure to a causative agent, who then develop clinical disease.

Population. The total number of inhabitants of a given area or country. In sampling, the population may refer to the units from which the sample is drawn, not necessarily the total population of people.

Prevalence. The number or proportion of cases or events or conditions in a given population.

Proportion. A type of ratio in which the numerator is included in the denominator. The ratio of a part to the whole, expressed as a "decimal fraction" (e.g., 0.2), as a fraction (1/4), or, loosely, as a percentage (22 percent).

Random Sample. A sample derived by selecting individuals such that each individual has the same probability of selection.

Range. In statistics, the difference between the largest and smallest values in a distribution. In common use, the span of values from smallest to largest.

Rate. An expression of the frequency with which an event occurs in a defined population.

Relative Risk. A comparison of the risk of some health-related event such as disease or death in two groups.

Representative Sample. A sample whose characteristics correspond to those of the original population or reference population.

Reservoir. The habitat in which an infectious agent normally lives, grows, and multiplies; reservoirs include human reservoirs, animal reservoirs, and environmental reservoirs.

Risk. The probability that an event will occur, e.g., that an individual will become ill or die within a stated period of time or age.

Risk Factor. An aspect of personal behavior or lifestyle, an environmental exposure, or an inborn or inherited characteristic that is associated with an increased occurrence of disease or other health-related event or condition.

Risk Ratio. A comparison of the risk of some health-related event such as disease or death in two groups.

Sample. A selected subset of a population. A sample may be random or non-random and it may be representative or non-representative.

Scatter Diagram. A graph in which each dot represents paired values for two continuous variables, with the x-axis representing one variable and the y-axis representing the other; it is used to display the relationship between the two variables; it is also called a scattergram.

Sensitivity. The ability of a system to detect epidemics and other changes in disease occurrence. The proportion of persons with disease who are correctly identified by a screening test or case definition as having disease.

Skewed. A distribution that is asymmetrical.

Specificity. The proportion of persons without disease who are correctly identified by a screening test or case definition as not having disease.

Spot Map. A map that indicates the location of each case of a rare disease or outbreak by a place that is potentially relevant to the health event being investigated, such as where each case lived or worked.

Standard Deviation. The most widely used measure of dispersion of a frequency distribution, equal to the positive square root of the variance.

Transmission of Infection. Any mode or mechanism by which an infectious agent is spread through the environment or to another person.

Trend. A long-term movement or change in frequency, usually upward or downward.

Validity. The degree to which a measurement actually measures or detects what it is supposed to measure.

Variable. Any characteristic or attribute that can be measured.

Variance. A measure of the dispersion shown by a set of observations, defined by the sum of the squares of deviations from the mean, divided by the number of degrees of freedom in the set of observations.

Vector. An absolute intermediary in the indirect transmission of an agent that carries the agent from a reservoir to a susceptible host.

Vehicle. An inanimate intermediary in the indirect transmission of an agent that carries the agent from a reservoir to a susceptible host.

Virulence. The proportion of persons with clinical disease, who after becoming infected, become severely ill or die.

Years of Potential Life Lost. A measure of the impact of premature mortality on a population, calculated as the sum of the differences between some predetermined minimum or desired life span and the age of death for individuals who died earlier than that predetermined age.

Zoonoses. An infectious disease that is transmissible under normal conditions from animals to humans.

THE BOTTOMLINE

Epidemiology, including wastewater-based epidemiology, is the scientific, systematic, data driven study of the distribution (frequency, pattern) and determinants (causes, risk factors) of health-related states and events (not just diseases) in specified populations (patient is the community, individuals viewed collectively), and the application of (since epidemiology is a discipline within public health) this study to the control of health problems (CDC, 2012).

REFERENCES

CDC, 2012. *Principles of Epidemiology in Public Heath Practice*, 3rd ed. Atlanta, GA: Centers for Disease Control and Prevention.

Greenwood, M., 1935. *Epidemics and Crowed Diseases: An Introduction to the Study of Epidemiology*. New York: Oxford University Press.

WEF (Water Environment Federation), 2020. *CDC Scientists Discuss Wastewater-Based Epidemiology*. WEF Highlights. Arlington, VA.

Part 2

Statistical Methods

3 Statistics and Biostatistics Review

There are three kinds of lies: lies, damned lies, and statistics.

—Benjamin Disraeli

To the uninitiated it may often appear that the statistician's primary function is to prevent or at least impede the progress of research. And even those who suspect that statistical methods may be more boon than bane are at times frustrated in their efforts to make use of the statistician's wares.

—Frank Freese

No aphorism is more frequently repeated in connection with field trials, than that we must ask Nature few questions, or, ideally, one question, at a time. The writer is convinced that this view is wholly mistaken.

—Ronald Fisher

EXTRACTION OF DATA

Suppose that you work in a city health department and are faced with a couple of challenges. First, a case of COVID-19 is reported to the health department. The patient, a 70-year-old man, denies having any of the common risk factors for the disease: he has never had any disease; he is Caucasian, thus race is not an issue; he has not used any form of medication other than an aspirin per day beginning in 1979 to the present; he has not been exposed to overcrowding or poverty; he served as an accountant and thus not exposed via occupation. However, he recounts going to dancing lessons two months earlier, where he was in a room with dozens of other people. COVID-19 has been transmitted between groups and patients, particularly before the dancers wore protective personal protection equipment (PPE) such as masks or gloves.

Question*: What proportion of other persons with new onset of COVID-19 reported recent exposure to the same dance instruction session(s), or to any dance instruction during their likely period of exposure?*

Subsequently, in the following week, the health department receives 46 death certificates. A new employee in the local Vital Statistics office wonders how many death certificates the local health department usually receives each week.

Question*: What is the average number of death certificates the local health department receives each week? By how much does this number differ? What is the range over the past year?*

Given the appropriate raw data, would you be able to answer these questions with assurance? Well, if not, this lesson provides a template, outline, model, and/or statistical procedure to allow you to do so—and much more.

IT'S ALL ABOUT MEASUREMENT

Earlier it was stated that epidemiology is concerned with measurement within a population. To properly perform this vital function, it is important to understand the tools inside the epidemiologist's toolbox, including frequency distribution: calculating and interpreting four measures of central location (mode, median, arithmetic mean, and geometric mean), the ability to apply the most appropriate measure of central location for a frequency distribution; and applying and interpreting four measures of spread, range, interquartile range, standard deviation, and confidence interval (for mean)

In order to measure anything, data must be available. Measurement begins with collecting data and then organizing it. This is the case if you are conducting routine surveillance, investigating an outbreak, or conducting a study. One common method is to create a *line list* or *line listing.*

LINE LIST/LISTING

The Centers for Disease Control and Prevention (CDC) has provided a line list template for the investigation of outbreaks of unexplained illness. The line list template is shown below. Upon notification of a potential cluster or outbreak, the line list template can be used to collect and organize preliminary information on cases. Note that each case should be assigned a unique identifier (CaseID) and his/her respective information should be added to a single row in the spreadsheet. The data fields contained on this spreadsheet are suggestions and should be modified to reflect the needs of the current outbreak situation. Definitions and suggested values for the data fields are also included below.

SAMPLE LINE LIST TEMPLATE*

Reporting City, County or State: ________ Date of Initial Report: ____________

CaseID	Case initials	Age	Sex	Onset date	Current status	Location	Case category	Epi links	Underlying conditions

(Continued)

* From CDC Atlanta, Georgia accessed 09/20/2020 at https://www.cdc.gov/urdo/downloaeds/linelisttemplate.

CaseID	Case initials	Age	Sex	Onset date	Current status	Location	Case category	Epi links	Underlying conditions

Explanations:

CaseID:	Unique identifier assigned to each case-patient for this investigation
Case initials:	Case-patient initials
Age:	Age in years
Sex:	Male, female or unknown
Onset date:	Date of symptom onset, mm/dd/yy
Current status:	Outpatient, inpatient, inpatient ICU, discharge, died
Location:	Hospital, city, county
Case category:	Confirmed, probable, suspect
Epi Links:	Known exposures, affiliations or connections to other cases
Underlying conditions:	Significant immunodeficiencies, medications or other conditions that may alter the patient's susceptibility or course

DID YOU KNOW?

In the line list sample shown above, each row is called a record or observation and represents one person or case of disease. Each column is called a variable and contains information about one characteristic of the individual, such as race or date of birth.

DID YOU KNOW?

A variable can be any characteristic that differs from person to person, such as height, sex, vaccination status, or physical activity pattern (CDC. 2012).

VARIABLES

The type of values influences the way in which the variables can be summarized. Variables can be classified into one of four types, depending on the type of scale used to characterize their values.

- A *nominal-scale variable* is one whose values are categories without any numerical ranking, such as county or residence. In epidemiology, nominal variables with only two categories are very common: alive or dead, ill or well, vaccinated or unvaccinated, or did or did not eat the macaroni salad. Note that a nominal variable with two mutually exclusive categories is sometimes called a dichotomous variable (i.e., a variable that contains precisely two distinct values)—considered a qualitative or categorical variable.
- An *ordinal-scale variable* has values that can be ranked but are not necessarily evenly spaced, such as stage of cancer or the socio-economic status of families. We know "upper middle" is higher than middle but can't say "how much higher"—considered a qualitative or categorical variable.
- An *interval-scale variable* is measured on a scale of equally spaced units, but without a true zero point, such as date of birth—considered quantitative or continuous variable.
- A *ratio-scale variable* is an interval variable with a true zero point, such as height in centimeters or duration of illness—considered quantitative or continuous variable.

After collecting pertinent data and listing it in logical and workable order, what is next? What is next is measurement and this is where statistical methods come into play.

REFERENCE

CDC, 2012. *Principles of Epidemiology in Public Heath Practice*, 3rd ed. Atlanta, GA: Centers for Disease Control and Prevention.

4 Basic Concepts

INTRODUCTION

Epidemiological practitioners use statistical analysis of the results of their investigations; it is one of the main tools inside the epidemiologist's toolbox. The principal concept of statistics is that of variation. In conducting typical environmental health functions statistics and biostatistics, there is a wide range in the application of statistics to an even wider range of topics in biology—such as toxicological or biological sampling protocol for air contamination, epidemics, health issues, and other environmental functions. This chapter provides a survey of the basic statistical and data-analysis techniques that can be used to address many of the problems that he or she will encounter on a daily basis. It covers the data-analysis process, from research design to data collection, analysis, reaching of conclusions of, and, most importantly, presentation of findings.

Finally, it is important to point out that statistics can be used to justify the implementation of a program, identify areas that need to be addressed, or justify the impact that various environmental health and safety programs have on losses and accidents. A set of occupational health and safety data (or other data) is only useful if it is analyzed properly. Better decisions can be made when the nature of the data is properly characterized. For example, the importance of using statistical data in selling the environmental health and other environmental operations, the innovation, the winning over of those who control the purse strings cannot be overemphasized.

With regard to understanding and grasping the principles of statistics and biostatistics, much of the difficulty is due to not understanding the basic objectives of statistical methods. We can boil these objectives down to two:

1. The estimation of population parameters (values that characterize a particular population)
2. The testing of hypotheses about these parameters

A common example of the first is the estimation of the coefficients a and b in the linear relationship. $Y = a + bX$, between the variables Y and X. To accomplish this objective, one must first define the population involved and specify the parameters to be estimated. This is primarily the research worker's job. Statistics helps devise efficient methods of collecting data and calculating the desired estimates.

Unless the whole population is examined, an estimate of a parameter is likely to differ to some degree from the population value. The unique contribution of statistics to research is that it provides ways of evaluating how far off the estimate may be. This is ordinarily done by computing confidence limits, which have a known probability of including the true value of the parameter. Thus, the mean diameter of the trees in a pine plantation may be estimated from a sample as 9.2 inches, with 95-percent confidence limits of 8.8 and 9.6 inches. These limits (if properly obtained) tell us that, unless a 1-in-20 chance has occurred in sampling, the true mean diameter is somewhere between 8.8 and 9.6 inches.

The second basic objective in statistics is to test some hypotheses about the population parameters. A common example is a test of the hypothesis that the regression coefficient in the linear model has some specified value (say zero).

$$Y = a + bX$$

Another example is a test of the hypothesis that the difference between the means of two populations is zero.

Again, it is the research worker who should formulate meaningful hypotheses to be tested, not the statistician. This task can be tricky. The novice would do well to work with the statistician to be sure that the hypothesis is put in a form that can be tested. Once the hypothesis is set, it is up to the statistician to work out ways of testing it and to devise efficient procedures for obtaining the data (Freese, 1969; Spellman, 2020).

PROBABILITY AND STATISTICS

Those who work with probabilities are commonly thought to have an advantage when it comes to knowing, for example, the likelihood of tossing coins heads up six times in a row, or the chances of a crapshooter making several consecutive winning throws ("passes"), and other such useful bits of information. It is fairly well known that statisticians work with probabilities; thus, they are often associated with having the upper hand, so to speak, on predicting outcomes in games of chance. However, statisticians also know that this assumed edge they have in games of chance is often dependent on other factors.

The fundamental role of probability in statistical activities is often not appreciated. In putting confidence limits on an estimated parameter, the part played by probability is fairly obvious. Less apparent to the neophyte is the operation of probability in the testing of hypotheses. Some of them say with derision, "You can prove anything with statistics"—remember what Disraeli said about statistics. Anyway, the truth is, you can prove nothing; you can at most compute the probability of something happening and let the researcher draw his or her own conclusions.

Let's return to our game of chance to illustrate this point. In the game of craps, the probability of the shooter winning (making a pass) is approximately 0.493—assuming, of course, a perfectly balanced set of dice and an honest shooter. Suppose now that you run up against a shooter who picks up the dice and immediately makes seven passes in a row! It can be shown that if the probability of making a single pass is really 0.493, then the probability of seven or more consecutive passes is about 0.007 (or 1 in 141). This is where statistics ends; you draw your own conclusions about the shooter. If you conclude that the shooter is pulling a fast one, then in statistical terms you are rejecting the hypothesis that the probability of the shooter making a single pass is 0.493.

In practice, most statistical tests are of this nature. A hypothesis is formulated, and an experiment is conducted, or a sample is selected to test it. The next step is to compute the probability of the experimental or sample results occurring by chance if the hypothesis is true. If this probability is less than some preselected value (perhaps 0.05 or 0.01), the hypothesis is rejected. Note that nothing has been proved—we haven't even proved that the hypothesis is false. We merely inferred this because of the low probability associated with the experiment or sample results.

Our inferences may be incorrect if we are given inaccurate probabilities. Obviously, reliable computation of these probabilities requires knowledge of how the variable we are dealing with is distributed (that is, what the probability is of the chance occurrence of different values of the variable). Accordingly, if we know that the number of beetles caught in light traps follows what is called the Poisson distribution, we can compute the probability of catching X or more beetles. But, if we assume that this variable follows the Poisson when it actually follows the negative binomial distribution, our computed probabilities may be in error.

Even with reliable probabilities, statistical tests can lead to the wrong conclusions. We will sometimes reject a hypothesis that is true. If we always test at the 0.05 level, we will make this mistake on the average of 1 time in 20. We accept this degree of risk when we select the 0.05 level of testing. If we're willing to take a bigger risk, we can test at the 0.10 or the 0.25 level. If we're not willing to take this much risk, we can test at the 0.01 or 0.001 level.

Researchers can make more than one kind of error. In addition to rejecting a hypothesis that is true (a Type 1 error), he or she can make the mistake of not rejecting a hypothesis that is false (a Type II error). In crap shooting, it is a mistake to accuse an honest shooter of cheating (Type I error—rejecting true hypothesis), but it is also a mistake to trust a dishonest shooter (Type II error—failure to reject a false hypothesis).

The difficulty is that for a given set of data, reducing the risk of one kind of error increases the risk of the other kind. If we set 15 straight passes as the critical limit for a crap shooter, then we greatly reduce the risk of making a false accusation (probability about 0.00025). But in so doing we have dangerously increased the probability of making a Type II error—failure to detect a phony. A critical step in designing experiments is the attainment of an acceptable level of probability of each type of error. This is usually accomplished by specifying the level of testing (i.e., probability of an error of the first kind) and then making the experiment large enough to attain an acceptable level of probability for errors of the second kind.

It is beyond the scope of this book to go into basic probability computations, distribution theory, or the calculation of Type II errors. But anyone who uses statistical methods should be fully aware that he or she is dealing primarily with probabilities (not necessarily lies or damnable lies) and not with immutable absolutes. Remember, 1-in-20 chances do actually occur—about one time out of twenty.

MEASURE OF CENTRAL TENDENCY

As mentioned above, when talking statistics, it is usually because we are estimating something with incomplete knowledge. Maybe we can only afford to test 1 percent of the items we are interested in, and we want to say something about what the properties of the entire lot are; possibly because we destroy the sample by testing it. In that case, 100 percent sampling is not feasible if someone is supposed to get the items after we are done with them.

The questions we are usually trying to answer are, "What is the central tendency of the item of interest?" and "How much dispersion about this central tendency can we expect?"

Simply, the average or averages that can be compared are measures of *central tendency* or central location of the data.

SYMBOLS, SUBSCRIPTS, BASIC STATISTICAL TERMS, AND CALCULATIONS

In statistics, *symbols* such as X, Y, and Z are used to represent different sets of data. Hence, if we have data for five companies, we might let

X = company income
Y = company materials expenditures
Z = company savings

Subscripts are used to represent individual observations within these sets of data. Thus, X_i represents the income of the company, where *i* takes on values 1, 2, 3, 4, and 5. In this notation X_1, X_2, X_3, X_4, and X_5 stand for the incomes of the first company, the second company, and so on. The data are arranged in some order, such as by size of income, the order in which the data were gathered, or any other way suitable to the purposes or convenience of the investigator.

The subscript *i* is a variable used to index the individual data observations. Therefore, X_i, Y_i and Z_i represent the income, materials expenditures, and savings of the *i*th company. For example, X_2 represents the income of the second company, Y_2 materials expenditures of the second (same) company, and Z_5 the savings of the fifth company.

Suppose that we have data for two different samples, the net worth of 100 companies and the test scores of 30 students. To refer to individual observations in these samples, we can let X_i denote the net worth of the *i*th company, where *i* assumes values from 1 to 100. (This later idea is indicated by the notation I = 1, 2, 3, ... 100.) We can also let Y_j denote the test score of the *j*th student, where j = 1, 2, 3, ... 20. The different subscript letters make it clear that different samples are involved. Letters such as *X*, *Y*, and *Z* generally represent the different variables or types of measurements involved, whereas subscripts such as *i*, *j*, *k*, and *l* designated individual observations (Hamburg, 1987).

Next, we turn our attention to the method of expressing summations of sets of data. Suppose we want to add a set of four observations, denoted X_1, X_2, X_3, and X_4. A convenient way of designating this addition is

$$\sum_{i=1}^{4} X_i = X_1 + X_2 + X_3 + X_4$$

Where the symbol Σ (Greek capital "sigma") means "the sum of." Thus, the symbol

$$\sum_{i=1}^{4} X_i$$

is read "the sum of the X_i's, i going from 1 to 4." For example, if $X_1 = 5$, $X_2 = 1$, $X_3 = 8$, and $X_4 = 6$,

$$\sum_{i=1}^{4} X_i = 5 + 1 + 8 + 6 = 20$$

In general, if there are n observations, we write

$$\sum_{i=1}^{n} X_i = X_1 + X_2 + X_3 + \cdots + X_n$$

Basic statistical terms include mean or average, median, mode, and range. The following is an explanation of each of these terms.

THE MEAN

Mean is one of the most familiar and commonly estimated population parameters; it is the total of the values of a set of observations divided by the number of observations. Given a random sample, the population mean is estimated by

$$\bar{X} = \frac{\sum_{i=1}^{n} x_i}{n}$$

where:

X_i = observed value of the ith unit in the sample

N = number of units in the sample

$\sum_{i=1}^{n} x_i$ means to sum up all n of the X-values in the sample.

If there are N units in the population, the total of the X-values over all units in the population would be estimated by

$$\hat{T} = N\bar{X}$$

The circumflex ($\hat{T}$) over the T is frequently used to indicate an estimated value as opposed to the true but unknown population value. It should be noted that this estimate of the mean is used for a simple random sample. It may not be appropriate if the units included in the sample are not selected entirely at random.

MEDIAN

The median is the value of the central item when the data are arrayed in size.

MODE

The mode is the observation that occurs with the greatest frequency and thus is the most "fashionable" value.

RANGE

The range is the difference between the values of the highest and lowest terms.

Example 4.1

PROBLEM:

Given the following laboratory results for the measurement of dissolved oxygen (DO) in water, find the mean, median, mode, and range.

Data:

6.5 mg/L, 6.4 mg/L, 7.0 mg/L, 6.9 mg/L, 7.0 mg/L

SOLUTION:

To find the Mean:

$$\bar{X} = \frac{\sum_{i=1}^{n} x_i}{n}$$

$$\text{Mean } (\bar{X}) = \frac{(6.5 \text{ mg/L} + 6.4 \text{ mg/L} + 7.0 \text{ mg/L} + 6.0 \text{ mg/L} + 7.0 \text{ mg/L})}{5}$$

$$= 6.58 \text{ mg/L}$$

Mode = 7.0 mg/L (number that appears most often)
arrange in order: 6.4 mg/L, 6.5 mg/L, 6.9 mg/L, 7.0 mg/L, 7.0 mg/L
Median = 6.9 mg/L (central value)
Range = 7.0 mg/L - 6.4 mg/L = 0.6 mg/L

The importance of using statistically valid sampling methods cannot be overemphasized. Several different methodologies are available. A careful review of these methods (with the emphasis on designing appropriate sampling procedures) should be made before computing analytic results. Using appropriate sampling procedures along with careful sampling techniques will provide basic data that is accurate.

The need for statistics in environmental practice is driven by the discipline itself. As mentioned, environmental studies often deal with entities that are variable. If there were no variation in collected data, there would be no need for statistical methods.

Over a given time interval there will always be some variation in sampling analyses. Usually, the average and the range yield the most useful information. For example, in evaluating the indoor air quality (IAQ) in a factory, a monthly summary of air flow measurements, operational data, and laboratory tests for the factory would be used. Another example is seen in evaluating a work center or organization's monthly on-the-job report of accidents and illnesses, a monthly summary of reported injuries, lost-time incidents and work-caused illnesses for the entity would be used.

In the preceding section, we used the term "sample" and the scenario sampling to illustrate the use and definition of mean, mode, median and range. Though these terms are part of the common terminology used in statistics, the term "sample" in statistics has its own unique meaning. There is an exact delineation between both the term "sample" and the term "population." In statistics we most often obtain data from a sample and use the results from the *sample* to describe an entire population. The *population* of a sample signifies that one has measured a characteristic for everyone or everything that belongs in a particular group. For example, if one wishes to measure a certain characteristic of the population defined as environmental professionals, one will have to obtain a measure of that characteristic for every environmental professional possible. Measuring a population is difficult, if not impossible.

We use the term *subject* or *case* to refer to a member of a population or sample. There are statistical methods for determining how many cases must be selected in order to have a credible study. *Data* is another important term: it consists of the measurements taken for the purposes of statistical analysis. Data can be classified as either qualitative or quantitative. *Qualitative* data deal with characteristics of the individual or subject (e.g., gender of a person or the color or a car). *Quantitative* data describe a characteristic in terms of a number (e.g., the age of a horse or the number of lost-time injuries an organization had over the previous year).

Along with common terminology, the field of statistics also generally uses some common symbols. Statistical notation uses Greek letters and algebraic symbols to convey meaning about the procedures that one should follow in order to complete a particular study or test. Greek letters are used as statistical notation for a population, while English letters are used for statistical notation for a sample. Table 4.1 summarizes some of the more common statistical symbols, terms, and procedures used in statistical operations.

TABLE 4.1
Commonly Used Statistical Symbols and Procedures

Term or procedure	Symbol	
	Population symbol	Sample notation
Mean	$\bar{\mu}$	$\bar{x}$
Standard deviation	σ	s
Variance	σ^2	s^2
Number of cases	N	n
Raw number or value	X	X
Correlation coefficient	R	R
Procedure	**Symbol**	
Sum of	Σ	
Absolute value of x	$\lvert x \rvert$	
Factorial of n	$n!$	

REFERENCES

Freese, F., 1969. *Elementary Statistical Methods for Foresters.* Washington, DC: USDA.
Hamburg, J.M., 1987. *Institut fur Asienkude.* Germany: Frankfurt.
Spellman, F.R., 2020. *Environmental Engineers Mathematics Handbook.* Boca Raton, FL: CRC Press.

5 Distribution

THE 411 ON DISTRIBUTION

If an epidemiologist or any environmental professional wishes to conduct a research study, and the data is collected and a group of raw data is obtained, in order to make sense out of that data it must be organized into a meaningful format.

The formatting begins by putting the data into some logical order, then grouping the data. Before the data can be compared to other data, it must be organized. Organized data are referred to as *distributions*.

When confronted with masses of ungrouped data (i.e., listings of individual figures), it's difficult to generalize about the information these masses contain. However, if a *frequency distribution* of the figures is formed, many features become readily discernible. A frequency distribution records the number of cases that fall in each class of the data; it displays the values a variable can take and the number of person or records with each value.

Example 5.1

An environmental health and safety professional gathers data on the medical costs of 24 on-the-job injury claims for a given year. The raw data collected is shown in Table 5.1. To develop a frequency distribution, the investigator takes the values of the claims and places them in order. Then the investigator counts the frequency of occurrences for each value as shown in Table 5.2.

In order to develop a frequency distribution, groupings are formed using the values in the table above, ensuring that each group has an equal range. The environmental professional grouped the data into ranges of 1,000. The lowest range and highest range are determined by the data. Because it was decided to group by thousands, values will fall in the ranges of $0 to $4,999, and the distribution will end with this. The frequency distribution for this data appears in Table 5.3.

The data in a frequency distribution can be graphed. We call this type of graph a histogram. Figure 5.1 is a graph of the number of outbreak-related salmonellosis cases by illness onset.

NORMAL OR GAUSSIAN DISTRIBUTION

When large amounts of data are collected on certain characteristics, the data and subsequent frequency can follow a distribution that is bell-shaped in nature—the *normal distribution*. The normal distributions are an especially important class of statistical distributions. As stated, all normal distributions are symmetric and have bell-shaped curves with a single peak (see Figure 5.2).

TABLE 5.1
Raw Data on Medical Costs

$60	$1,500	$85	$120
$110	$150	$110	$340
$2,000	$3,000	$550	$560
$4,500	$85	$2,300	$200
$120	$880	$1,200	$150
$650	$220	$150	$4,600

To speak specifically of any normal distribution, two quantities have to be specified: the mean μ (pronounced mu), where the peak of the density occurs, and the standard deviation σ (sigma). Different values of μ and σ yield different normal density curves and hence different normal distributions.

Although there are many normal curves, they all share an important property that allows us to treat them in a uniform fashion. All normal density curves satisfy the following property, which is often referred to as the *Empirical Rule*.

TABLE 5.2
Frequency of Occurrences

Value	Frequency
$60	1
$85	2
$110	2
$120	2
$150	3
$200	1
$220	1
$340	1
$550	1
$560	1
$650	1
$880	1
$1,200	1
$1,500	1
$2,000	1
$2,300	1
$3,000	1
$4,500	1
$4,600	1
Total	24

TABLE 5.3
Frequency Distribution Data

Range	Frequency
$0 - $999	17
$1,000 – 1,999	2
$2,000 – 2,999	2
$3,000 – 3,999	1
$4,000 – 4,999	2
Total	24

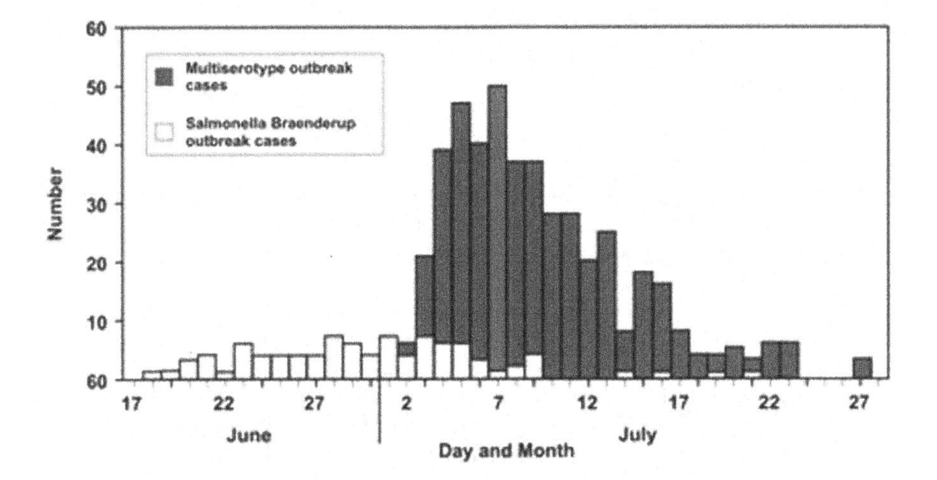

FIGURE 5.1 Number of outbreak-related Salmonellosis cases by date of onset of illness—United States, June–July 2004. *Source*: CDC, 2012.

- **68 percent** of the observations fall within **1 standard deviation** of the **mean**, that is, between $\mu - \sigma$ and $\mu + \sigma$.
- **95 percent** of the observations fall within **2 standard deviations** of the **mean**, that is, between $\mu - 2\sigma$ and $\mu + 2\sigma$.
- **98 percent** of the observations all fall within **3 standard deviations** of the **mean**, that is, between $\mu - 3\sigma$ and $\mu + 3\sigma$.

Thus, for a normal distribution, almost all values lie within 3 standard deviations of the mean (see Figure 5.2). It is important to stress that the rule applies to all normal distributions. Also remember that it applies *only* to normal distributions.

Note: Before applying the empirical rule, it is a good idea to identify the data being described, and the value of the mean and standard deviation. A sketch of a graph summarizing the information provided by the empirical rule should also be made.

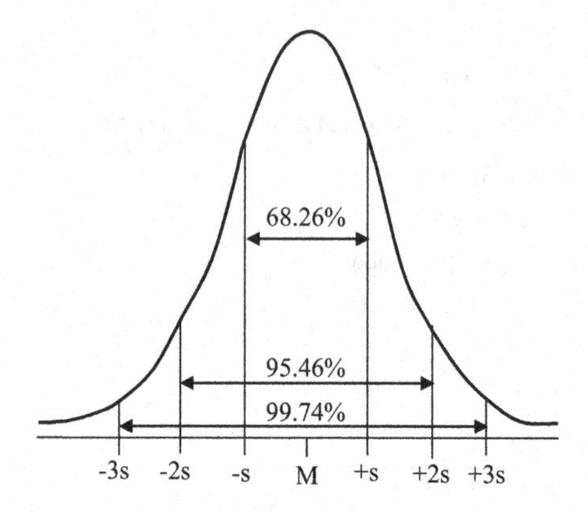

FIGURE 5.2 The normal distribution curve showing the frequency of a measurement.

Example 5.2

PROBLEM:

The scores for all high school seniors taking the math section of the Scholastic Aptitude Test (SAT) in a particular year had a mean of 490 and a standard deviation of 100. The distribution of SAT scores is bell-shaped.

1. What percentage of seniors scored between 390 and 590 on this SAT test?
2. One student scored 795 on this test. How did this student do compared to the rest of the scores?
3. A rather exclusive university only admits students who were among the highest 16 percent of the scores on this test. What score would a student need on this test to be qualified for admittance to this university?

The data being described are the math SAT scores for all seniors taking the test in one year. Because this is describing a population, we denote the mean and standard deviation as $\mu = 490$ and $\sigma = 100$, respectively. A bell-shaped curve summarizing the percentages given by the empirical rule is shown in Figure 5.3.

1. From Figure 5.3, about 68 percent of seniors scored between 390 and 590 on this SAT test.
2. Because about 99.7 percent of the scores are between 190 and 790, a score of 795 is excellent. This is one of the highest scores on this test.
3. Because about 68 percent of the scores are between 390 and 590, this leaves 32 percent of the scores outside the interval. Because a bell-shaped curve is symmetric, one-half of the scores, or 16 percent, are on each end of the distribution.

MEASURES OF SPREAD AND STANDARD DEVIATION

Spread is dispersion and dispersion is spread. Anyway, by any other name spread is an important feature of frequency distribution. Just as measure of central location

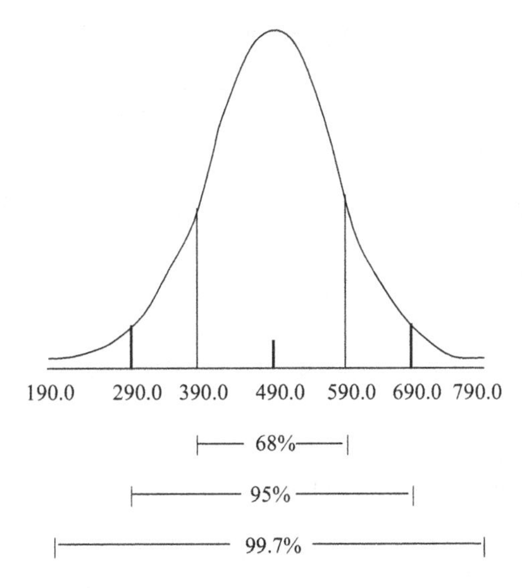

FIGURE 5.3 Sample Scholastic Aptitude Test (SAT) math percentages given by the empirical rule (CDC, 2012).

describes where the peak is located, measures of spread describe the dispersion (or variation) of values from that peak in the distribution. Range, interquartile range, and standard deviation are measures of spread.

Range

The *range* of a set of data is the difference between its largest (maximum) value and its smallest (minimum) value. Stated more simply, range is a measure between the smallest to the largest. In the statistical world, the range is reported as a single number and is the result of subtracting the maximum from the minimum value. In the epidemiologic community, the range is usually reported as "from (the minimum) to the maximum," that is, as two numbers rather than one. Note that range is measured in the same units as the data (CDC, 2020).

How to Identify the Range

There are two basic steps involved with identifying the range:

Step 1: Identify the smallest (minimum) observation and the largest (maximum) observation.

Step 2: From an epidemiology point of view, report the minimum and maximum values. Statistically, subtract the minimum from the maximum value.

Example 5.3

PROBLEM:

Find the range of the following incubation periods for hepatitis A: 28, 32, 15, 30, and 21 days.

SOLUTION:

Step 1: Identify the minimum and maximum values.

Minimum = 15, maximum = 32

Step 2: Subtract the minimum from the maximum value.

Range = 32 – 15 = 17 days

For an epidemiologic or lay audience, you could report that "incubation periods ranged from 15 to 32 days." Statistically, that range is 17 days.

Percentiles

Dividing the group data in a distribution into 100 equal parts is known as *percentiles*. The *P*th percentile (*P* ranging from 0 to 100) is the value that has *P* percent of the observations falling at or below it. To put it in other words, the 90th percentile has 90 percent of the observations at or below it. The median, the halfway point of the distribution, is the 50th percentile. The maximum value is the 100th percentile because all values fall at or below the maximum.

Qaurtiles

It is not uncommon for epidemiologists to group data into four equal parts, or *quartiles*. Each quartile includes 25 percent of the data. The cut-off for the first quartile is the 25th percentile. The cut-off for the second quartile is the 50th percentile, which is the median. The cut-off for the third quartile is the 70th percentile. And the cut-off for the fourth quartile is the 100th percentile, which is the maximum.

Interquartile Range

The measure of spread used most commonly with the median is known as the *interquartile range* (IQR), also called the midspread, middle 50 percent, or the H-spread. It represents the middle portion of the distribution, from the 25th percentile to the 75th percentile. In other words, the interquartile range includes the second and third quartiles of a distribution. Thus, the interquartile range includes approximately one-half of the observations in the set, leaving one-quarter of the observations on each side.

Example 5.4

Problem: Determine the interquartile range

Step 1: Arrange the observations in increasing order.

Step 2: Find the position of the 1st and 3rd quartiles with the following formulas. Divide the sum of the number of observations.

Position of 1st quartile (Q_1) = 25th percentile = (n + 1)/4
Position of 3rd quartile (Q_3) = 75th percentile = 3(n + 1)/4 = $3Q_1$

Step 3: Identify the value of the 1st and 3rd quartiles.

a. If a quartile lies *on an observation* (i.e., if its position is a whole number), the value of the quartile is the value of the observation. For example, if the position of a quartile is 20, its value is the value of the 20th observation.
b. If a quartile lies *between observations,* the value of the quartile is the value of the lower observation plus the specified fraction of the difference between the observations. For example, if the position of a quartile is 20¼, it lies between the 20th and 21st observations, and its value is the value of the 20th observation, plus ¼ the difference between the value of the 20th and 21st observations (CDC, 2012; Spellman, 2020).

Step 4: In epidemiology we report the values at Q_1 and Q_3. In statistics, calculate the interquartile range as Q_3 minus Q_1.

DID YOU KNOW?

The interquartile range is generally used in conjunction with the median. Together, they are useful for characterizing the central location and spread of any frequency distribution, particularly those that are skewed.

Standard Deviation

The standard deviation, s or σ (sigma), is often used as an indicator of precision. The *standard deviation* is a measure of the variation (the spread in a set of observations) in the results; that is, it gives us some idea whether most of the individuals in a population are close to the mean or spread out. In order to gain better understanding and perspective of the benefits derived from using statistical methods in epidemiology, it is appropriate to consider some of the basic theories of statistics. In any set of data, the true value (mean) will lie in the middle of all the measurements taken. This is true, provided the sample size is large and only random error is present in the analysis. In addition, the measurements will show a normal distribution as shown in Figure 5.1.

Figure 5.1 shows that 68.26 percent of the results fall between M + s and M – s, 95.46 percent of the results lie between M + 2s and M – 2s, and 99.74 percent of the results lie between M + 3s and M – 3s. Therefore, if precise, then 68.26 percent of all the measurements should fall between the true value estimated by the mean, plus the standard deviation and the true value minus the standard deviation. The following equation is used to calculate the sample standard deviation:

$$s = \frac{\sqrt{\Sigma(X - \bar{X})^2}}{n-1}$$

where:

s = standard deviation
n = number of samples
X = the measurements from X to Xn
$\bar{X}$ = the mean
Σ = means to sum the values from X to Xn

Example 5.5

PROBLEM:

Calculate the standard deviation, σ, of the following dissolved oxygen values:

9.5, 10.5, 10.1, 9.9, 10.6, 9.5, 11.5, 9.5, 10.0, 9.4

$$\bar{X} = 10.0$$

X	$X - \bar{X}$	$(X - \bar{X})^2$
9.5	–0.5	0.25
10.5	0.5	0.25
10.1	0.1	0.01
9.9	–0.1	0.01
10.6	0.6	0.36
9.5	–0.5	0.25
11.5	1.5	2.25
9.5	–0.5	0.25
10.0	0	0
9.4	–0.6	0.36
		3.99

$$\sigma = \sqrt{\frac{1}{(10-1)}}(3.99)$$

COEFFICIENT OF VARIATION

In nature, populations with large means often show more variation than populations with small means. The coefficient of variation (C) facilitates comparison of variability about different sized means. It is the ratio of the standard deviation to the mean. A standard deviation of 2 for a mean coefficient of variation would be 0.20 or 20 percent in each case. If we have a standard deviation of 1.414 and a mean of 9.0, the coefficient of variation would be estimated by

$$C = \frac{s}{\overline{X}} = \frac{1.414}{9.0} = 0.157, \text{ or } 15.7 \text{ percent}$$

STANDARD ERROR OF THE MEAN

As mentioned, there is usually variation among the individual units of a population. Again, the standard deviation is a measure of this variation. Because the individual units vary, variation may also exist among the means (or any other estimates) computed from samples of these units. Take, for example, a population with a true mean of 10. If we were to select four units at random, they might have a sample mean of 8. Another sample of four units from the same population might have a mean of 11, another 10.5, and so forth. Clearly, it would be desirable to know the variation likely to be encountered among the means of samples from this population. A measure of the variation among sample means is the standard error of the mean. It can be thought of as a standard deviation among sample means; it is a measure of the variation among sample means, just as the standard deviation is a measure of the variation among individuals. The standard error of the mean may be used to compute confidence limits for a population mean.

The computation of the standard error of the mean (often symbolized by $S_{\bar{x}}$) depends on the manner in which the sample was selected. For simple random sampling without replacement (i.e., a given unit cannot appear in the sample more than once) from a population having a total of N units the formula for the estimated standard error of the mean is

$$S_{\bar{x}} = \sqrt{S^2/n}(1 - n/N)$$

In a forestry example, if we had $n = 10$ and found that $s = 1.414$ or $s^2 = 2$ in the population that contains 1,000 trees, the estimated mean diameter ($X = 9.0$) would have a standard error of

$$S_{\bar{x}} = \sqrt{2/10\,(1 - 10/1{,}000)} = \sqrt{0.198}$$
$$= 0.445$$

Note: The term (1 – n/N) is called the *finite population correction* or fpc. The fpc is the finite population correct and is used when the sampling fraction (the number of elements or respondents sampled relative to the population) becomes large. The fpc is used in the calculation of the standard error of the estimate. If the value of the fpc is close to 1, it will have little impact and can be safely ignored.

COVARIANCE

Very often, each unit of a population will have more than a single characteristic. In forest practice, for example, trees may be characterized by their height, diameter, and form class (i.e., the amount of taper). The covariance is a measure of the association between the magnitudes of two characteristics. If there is little or no association, the covariance will be close to zero. If the large values of one characteristic tend to be associated with the small values of another characteristic, the covariance will be negative. If the large values of one characteristic tend to be associated with the large values of another characteristic, the covariance will be positive. The population covariance of X and Y is often symbolized by σ_{xy}; the sample estimate by s_{xy}.

Let's return to a forestry practice example. Suppose that the diameter (inches) and age (years) have been obtained for a number of randomly selected trees. If we symbolize diameter by Y and the age by X, the sample covariance of diameter and age is given by

$$s_{xy} = \frac{\Sigma XY - \frac{(\Sigma X)(\Sigma Y)}{n}}{(n-1)}$$

This is equivalent to the formula

$$s_{xy} = \frac{\Sigma(X-\bar{X})(Y-\bar{Y})}{(n-1)}$$

If n = 12 and the Y and X values were as follows:

												Sums
Y| 4	9	7	7	5	10	9	6	8	6	4	11|	86
X| 20	40	30	45	25	45	30	40	20	35	25	40|	395

Then

$$s_{xy} = \frac{(4)(20)+(9)(40)+\cdots+(11)(40)-\frac{(86)(395)}{12}}{12-1}$$

$$= \frac{2,960-2,830.83}{11} = 11.74$$

The positive covariance is consistent with the well-known and economically unfortunate fact that the larger diameters tend to be associated with the older ages.

SIMPLE CORRELATION COEFFICIENT

The magnitude of the covariance, like that of the standard deviation, is often related to the size of the variables themselves. Units with large X and Y values tend to have larger covariances than units with small X and Y values. Also, the magnitude of the covariance depends on the scale of measurement; in the previous example, had diameter been expressed in millimeters instead of inches, the covariance would have been 298.196 instead of 11.74.

The simple correlation coefficient, a measure of the degree of linear association between two variables, is free of the effects of scale of measurement. It can vary from −1 and +1. A correlation of 0 indicates that there is no linear association (there may be a very strong nonlinear association, however). A correlation of +1 or −1 would suggest a perfect linear association. As for the covariance, a positive correlation implies that the large values of X are associated with the large values of Y. If the large values of X are associated with the small values of Y, the correlation is negative.

The population correlation coefficient is commonly symbolized by ρ (rho), and the sample-based estimate r. The population correlation coefficient is defined to be

$$\rho = \frac{\text{Covariance of } X \text{ and } Y}{\sqrt{(\text{Variance of } X)(\text{Variance of } Y)}}$$

For a simple random sample, the sample correlation coefficient is computed as follows:

$$r = \frac{s_{xy}}{s_x \cdot s_y} = \frac{\Sigma xy}{\sqrt{(\Sigma x^2)(\Sigma y^2)}}$$

where:

s_{xy} = Sample covariance of X and Y
s_x = Sample standard deviation of X
s_y = Sample standard deviation of Y
$*xy$ = Corrected sum of XY products

$$= \Sigma XY - \frac{(\Sigma X)(\Sigma Y)}{n}$$

$*x^2$ = Corrected sum of squares for X

$$= \Sigma X^2 - \frac{(\Sigma X)^2}{n}$$

S_y^2 = Corrected sum of squares for Y

$$= \Sigma Y^2 - \frac{(\Sigma Y)^2}{n}$$

For the values used to illustrate the covariance we have:

$$\Sigma xy = (4)(20) + (9)(40) + \cdots + (11)(40) - \frac{(86)(395)}{12} = 129.1667$$

$$\Sigma y^2 = 4^2 + 9^2 + \cdots + 11^2 - \frac{86^2}{12} = 57.6667$$

$$\Sigma x^2 = 20^2 + 40^2 + \cdots + 40^2 - \frac{395^2}{12} = 922.9167$$

So,

$$R = \frac{129.1667}{\sqrt{(57.6667)(922.9167)}} = \frac{129.1667}{230.6980} = .56$$

DID YOU KNOW?

The computed value of a statistic such as the correlation efficient depends on which particular units were selected for the sample. Such estimates will vary from sample to sample. More important, they will usually vary from the population value which we try to estimate.

VARIANCE OF A LINEAR FUNCTION

Routinely, we combine variables or population estimates in a linear function. For example, in forestry, if the mean timber volume per acre has been estimated as $\overline{X}$, then the total volume on M acres with be $M\overline{X}$; the estimate of total volume is a linear function of the estimated-mean volume. If the estimate of cubic volume per acre in saw timber is $\overline{X}_1$ and of pulpwood above the saw timber top is $\overline{X}_2$, then the estimate of total cubic foot volume per acre is $\overline{X}_1 + \overline{X}_2$. If on a given tract the mean volume per half-acre is $\overline{X}_1$ for spruce and the mean volume per quarter-acre is $\overline{X}_2$ for yellow birch, then the estimated total volume per acre of spruce and birch would be $2\overline{X} + 4\overline{X}_2$.

In general terms, a liner function of three variables (say X_1, X_2, and X_3) can be written as

$$L = a_1X_1 + a_2X_2 + a_3X_3$$

where a_1, a_2, and a_3 are constants.

If the variances are S_1^2, S_2^2, and S_3^2 (for X_1, X_2, and X_3, respectively) and the covariances are $S_{1,2}$, $S_{1,3}$, and $S_{2,3}$, then the variance of L is given by

$$s_L^2 = a_1^2 s_1^2 + a_2^2 s_2^2 + a_3^2 s_3^2 + 2\left(a_2 a_2 s_{1,2} + a_1 a_3 s_{1,3} + a_2 a_3 s_{2,3}\right)$$

The standard deviation (or standard error) of L is simply the square root of this.

The extension of the rule to cover any number of variables should be fairly obvious.

Example 5.6

PROBLEM:

The sample mean volume per forested acre for a 10,000-acre tract is X = 5,680 board feet with a standard error of $s_x = 632$ (So $s_{\bar{x}}^2 = 399{,}424$). The estimated total volume is

$$L = 10{,}000(\bar{X}) = 56{,}800{,}000 \text{ board feet.}$$

The variance of this estimate would be

$$S_L^2 = (10{,}000)^2 (s\bar{x}^2) = 39{,}942{,}400{,}000{,}000$$

Since the standard error of an estimate is the square root of its variance, the standard error of the estimated total is

$$s_L = \sqrt{sL^2} = 6{,}320{,}000$$

Example 5.7

PROBLEM:

In 1995 a random sample of 40 one-quarter-acre circular plots was used to estimate the cubic foot volume of a stand of pine. Plot centers were monumented for possible relocation at a later time. The mean volume per plot was $\bar{X}_1 = 225 \text{ ft}^3$. The plot variance was $s_{x1}^2 = 8{,}281$ so that the variance of the mean was $s_{\bar{x}.1}^2 = 8{,}281/40 = 207.025$.

In 2000 a second inventory was made using the same plot centers. This time, however, the circular plots were only one-tenth acre. The mean volume per plot was $\bar{X}_2 = 122 \text{ ft}^3$. The plot variance was $s_{x2}^2 = 6{,}084$, so the variance of the mean was $s_{\bar{x}_2}^2 = 152.100$. The covariance of initial and final plot volumes was $s_{x1,x2} = 4{,}259$, making the covariance of the means $s_{\bar{x}1,\bar{x}2} = 4{,}259/40 = 106.475$.

The net periodic growth per acre would be estimated as

$$G = 10\bar{X}_2 - 4\bar{X}_1 = 10(122) - 4(225) = 320 \text{ cubic feet per acre.}$$

By the rule for linear functions the variance of G would be

$$\begin{aligned} S_G^2 &= (10)^2 s\bar{x}_2^2 + (-4)^2 s\bar{x}_1^2 + 2(1)(-4) s\bar{x}_1,\bar{x}_2 \\ &= 100(152.100) + 16(207.025) - 80(106.475) \\ &= 10{,}004.4 \end{aligned}$$

In this example there was a statistical relationship between the 2000 and 1995 means because the same plot locations were used in both samples. The covariance of the means ($s_{\bar{x}1,\bar{x}2}$) is a measure of this relationship. If the 2000 plots had been located at random rather than at the 1995 locations, the two means would have been considered statistically independent and their covariance would have been set at zero. In this case the equation for the variance of the net periodic growth per acre (*G*) would reduce to

$$S_{\bar{G}}^2 = (10)^2 s\bar{x}_2^2 + (-4)^2 s\bar{x}_1^2$$

$$= 100(152.100) + 16(207.025) = 18,522.4$$

REFERENCES

CDC, 2012. *Principles of Epidemiology in Public Heath Practice*, 3rd ed. Atlanta, GA: Centers for Disease Control and Prevention.

CDC, 2020. *Range Statistics: Data & Statistics*. Accessed 9/2/2020 @ https://www.cdc.gov/datastatistics

Spellman, F.R., 2020. *Environmental Engineers Mathematics Handbook*. Boca Raton, FL: CRC Press.

6 Sampling—Measurement Variables

Note: To ensure understanding of the concepts presented in this chapter, the author uses real-world examples and, for simplicity, the material used by modern foresters dealing with seedlings, seeds, trees, and undergrowth.

SIMPLE RANDOM SAMPLING

Epidemiologists are familiar with *simple random sampling.* As in any sampling system, the aim is to estimate some characteristic of a population without measuring all of the population units. In a simple random sample of size n, the units are selected so that every possible combination of n units has an equal chance of being selected. If sampling is without replacement, then at any stage of the sampling each unused unit should have an equal chance of being selected.

Sample estimates of the population mean and total—From a population of $N =$ 100 units, $n = 20$ units were selected at random and measured. Sampling was without replacement—once a unit had been included in the sample it could not be selected again. The unit values were:

10	9	10	9	11
16	11	7	12	12
11	3	5	11	14
8	13	12	20	10

Sum of all 20 random units = 214

From this sample we estimate the population mean as

$$\bar{X} = \frac{\Sigma X}{n} = \frac{214}{20} = 10.7$$

A population of N = 100 units having a mean of 10.7 would then have an estimated total of

$$\hat{T} = N\bar{X} = 100(10.7) = 1{,}070$$

Standard Errors

The first step in calculating a standard error is to obtain an estimate of the population variance (σ^2) or standard deviation (σ). As noted in a previous section, the standard deviation for a simple random sample is estimated by

$$S = \sqrt{\frac{\Sigma X^2 - \frac{(\Sigma X)^2}{n}}{n-1}} = \sqrt{\frac{10^2 + 16^2 + \cdots + 10^2 - \frac{214^2}{20}}{19}}$$

$$= \sqrt{13.4842} = 3.672$$

For sampling without replacement, the standard error of the mean is

$$s\bar{x} = \sqrt{\frac{s^2}{n}\left(1 - \frac{n}{N}\right)} = \sqrt{\frac{13.4842}{20}\left(1 - \frac{20}{100}\right)}$$

$$= \sqrt{0.539368} = 0.734$$

From the formula for the variance of a linear function we find that the variance of the estimated total is

$$S\hat{i}^2 = N^2 s\bar{x}^2$$

The standard error of the estimated total is the square root of this, or

$$S\hat{i} = N s\bar{x} = 100(0.734) = 73.4$$

Confidence Limits

Sample estimates are subject to variation. How much they vary depends primarily on the inherent variability of the population (Var^2) and on the size of the sample (n) and of the population (N).

The statistical way of indicating the reliability of an estimate is to establish *confidence* limits. For estimates made from normally distributed populations, the confidence limits are given by

$$(\text{Estimate}) \pm (t)\ (\text{Standard Error})$$

For setting confidence limits on the mean and total we already have everything we need except for the value of t, and that can be obtained from a table of the t distribution.

In the previous example the sample of $n = 20$ units had a mean of $\bar{X} = 10.7$ and a standard error of $S\bar{x} = 0.734$. For 95-percent confidence limits on the mean we would use a t value (from t table) of 0.05 and (also from a t table) 19 degrees of freedom. As $t_{.05} = 2.093$, the confidence limits are given by

$$\bar{X} \pm (t)(s\bar{x}) = 10.7 \pm (2.093)(0.734) = 9.16 \text{ to } 12.24$$

This says that unless a 1-in-20 chance has occurred in sampling, the population mean is somewhere between 9.16 and 12.24. It does not say where the mean of future samples from this population might fall. Nor does it say where the mean may be if mistakes have been made in the measurements.

For 99 percent confidence limits we find $t_{0.01} = 2.861$ (with 19 degrees of freedom), and so the limits are

$$10.7 \pm (2.861)(0.734) = 8.6 \text{ to } 12.8$$

These limits are wider, but they are more likely to include the true population mean.

For the population total the confidence limits are:

95 percent limits: 1,070 ± (2.093) (73.4) = 916 to 1,224
99 percent limits: 1,070 ± (2.861) (73.4) = 860 to 1,280

For large samples ($n > 60$) the 95 percent limits are closely approximated by

Estimate ± (2) (Standard Error)

and the 99 percent limits by

Estimate ± (2.6) (Standard Error)

Sample Size

Samples cost money. So do errors. The aim in planning a survey should be to take enough observations to obtain the desired precision—no more, no less.

The number of observations needed in a simple random sample will depend on the precision desired and the inherent variability of the population being sampled. Since sampling precision is often expressed in terms of confidence interval on the mean, it is not unreasonable in planning a survey to say that in the computed confidence interval

$$\bar{X} \pm ts\bar{x}$$

we would like to have the $ts\bar{x}$ equal to or less than some specified value E, unless a 1-in-20 (or 1-in-100) chance has occurred in sample. That is, we want

$$ts\bar{x} = E$$

or, since $s\bar{x} = \frac{s}{\sqrt{n}}$, we want

$$t = \left(\frac{s}{\sqrt{n}}\right) = E$$

Solving this for n gives the desired sample size.

$$n = \frac{t^2 s^2}{E^2}$$

To apply this equation, we need to have an estimate (s^2) of the population variance and a value for student's t at the appropriate level of probability.

The variance estimate can be a real problem. One solution is to make the sample survey in two stages. In the first state, n_1 random observations are made and from these an estimate (s^2) of the variance is computed. Then this value is plugged into the sample size equation

$$n = \frac{t^2 s^2}{E^2}$$

where t has $n_1 - 1$ degrees of freedom and is selected from the appropriate table. The computed value of n is the total size of sample needed. As we have already observed n_1 units, this means that we will have to observe $(n - n_1)$ additional units.

If pre-sampling as described above is not feasible, then it will be necessary to make a guess at the variance. Assuming our knowledge of the population is such that the guessed variance (s^2) can be considered fairly reliable, then the size of sample (n) needed to estimate the mean to within $\pm E$ units is approximately

$$n = \frac{4s^2}{E^2}$$

for 95 percent confidence, and

$$n = \frac{20(s^2)}{3E^2}$$

for 99 percent confidence.

Less reliable variance estimates could be doubled (as a safety factor) before applying these equations. In many cases the variance estimate may be so poor as to make the sample size computation just so much statistical window dressing.

When sampling is without replacement (as it is in most forest sampling situations) the sample size estimates given above apply to populations with an extremely large number (N) of units so that the sampling fraction (n/N) is very small. If the sampling fraction is not small (say $n/N = 0.05$) then the sample size estimates should be adjusted. This adjusted value of n is

$$n_a = \frac{n}{1 + n/N}$$

DID YOU KNOW?

It is important that the specified error (E) and the estimated variance (s^2) be on the same scale of measurement. We could not, for example, use a board-foot variance in conjunction with an error expressed in cubic feet. Similarly, if the error is expressed in volume per acre, the variance must be put on a per acre basis.

Suppose that we plan to use quarter-acre plots in a survey and estimate the variance among plot volumes to be $s^2 = 160{,}000$. If the error limit is $E = 5{,}000$ ft per acre, we must convert the variance to an acre basis or the error to a quarter-acre basis. To convert a quarter-acre volume to a per-acre basis we multiply by 4, and to convert a quarter-acre variance to an acre variance we multiply by 16. Thus, the variance would be 2,560,000 and the sample size formula would be

$$n = \frac{t^2(2{,}560{,}000)}{(500)^2} = t^2(10.24)$$

Alternatively, we can leave the variance alone and convert the error statement from an acre to a quarter-acre basis; $E = 125$. Then the sample size formula is

$$n = \frac{t^2(160{,}000)}{(125)^2} = t^2(10.24), \text{ as before.}$$

STRATIFIED RANDOM SAMPLING

In *stratified sampling*, a population is divided into subpopulations (strata) of known size, and a simple random sample of at least two units is selected in each subpopulation. This approach has several advantages. For one thing, if there is more variation between subpopulations than within them, the estimate of the population mean will be more precise than that given by a simple random sample of the same size. Also, it may be desirable to have separate estimates for each subpopulation (e.g., administrative subunits). And it may be administratively more efficient to sample by subpopulations.

SAMPLE ALLOCATION

If a sample of n units is taken, how many units should be selected in each stratum? Among several possibilities, the most common procedure is to allocate the sample in proportion to the size of the stratum; in a stratum having two fifths of the units of the population, we would take two fifths of the samples. In the population discussed in the previous example the proportional allocation of the 55 sample units would have been (and was) as follows:

Stratum	Relative size (N_h/N)	Sample allocation
1	0.54	29.7 or 30
2	0.28	15.4 or 15
3	0.18	9.9 or 10
Sums	1.00	55

Some other possibilities are equal allocation, allocation proportional to estimated value, and optimum allocation. In optimum allocation an attempt is made to get the smallest standard error possible for a sample of n units. This is done by sampling more heavily in the state having a larger variation. The equation for optimum allocation is

$$n_h = \left(\frac{N_h s_h}{\Sigma N_h s_h} \right) n$$

Optimum allocation obviously requires estimates of the within-stratum variances—information that may be difficult to obtain. A refinement of optimum allocation is to take sampling cost differences into account and allocate the sample so as to get the most information per dollar. If the cost per sampling unit in stratum h is c_h, the equation is

$$n_h = \left(\frac{\frac{N_h s_h}{\sqrt{c_h}}}{\Sigma \left(\frac{N_h s_h}{\sqrt{c_h}} \right)} \right) n$$

Sample Size

To estimate the size of sample to take for a specified error at a given level of confidence, it is first necessary to decide on the method of allocation. Ordinarily, proportional allocation is the simplest and perhaps the best choice. With proportional allocation, the size of sample needed to be within $\pm E$ units of the true value at the 0.05 probability level can be approximated by

$$n = \frac{N(\Sigma N_h s_h^2)}{\frac{N^2 E^2}{4} + \Sigma N_h s_h^2}$$

For the 0.01 probability level, use 6.76 in place of 4.

SAMPLING—DISCRETE VARIABLES

Random Sampling

The sampling methods discussed in the previous sections apply to data that are on a continuous or nearly continuous scale of measurement. These methods may not be applicable if each unit observed is classified as alive. This type may follow what is known as the binomial distribution. They require slightly different statistical techniques.

As an illustration, suppose that a sample of 1,000 seeds was selected at random and tested for germination. If 480 of the seeds germinated, the estimated viability for the lot would be

$$\bar{p} = \frac{480}{1,000} = 0.48, \text{ or } 48 \text{ percent}$$

For large samples (say $n > 250$) with proportions greater than 0.20 but less than 0.80, approximate confidence limits can be obtained by first computing the standard error of $\bar{p}$ by the equation

$$S\bar{p} = \sqrt{\frac{\bar{p}(1-\bar{p})}{(n-1)}\left(1-\frac{n}{N}\right)}$$

Then, the 95 percent confidence limits are given by

$$95\text{-percent confidence interval} = \bar{p} \pm [2(s\bar{p}) + 1/2n]$$

Applying this to the above example we get

$$S\bar{p} = \sqrt{\frac{(0.48)(0.52)}{999}} \text{ (fpc ignored)}$$

$$= 0.0158$$

And,

$$95\text{-percent confidence interval} = .48 \pm \left[2(0.0158) + \frac{1}{2(1,000)}\right]$$

$$= .448 \text{ to } .512$$

The 99 percent confidence limits are approximated by

$$99\text{-percent confidence interval} = \bar{p} \pm [2.6\,s\bar{p} + 1/2n]$$

SAMPLE SIZE

An appropriate table can be used to estimate the number of units that would have to be observed in a simple random sample in order to estimate a population proportion with some specified precision.

Suppose, for example, that we wanted to estimate the germination percent for a population to within plus or minus 10 percent (or 0.10) at the 95 percent confidence level. The first step is to guess what the proportion of seed germination will be. If a good guess is not possible, then the safest course is to guess $\overline{p} = 0.59$ as this will give the maximum sample size.

Next, pick any of the sample sizes given in the appropriate table (e.g., 10, 15, 20, 30, 50, 100, 250, and 1,000) and look at the confidence interval for the specified value of $\overline{p}$. Inspection of these limits will tell whether or not the precision will be met with a sample of this size or if a larger or smaller sample would be more appropriate.

Thus, if we guess $\overline{p} = 0.2$, then in a sample of $n = 50$ we would expect to observe (0.2) (50) = 10, and the table says that the 95 percent confidence limits on $\overline{p}$ would be 0.10 and 0.34. Since the upper limit is not within 0.10 of $\overline{p}$, a larger sample would be needed. For a sample of $n = 100$ the limits are 0.13 to 0.29. Since both of these values are within 0.10 of $\overline{p}$, a sample of 100 would be adequate.

If the table indicates the need for a sample of over 250, the size can be approximated by

$$n = \frac{4(\overline{p})(1-\overline{p})}{E^2}, \text{ for 95-percent confidence}$$

or,

$$n = \frac{20(\overline{p})(1-\overline{p})}{3E^2}, \text{ for 99-percent confidence}$$

where E = The precision with which $\overline{p}$ is to be estimated (expressed in same for as $\overline{p}$, either percent or decimal).

CLUSTER SAMPLING FOR ATTRIBUTES

Simple random sampling of discrete variables is often difficult or impractical. In estimating tree plantation survival, for example, we could select individual trees at random and examine them, but it wouldn't make much sense to walk down a row of planted trees in order to observe a single member of that row. It would usually be more reasonable to select rows at random and observe all of the trees in the selected row.

Seed viability is often estimated by randomly selecting several lots of 100 or 200 seeds each and recording for each lot the percentage of the seeds that germinate.

These are examples of cluster sampling; the unit of observation is the cluster rather than the individual tree or single seed. The value attached to the unit is the proportion having a certain characteristic rather than the simple fact of having or not having that characteristic.

If the clusters are large enough (say over 100 individuals per cluster) and nearly equal in size, the statistical methods that have been described for measurement variables can often be applied. Thus, suppose that the germination percent of a seed lot is estimated by selecting $n = 10$ sets of 200 seed each and observing the germination percent for each set.

Set	1	2	3	4	5	6	7	8	9	10	Sum
Germination Percent (p)	78.5	82.0	86.0	80.5	74.5	78.0	79.0	81.0	80.5	83.5	803.5

then the mean germination percent is estimated by

$$\bar{p} = \frac{\Sigma p}{n} = \frac{8.03.5}{10} = 80.35 \text{ percent}$$

The standard deviation of p is

$$S_p = \frac{\sqrt{\Sigma p^2 - \dfrac{(\Sigma p)^2}{n}}}{n-1} = \sqrt{\frac{78.5^2 + \cdots + 83.5^2 = (803.5)^2/10}{9}}$$

$$= \sqrt{10.002778} = 3.163$$

And the standard error for $\bar{p}$ is

$$S\bar{p} = \sqrt{\frac{S_p{}^2}{n}\left(1 - \frac{n}{N}\right)}$$

$$= \sqrt{\frac{10.002778}{10}} = 1.000 \text{ (fpc ignored)}$$

Note that n and N in these equations refer to the number of clusters, not to the number of individuals.

The 95-percent confidence interval, computed by the procedure for continuous variables:

$$= \bar{p} \pm (t_{0.5})(S\bar{p}), \ (t \text{ has } (\text{n} - 1) = 9\,\text{df})$$

$$= 80.35 \pm 2.262\ (1.000) = 78.1 \text{ to } 82.6$$

Transformations

The above method of computing confidence limits assumes that the individual percentages follow something close to a normal distribution with homogenous variance (i.e., same variance regardless of the size of the percent). If the clusters are small (say less than 100 individuals per cluster) or some of the percentages are greater than 80 or less than 20, the assumptions may not be valid and the computed confidence limits will be unreliable.

In such cases it may be desirable to compute the transformation

$$y = \text{arc sine}\sqrt{\text{percent}}$$

and to analyze the transformed variable.

CHI-SQUARE TESTS

Test of Independence

Individuals are often classified according to two (or more) distinct systems. A tree can be classified as to species, and at the same time, according to whether or not it is infected with some disease. A plot can be classified as to whether or not it is stocked with adequate reproduction and whether it is shaded or not shaded. Given such a cross-classification, it may be desirable to know whether the classification of an individual according to one system is independent of its classification by the other system. In the species-infection classification, for example, independence of species and infection would be interpreted to mean that there is no difference in infection rate between species (i.e., infection rate does not depend on species).

The hypothesis that two or more systems of classification are independent can be tested by chi-square. The procedure can be illustrated by a test of three termite repellents. A batch of 1,500 wooden stakes was divided at random into three groups of 500 each, and each group received a different termite-repellent treatment. The treated stakes were driven into the ground, with the treatment at any particular stake location being selected at random. Two years later the stakes were examined for termites. The number of stakes in each classification is shown in the following 2 by 3 (two rows and three columns) contingency table:

	Group I	Group II	Group III	Subtotals
Attacked by termites	193	148	210	551
Not attacked	307	352	390	949
Subtotals	500	500	500	1,500

The data in the table is symbolized as shown below:

	I	II	III	
Attacked	a_1	a_2	a_3	A
Not attacked	b_1	b_2	b_3	B
	T_1	T_2	T_3	G

the test of independence is made by computing

$$x^2 = \frac{1}{(A)(B)} \sum_{t=1}^{3} \left(\frac{[a_i B - b_i A]^2}{T_i} \right)$$

$$= \frac{1}{(551)(949)} \left[\frac{(193)(949)\quad(307)(551)^2}{500} + \cdots + \frac{(210)(949)\quad(290)(551)^2}{500} \right]$$

$$= 17.66$$

The result is compared to the appropriate tabular accumulative distribution of chi-square value of x^2 with $(c - 1)$ degrees of freedom, where c is the number of columns in the table of data. If the computed value exceeds the tabular value given in the 0.05 column, the difference among treatments is said to be significant at the 0.05 level (i.e., we reject the hypothesis that attack classification is independent of termite-repellent treatment) (Spellman, 2020; Freese, 1969).

For illustrative purposes, in this example, we say that the computed value of 17.66 (2 degrees of freedom) exceeds the tabular value in the 0.01 column, and so the difference in rate of attack among treatments is said to be significant at the 1-percent level. Examination of the data suggests that this is primarily due to the lower rate of attack on the Group II stakes.

Test of a Hypothesized Count

A geneticist hypothesized that, if a certain cross were made, the progeny would be of four types, in the proportions

$$A = 0.48 \quad B = 0.32 \quad C = 0.12 \quad D = 0.08$$

The actual segregation of 1,225 progeny is shown below, along with the numbers expected according to the hypothesis.

Type	***A***	***B***	***C***	***D***	**Total**
Number (X_i)	542	401	164	118	1,225
Expected (m_i)	588	392	147	98	1,225

As the observed counts differ from those expected, we might wonder if the hypothesis is false. Or can departures as large as this occur strictly by chance?

The chi-square test is

$$X^2 = \sum_{i=1}^{k} \left(\frac{(X_i - m_i)^2}{m_i} \right), \text{ with } (k-1) \text{ degrees of freedom}$$

where

k = number of groups recognized
X_i = observed count for the ith
M_i = count expected in the ith group if the hypothesis is true

For the above data,

$$X^2_{3\,\text{df}} = \frac{(542-588)^2}{588} + \frac{(401-392)^2}{392} + \frac{(164-147)^2}{147} + \frac{(118-98)^2}{98} = 9.85$$

This value exceeds the tabular x^2 with 3 degrees of freedom at the 0.05 level (i.e., it is greater than 7.81). Hence the hypothesis would be rejected (if the geneticist believed in testing at the 0.05 level).

BARTLETT'S TEST OF HOMOGENEITY OF VARIANCE

Many of the statistical methods described later are valid only if the variance is homogenous (i.e., variance within each of the populations is equal). The t test of the following section assumes that the variance is the same for each group, and so does the analysis of variance. The fitting of an unweighted regression as described in the last section also assumes that the dependent variable has the same degree of variability (variance) for all levels of the independent variables.

Bartlett's test offers a means of evaluating this assumption. Suppose that we have taken random samples in each of four groups and obtained variances (s^2) of 84.2, 63.8, 88.6, and 72.1 based on samples of 9, 21, 5, and 11 units, respectively. We would like to know if these variances could have come from populations all having the same variance. The quantities needed for Bartlett's test are tabulated here:

Group	Variance (s^2)	(n−1)	Corrected sum of squares SS	1/n−1	log s^2	(n−1) (logs^2)
1	84.2	8	673.6	0.125	1.92531	15.40248
2	63.8	20	1,276.0	0.050	1.80482	36.09640
3	88.6	5	443.0	0.200	1.94743	9.73715
4	72.1	10	721.0	0.100	1.85794	18.57940
k = 4 groups		Sums 43	3,113.6	0.475		79.81543

where k = number of groups (= 4).

$$\text{SS} = \text{The corrected sum of squares} = \left(\Sigma X^2 - \frac{(\Sigma X)^2}{n}\right) = (n-1)\,s^2$$

From this we compute the pooled within-group variance

$$\bar{s}^2 = \frac{\Sigma SS_i}{\Sigma(n-1)} = \frac{3113.6}{43} = 72.4093$$

and

$$\log \bar{s}^2 = 1.85979$$

Then the test for homogeneity is

$$X^2_{(k-1)\,\text{df}} = (2.3026^*)[(1.85979)(43) - 79.81543]$$

$$= 0.358$$

This value of x^2 is now compared with the value of x^2 in an accumulative distribution of chi-square for the desired probability level. Note that a value greater than that given in the table would lead us to reject the homogeneity assumption.

****Note***: The original form of this equation used natural logarithms in place of the common logarithms shown here. The natural log of any number is approximately 2.3026 times its common log—hence the constant of 2.3026 in the equation. In computations, common logarithms are usually more convenient than natural logarithms.

The x^2 value given by the above equation is biased upward. If x^2 is nonsignificant, the bias is not important. However, if the computed x^2 is just a little above the threshold value for significance, a correction for bias should be applied. The correction is:

$$C = \frac{3(k-1) + \left[\Sigma\left(\frac{1}{N_i - 1}\right) - \frac{1}{\Sigma(n_i - 1)}\right]}{3(k-1)}$$

$$= \frac{3(4-1) + (0.475 - 1/43)}{3(4-1)}$$

$$= 1.0502$$

The corrected value of x^2 is then

$$\text{Corrected } x^2 = \frac{\text{Uncorrected } x^2}{C} = \frac{0.358}{1.0502} = 0.341$$

COMPARING TWO GROUPS BY THE *T* TEST

THE *T* TEST FOR UNPAIRED PLOTS

An individual unit in a population may be characterized in a number of different ways. A single tree, for example, can be described as alive or dead, hardwood or softwood, infected or uninfected, and so forth. When dealing with observations of this type, we usually want to estimate the proportion of a population having a certain attribute. Or, if there are two or more different groups, we will often be interested in testing whether or not the groups differ in the proportions of individuals having the specified attribute. Some methods of handling these problems have been discussed in previous sections.

Alternatively, we might describe a tree by a measurement of some characteristic, such as its diameter, height, or cubic volume. For this measurement type of observation, we may wish to estimate the mean for a group as discussed in the section on sampling for measurement variables. If there are two or more groups, we will frequently want to test whether or not the group means are different. Often the groups will represent types of treatment which we wish to compare. Under certain conditions, the t or F tests may be used for this purpose.

Both of these tests have a wide variety of applications. For the present we will confine our attention to tests of the hypothesis that there is no difference between treatment (or group) means. The computational routine depends on how the observations have been selected or arranged. The first illustration of a t test of the hypothesis that there is no difference between the means of two treatments assumes that the treatments have been assigned to the experimental units completely at random. Except for the fact that there are usually (but not necessarily) an equal number of units or "plots" for each treatment, there is no restriction on the random assignment of treatments.

DID YOU KNOW?

According to Ernst Mayr (1970; 2002), *races* are distinct generally divergent populations within the same species with relatively small morphological and genetic differences. The populations can be described as ecological races if they arise from adaptation to different local habitats or geographic races when they are geographically isolated. If sufficiently different, two or more races can be identified as subspecies, which is an official biological taxonomy unit subordinate to species. If not, they are denoted as races, which means that a formal rank should not be given to the group, or taxonomists are unsure whether or not a formal rank should be given. Again, according to Mayr, "a subspecies is a geographical race that is sufficiently different taxonomically to be worthy of a separate name" (pp 89–94).

In this example the "treatments" were two races of white pine which were to be compared on the basis of their volume production over a specified period of time. Twenty-two square one-acre plots were staked out for the study. Eleven of these were selected entirely at random and planted with seedlings of race A. The remaining eleven were planted with seedlings of race B. After the prescribed time period the pulpwood volume (in cords—a stack of wood 4 ft wide by 4 ft high by 8 ft in length) was determined for each plot. The results were as follows:

Race *A*			Race *B*		
11	5	9	9		69
8	10	11	9	13	8
10	8	11	6		56
8	8		10	7	
Sum = 90			Sum = 88		
Average = 9.0			Average = 8.0		

To test the hypothesis that there is no difference between the race means (sometimes referred to as a null hypothesis—general or default position) we compute

$$t = \frac{\bar{X}_A - \bar{X}_B}{\sqrt{\dfrac{s^2(n_A + n_B)}{(n_A)(n_B)}}}$$

where

X_A and X_B = arithmetic means for groups *A* and *B*.

n_A and n_B = *The* number of observations in groups *A* and *B* (n_A and n_B do not have to be the same).

s^2 = pooled within-group variance (calculation shown below).

To compute the pooled within-group variance, we first get the corrected sum of squares (SS) within the group.

$$SS_A = \Sigma X^2{}_A \frac{(\Sigma X_A)^2}{n_A} = 11^2 + 8^2 + \cdots + 11^2 - \frac{(99)^2}{11} = 34$$

$$SS_B = \Sigma X^2{}_B \frac{(\Sigma X_B)^2}{n_B} = 9^2 + 9^2 + \cdots + 6^2 - \frac{(88)^2}{11} = 54$$

Then the pooled variance is

$$s^2 = \frac{SS_A + SS_B}{(n_A - 1) + (n_B - 1)} = \frac{88}{20} = 4.4$$

Hence,

$$t = \frac{9.0 - 8.0}{\sqrt{4.4\left(\dfrac{11+11}{(11)(11)}\right)}} = \frac{1.0}{\sqrt{.800000}} = 1.118$$

This value of t has $(n_A-1) + (n_B-1)$ degrees of freedom. If it exceeds the tabular value (from a distribution of t table) at a specified probability level, we would reject the hypothesis. The difference between the two means would be considered significant (larger than would be expected by chance if there is actually no difference).

In this case, tabular t with 20 degrees of freedom at the 0.05 level is 2.086. Since our sample value is less than this, the difference is not significant at the 0.05 levels.

Requirements—One of the unfortunate aspects of the t test and other statistical methods is that almost any kind of numbers can be plugged into the equations. But if the numbers and methods of obtaining them do not meet certain requirements, the result may be a fancy statistical facade with nothing behind it. In a handbook of this scope it is not possible to make the reader aware of all of the niceties of statistical usage, but a few words of warning are certainly appropriate.

A fundamental requirement in the use of most statistical methods is that the experimental material be a random sample of the population to which the conclusions are to be applied. In the t test of white pine races, the plots should be a sample of the sites on which the pine are to be grown, and the planted seedlings should be a random sample representing the particular race. A test conducted in one corner of an experimental forest may yield conclusions that are valid only for that particular area or sites that are about the same. Similarly, if the seedlings of a particular race are the progeny of a small number of parents, their performance may be representative of those parents only, rather than of the race.

In addition to assuming that the observations for a given race are a valid sample of the population of possible observations, the t test described above assumes that the population of such observations follows the normal distribution. With only a few observations, it is usually impossible to determine whether or not this assumption has been met. Special studies can be made to check on the distribution, but often the question is left to the judgment and knowledge of the research worker.

Finally, the t test of unpaired plots assumes that each group (or treatment) has the same population variance. Since it is possible to compute a sample variance for each group, this assumption can be checked with Bartlett's test for homogeneity of variance. Most statistical textbooks present variations of the t test that may be used if the group variances are unequal.

Sample Size

If there is a real difference of D feet between the two races of white pine, how many replicates (plots) would be needed to show that it is significant? To answer this, we first assume that the number of replicates will be the same for each group ($n_A = n_B = n$). The equation for t can then be written

$$t = \frac{D}{\sqrt{\frac{2s^2}{n}}} \quad \text{or} \quad n = \frac{2t^2s^2}{D^2}$$

Next, we need an estimate of the within-group variance, s^2. As usual, this must be determined from previous experiments, or by a special study of the populations.

Example 6.1

PROBLEM:

Suppose that we plan to test at the 0.05 level and wish to detect a true difference of $D = 1$ cord if it exists. From previous tests we estimate $s^2 = 5.0$. Thus, we have

$$n = \frac{2t^2s^2}{D^2} 2t^2\left(\frac{5.0}{1.0}\right)$$

Here we hit a snag. In order to estimate n we need a value for t, but the value of t depends on the number of degrees of freedom, which depends on n. The situation calls for an iterative solution—a mathematical procedure that generates a sequence of improving approximate solutions for a class of problems, in other words, a fancy name for trial and error. We start with a guessed value for n, say $n_o = 20$. As t has $(n_A - 1) + (n_B - 1) = 2\ (n - 1)$ degrees of freedom, we'll use $t = 2.025(= t_{.05}$ with 38 df) and compute

$$n_1 = 2(2.025)^2\left(\frac{5.0}{1.0}\right) = 41$$

The proper value of n will be somewhere between n_o and n_1—much closer to n_1 than to n_o. We can now make a second guess at n and repeat the process. If we try $n_2 = 38$, t will have $2(n - 1) = 74$ df and $t_{.05} = 1.992$. Hence,

$$n_2 = 2(1.992)^2\left(\frac{5.0}{1.0}\right) = 39.7$$

Thus, n appears to be over 39 and we will use $n = 40$ plots for each group or a total of 80 plots.

The *t* Test for Paired Plots

A second test was made of the two races of white pine. It also had 11 replicates of each race, but instead of the two races being assigned completely at random over the 22 plots, the plots were grouped into 11 pairs and a different race was randomly assigned to each member of a pair. The cordwood volumes at the end of the growth period were

Plot pair	1	2	3	4	5	6	7	8	9	10	11	Sum	Mean
Race *A*	12	8	8	11	10	9	11	11	13	10	7	110	10.0
Race *B*	10	7	8	9	11	6	10	11	10	8	9	99	9.0
$D_i = A_i - B_i$	2	1	0	2	−1	3	1	0	3	2	−2	11	1.0

As before, we wish to test the hypothesis that there is no real difference between the race means.

The value of *t* when the plots have been paired is

$$t = \frac{\bar{X}_A - \bar{X}_B}{\sqrt{\frac{s^2 d}{n}}} = \frac{\bar{d}}{\sqrt{\frac{s^2}{d}}}, \text{ with } (n-1) \text{ degress of freedom}$$

where

n = number of pairs of plots

$s^2{}_{d\,=}$ variance of the individual differences between *A* and *B*

$$s^2 d = \frac{\Sigma d^2{}_i - \frac{(\Sigma d_i)^2}{n}}{n-1} = \frac{2^2 + 1^2 + \cdots + (-2)^2 - \frac{11^2}{11}}{10}$$

$$= 2.6$$

So, in this example we find

$$t_{10} dt = \frac{10.0 - 9.0}{\sqrt{2.6/11}} = 2.057$$

When this value of 2.057 is compared to the tabular value of *t* in a distribution of *t* table ($t_{0.05}$ with 10df = 2.228), we find that the difference is not significant at the 0.05 level. That is, a sample means difference of 1 cord or more could have occurred by chance more than 1 time in 20, even if there is no real difference between the race means. Usually such an outcome is not regarded as sufficiently strong evidence to reject the hypothesis.

> The method of paired observations is a useful technique. Compared with the standard two-sample *t* test, in addition to the advantage that we do not have to assume that the two samples are independent, we also need not assume that the variances of the two samples are equal.
>
> **(Hamburg, 1987, p. 304)**

Moreover, the paired test will be more sensitive (capable of detecting smaller real differences) than the unpaired test whenever the experimental units (plots in this case) can be grouped into pairs such that the variation between pairs is appreciably larger than the variation within pairs. The basis for pairing plots may be geographic proximity or similarity in any other characteristic that is expected to affect the performance of the plot. In animal-husbandry studies, litter mates are often paired, and where patches of human skin are the plots, the left and right arms may constitute the pair. If the experimental units are very homogenous, there may be no advantage in pairing.

Number of Replicates

The number (n) of plot pairs needed to detect a true mean difference of size D is

$$n = \frac{t^2 s_d^{\ 2}}{D^2}$$

COMPARISON OF GROUPS BY ANALYSIS OF VARIANCE

Complete Randomization

A planter wanted to compare the effects of five site-preparation treatments on the early height growth of planted pine seedlings. He laid out 25 plots and applied each treatment to 5 randomly selected plots. The plots were then hand-planted and at the end of five years the height for all pines was measured and an average height computed for each plot. The plot averages (in feet) were as follows:

	Treatments					
	A	*B*	*C*	*D*	*E*	
	15	16	13	11	14	
	14	14	12	13	12	
	12	13	11	10	12	
	13	15	12	12	10	
	13	14	10	11	11	
Sum	67	72	58	57	59	313
Treatment Means	13.4	14.4	11.6	11.4	11.8	12.52

Looking at the data we see that there are differences among the treatment means: *A* and *B* have higher averages than *C*, *D*, and *E*. Soils and planting stock are seldom completely uniform, however, and so we would expect some differences even if every plot had been given exactly the same site-preparation treatment. The question is, can differences as large as this occur strictly by chance if there is actually no difference among treatments? If we decide that the observed differences are larger than might be expected to occur strictly by chance, the inference is that the treatment means are not equal. Statistically speaking, we reject the hypothesis of no difference among treatment means.

Problems like this are neatly handled by an analysis of variance. To make this analysis, we need to fill in a table like the following:

Source of variation	Degrees of freedom	Sums of squares	Mean squares
Treatments	4		
Error	20		
Total	24		

Source of variation—There are a number of reasons why the height growth of these 25 plots might vary, but only one can be definitely identified and evaluated—those attributable to treatments. The unidentified variation is assumed to represent the variation inherent in the experimental material and is labeled error. Thus, total variation is being divided into two parts: one part attributable to treatments, and the other unidentified and called error.

Degrees of freedom—Degrees of freedom are hard to explain in non-statistical language. In the simpler analyses of variance, however, they are not difficult to determine. For the total, the degrees of freedom are one less than the number of observations: there are 25 plots, so the total has 24 df's. For the sources, other than error, the df's are one less than the number of classes or groups recognized in the source. Thus, in the source labeled treatments there are five groups (five treatments), so there will be four degrees of freedom for treatments. The remaining degrees of freedom ((24 – 4) = 20) are associated with the error term.

Sums of squares—There is a sum of squares associated with every source of variation. These SS are easily calculated in the following steps.

First, we need what is known as a "correction term" or C.T. This is simply

$$\text{C.T.} = \frac{\left(\sum^{n} X\right)^2}{n} = \frac{313^2}{25} = 3918.76$$

where $\sum^{n}$ = the sum of n items

Then the total sum of squares is

$$\text{Total } \underset{24\text{ df}}{\text{SS}} = \sum^{n} X^2 - \text{C.T.} = (15^2 + 14^2 + \cdots + 11^2) - \text{C.T.} = 64.24$$

The sum of squares attributable to treatments is

$$\underset{4\text{ df}}{\text{Treatment SS}} = \frac{\sum^{5}(\text{treatment totals}^2)}{\text{No. of plots per treatment}} - \text{C.T.}$$

$$= \frac{67^2 + 72^2 + \cdots + 59^2}{5} - \text{C.T.} = \frac{19767}{5} - \text{C.T.} = 34.64$$

Note that in both SS calculations, the number of items squared and added was one more than the number of degrees of freedom associated with the sum of squares. The number of degrees of freedom just below the SS and the numbers of items to be squared and added over the n value, provided a partial check as to whether the proper totals are being used in the calculation—the degrees of freedom must be one less than the number of items.

Note also that the divisor in the treatment SS calculation is equal to the number of individual items that go to make up each of the totals being squared in the numerator. This was also true in the calculation of total SS, but there the divisor was 1 and hence did not have to be shown. Note further that the divisor times the number over the summation sign ($5 \times 5 = 25$ for treatments) must always be equal to the total number of observations in the test—another check.

The sum of squares for error is obtained by subtracting the treatment SS from total SS. A good habit to get into when obtaining sums of squares by subtraction is to perform the same subtraction using df's. In the more complex designs, doing this provides a partial check on whether the right items are being used.

Mean squares—The mean squares are now calculated by dividing the sums of squares by the associated degrees of freedom. It is not necessary to calculate the mean square for the total.

The items that have been calculated are entered directly into the analysis table, which at the present stage would look like this:

Source	Df	SS	MS (mean square)
Treatment	4	34.64	8.66
Error	20	29.60	1.48
Total	24	64.24	

An F test of treatments (used to reject the null hypothesis) is now made by dividing the MS for treatments by the MS for error. In this case

$$F = \frac{8.66}{1.48} = 5.851$$

Fortunately, critical values of the F ratio have been tabulated for frequently used significance levels analogous to the x^2 distribution. Thus, the result, 5.851, is compared to the appropriate value of F in the table. The tabular F for significance at the 0.05 level is 2.87 and that for the 0.01 level is 4.43. As the calculated value of F exceeds 4.43, we conclude that the difference in height growth between treatments is significant at the 0.01 level. (More precisely, we reject the hypothesis that there is no difference in mean height growth between the treatments.) If F had been smaller than 4.43 but larger than 2.87, we would have said that the difference is significant at the 0.05 level. If F had been less than 2.87, we would have said that the difference between treatments is not significant at the 0.05 level. The researcher should select his own level of significance (preferably in advance of the study), keeping in mind that significance at the a (alpha) level (for example) means this: if there is actually no difference among treatments, the probability of getting chance differences as large as those observed is a or less.

The t test versus the analysis of variance—If only two treatments are compared, the analysis of variance of a completely randomized design and the t test of unpaired plots lead to the same conclusion. The choice of test is strictly one of personal preference, as may be verified by applying the analysis of variance to the data used to illustrate the t test of unpaired plots. The resulting F value will be equal to the square of the value of t that was obtained (i.e., $F = t^2$).

Like the t test, the F test is valid only if the variable observed is normally distributed and if all groups have the same variance.

Multiple Comparisons

In the example illustrating the completely randomized design, the difference among treatments was found to be significant at the 0.01 probability level. This is interesting as far as it goes, but usually we will want to take a closer look at the data, making comparisons among various combinations of the treatments.

Suppose, for example, that A and B involved some mechanical form of site preparation while C, D, and E were chemical treatments. Then we might want to test whether the average of A and B together differed from the combined average of C, D, and E. Or we might wish to test whether A and B differ significantly from each other. When the number of replications (n) is the same for all treatments, such comparisons are fairly easy to define and test.

The question of whether the average of treatments A and B differs significantly from the average of treatments C, D, and E is equivalent to testing whether the linear contrast

$$\hat{Q} = (3\bar{A} + 3\bar{B}) - (2\bar{C} + 2\bar{D} + 2\bar{E})$$

differs significantly from zero ($\bar{A}$ = the mean for treatment A, etc.). Note that the coefficients of this contrast sum to zero (3 + 3 − 2 − 2 − 2 − 0) and are selected so as to put the two means in the first group on an equal basis with the three means in the second group.

F Test with Single Degree of Freedom

A comparison specified in advance of the study (on logical grounds and before examination of the data) can be tested by an F test with single degree of freedom. For the linear contrast

$$\hat{Q} = a_1 \bar{X}_1 + a_2 \bar{X}_2 + a_3 \bar{X}_3 + \cdots$$

among means based on the same number (n) of observations, the sum of squares has one degree of freedom and is computed as

$$\underset{1\,\text{df}}{\text{SS}} = \frac{n\hat{Q}^2}{\Sigma a^2_i}$$

This sum of squares divided by the mean square for error provides an F test of the comparison.

Thus, in testing A and B versus C, D, and E we have

$$\hat{Q} = 3(13.4) + 3(14.4) - 2(11.6) - 2(11.4) - 2(11.8) = 13.8$$

and

$$\underset{1\,\text{df}}{\text{SS}} = \frac{5(13.8)^2}{3^2 + 3^2 + (-2)^2 + (-2)^2 + (-2)^2} = \frac{952.20}{30} = 31.74$$

Then dividing by the error mean square gives the F value for testing the contrast.

$$F = \frac{31.74}{1.48} = 21.446 \text{ with 1 and 20 degrees of freedom}$$

This exceeds the tabular value of F (4.35) at the 0.05 probability level. If this is the level at which we decided to test, we would reject the hypothesis that the mean of treatments A and B does not differ from the mean of treatments C, D, and E.

If Q is expressed in terms of the treatment totals rather than their means so that

$$\hat{Q}_T = a_1\left(\Sigma X_1\right) + a_2\left(\Sigma X_2\right) + \cdots$$

then the equation for the single degree of freedom sum of squares is

$$\underset{1\,\text{df}}{\text{SS}} = \frac{\hat{Q}_T^2}{n\Sigma a^2_i}$$

The results will be the same as those obtained with the means. For the test of A and B versus C, D, and E,

$$\hat{Q}_T = 3(67) + 3(72) - 2(58) - 2(57) - 2(59) = 69$$

And,

$$\underset{1\text{ df}}{\text{SS}} = \frac{69^2}{5[3^2+3^2+(-2)^2+(-2)^2+(-2)^2]} = \frac{4761}{150} = 37.74\text{, as before.}$$

Working with the totals saves the labor of computing means and avoids possible rounding errors.

Scheffe's Test

Quite often we will want to test comparisons that were not anticipated before the data were collected. If the test of treatments was significant, such unplanned comparisons can be tested by the method of Scheffe, or Scheffe's test. Named after the American statistician Henry Scheffe, the Scheffe test adjusts significant levels in a linear regress analysis to account for multiple comparisons. It is particularly useful in analysis of variance, and in constructing simultaneous bands for regressions involving basic functions. When there are n replications of each treatment, k degrees of freedom for treatment, and v degrees of freedom forever, any linear contrast among the treatment means

$$\hat{Q} = a_1\bar{X}_1 + a_2\bar{X}_2 \ldots$$

is tested by computing

$$F = \frac{nQ^2}{K(\Sigma a^2{}_i)\text{ (Error mean square)}}$$

This value is then compared to the tabular value of F with k and v degrees of freedom.

For example, to test treatment B against the means of treatments C and E we would have

$$\hat{Q}[2\bar{B} - (\bar{C} + \bar{E})] = [2(14.4) - 11.6 - 11.8] = 5.4$$

And

$$F = \frac{5(5.4)^2}{(4)[2^2+(-1)+(-1)^2](1.48)} = 4.105\text{, with 4 and 20 degrees of freedom}$$

This figure is larger than the tabular value of F (= 2.87), and so in testing at the 0.05 level we would reject the hypothesis that the mean for treatment B did not differ from the combined average of treatments C and E. For a contrast (Q_T) expressed in terms of treatment totals, the equation for F becomes

$$F = \frac{\hat{Q}_T^2}{nk(\Sigma a^2{}_i)\text{ (Error mean square)}}$$

Unequal Replication

If the number of replications is not the same for all treatments, then for the linear contrast

$$\hat{Q} = a_1 X_1 + a_2 \bar{X}_2 + \cdots$$

The sum of squares in the single degree of freedom F test is given by

$$SS = \frac{\hat{Q}^2}{(k)\left(\frac{a^2_i}{n_1} + \frac{a^2_2}{n_2} + \cdots\right)}$$

where n_i = the number of replications on which $\bar{X}_i$ is based.

With unequal replication, the F value in Scheffe's test is computed by the equation

$$F = \frac{\hat{Q}^2}{(k)\left(\frac{a^2_1}{n_1} + \frac{a^2_2}{n_2} + \cdots\right)(\text{Error mean square})}$$

Selecting the coefficients (a_i) for such contrasts can be tricky. When testing the hypothesis that there is no difference between the means of two groups of treatments, the positive coefficients are usually

$$\text{positive } a_i = \frac{n_i}{p}$$

where p = total number of plots in the group of treatments with positive coefficients.

The negative coefficients are

$$\text{negative } a_j = \frac{n_i}{m}$$

where m = total number of plots in the group of treatments with negative coefficients.

To illustrate, if we wish to compare the mean of treatments A, B, and C with the mean of treatments D and E and there are two plots of treatment A, three of B, five of C, three of D, and two of E, then p = 2 + 3 + 5 = 10, m = 3 + 2 = 5 and the contrast would be

$$\hat{Q} = \left(\frac{2}{10}\bar{A} + \frac{3}{10}\bar{B} + \frac{5}{10}\bar{C}\right) - \left(\frac{3}{5}\bar{D} + \frac{2}{5}\bar{E}\right)$$

RANDOMIZED BLOCK DESIGN

There are two basic types of the two-factor analysis of variance: the *completely randomized design* (discussed in the previous section) and the *randomized block design*. In the completely randomized design, the error mean square is a measure of the variation among plots treated alike. It is in fact an average of the within-treatment variances, as may easily be verified by computation. If there is considerable variation among plots treated alike, the error mean square will be large and the *F* test for a given set of treatments is less likely to be significant. Only large differences among treatments will be detected as real and the experiment is said to be insensitive.

DID YOU KNOW?

The term "block" derives from experimental design work in agriculture, in which parcels of land are referred to as blocks. In a randomized block design, treatments are randomly assigned to units within each block. In testing the yield of different fertilizers, for example, this design ensures that the best fertilizer is applied to all types of soil, not just the best soil (Hamburg, 1987).

Often the error can be reduced (thus giving a more sensitive test) by use of a randomized block design in place of complete randomization. In this design, similar plots or plots that are close together are grouped into blocks. Usually, the number of plots in each block is the same as the number of treatments to be compared, though there are variations having two or more plots per treatment in each block. The blocks are recognized as a source of variation that is isolated in the analysis. A general rule in randomized block design is to "block what you can, randomize what you can't." In other words, blocking is used to remove the effects of nuisance variables or factors. Nuisance factors are those that may affect the measured result but are not of primary interest. For example, in applying a treatment, nuisance factors might be the time of day the experiment was run, the room temperature, or might be the specific operator who prepared the treatment (Addelman, 1969; 1970).

As an example, a randomized block design with five blocks was used to test the height growth of cottonwood cuttings from four selected parent trees. The field layout looked like this:

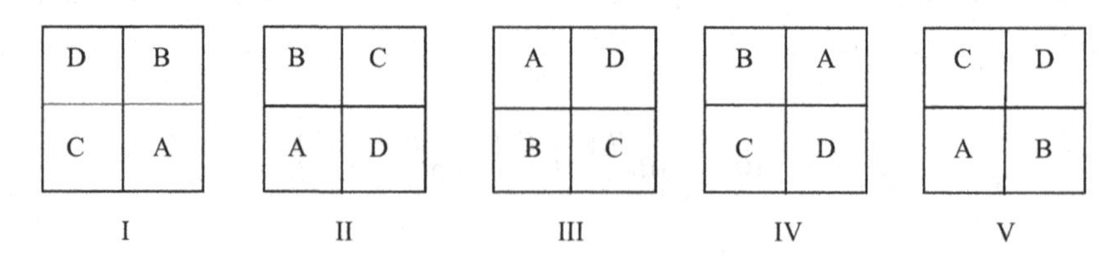

Each plot consisted of a planting of 100 cuttings of the clone assigned to that plot. When the trees were 5 years old the heights of all survivors were measured, and an average computed for each plot.

The plot averages (in feet) by clones and blocks are summarized below:

	Close				Block
Block	*A*	*B*	*C*	*D*	**total**
I	18	14	12	16	60
II	15	15	16	13	59
III	16	15	8	15	54
IV	14	12	10	12	48
V	12	14	9	14	49
Clone Totals	75	70	55	70	270
Clone Means	15	14	11	14	

The hypothesis to be tested is that clones do not differ in mean height.

In this design there are two identifiable sources of variation—that attributable to clones and that associated with blocks. The remaining portion of the total variation is used as a measure of experimental error. The outline of the analysis is therefore as follows:

Source of variation	df	Sums of squares	Mean squares
Blocks	4		
Clones	3		
Error	12		
Total	19		

The breakdown in degrees of freedom and computation of the various sums of squares follow the same pattern as in the completely randomized design. Total degrees of freedom (19) are one less than the total number of plots. Degrees of freedom for clones (3) are one less than the number of clones. With five blocks, there will be four degrees of freedom for blocks. The remaining 12 degrees of freedom are associated with the error term.

Sums-of-squares calculations proceed as follows:

1. The correction term

$$\text{C.T.} = \frac{\left(\sum^{20} X\right)^2}{N} = \frac{270^2}{20} = 3{,}645$$

2. $$\underset{19\text{ df}}{\text{Total SS}} = \sum^{20} X^2 - \text{C.T.} = (18^2 + 15^2 + \cdots + 14^2) - \text{C.T.}$$

$$= 3{,}766 - 3{,}645 = 121$$

3. $$\underset{3\text{ df}}{\text{Clone SS}} = \frac{\sum^{4}(\text{Clone totals}^2)}{\text{No. of plots per clone}} - \text{C.T.}$$

$$= \frac{75^2 + 70^2 + 55^2 + 70^2}{5} - \text{C.T.}$$

$$= 3{,}690 - 3{,}645 = 45$$

4. $$\underset{4\text{ df}}{\text{Block SS}} = \frac{\sum^{5}(\text{Block totals}^2)}{\text{No. of plots per block}} - \text{C.T.}$$

$$= \frac{60^2 + 59^2 + \cdots + 49^2}{4} - \text{C.T.}$$

$$= 3{,}675.5 - 3{,}645 = 30.5$$

5. $$\underset{12\text{ df}}{\text{Error SS}} = \underset{19\text{ df}}{\text{Total SS}} - \underset{3\text{ df}}{\text{Clone SS}} - \underset{4\text{ df}}{\text{Block SS}} = 45.5$$

Note that in obtaining the error SS by subtraction, we get a partial check on ourselves by subtracting clone and block df's from the total df to see if we come out with the correct number of error df. If these don't check, we have probably used the wrong sums of squares in the subtraction.

Mean squares are again calculated by dividing the sums of squares by the associated number of degrees of freedom.

Tabulating the results of these computations

Source	df	SS	MS
Blocks	4	30.5	7.625
Clones	3	45.0	15.000
Error	12	45.5	3.792
Total	19	121.0	

The F for clones is obtained by dividing clone MS by error MS. In this case F = 15.000/3,792 = 3.956. As this is larger than the tabular F of 3.49 (obtained from a distribution of F table) ($F_{0.05}$ with 3 and 12 degrees of freedom) we conclude that the difference between clones is significant at the 0.05 level. The significance appears to be due largely to the low value of *C* as compared to *A*, *B*, and *D*.

Comparisons among clone means can be made by the methods previously described. For example, to test the prespecified (i.e., before examining the data) hypothesis that there is no difference between the mean of clone *C* and the combined average *A*, *B*, and *D* we would have:

$$\underset{1\ \text{df}}{\text{SS}} \text{ for } \left(\text{A}+\text{B}+\text{D vs. C}\right) = \frac{5(3\bar{C}-\bar{A}-\bar{B}-\bar{D})^2}{[3^2+(-1)^2+(-1)^2+(-1)^2]}$$

Then,

$$F = \frac{41.667}{3.792} = 10.988$$

Tabular *F* at the 0.01 level with 1 and 12 degrees of freedom is 9.33. As calculated *F* is greater than this, we conclude that the difference between *C* and the average of *A*, *B*, and *D* is significant at the 0.01 level.

The sum of squares for this single-degree-of-freedom comparison (41.667) is almost as large as that for clones (45.0) with three degrees of freedom. This result suggests that most of the clonal variation is attributable to the low value of *C*, and that comparisons between the other three means are not likely to be significant.

There is usually no reason for testing blocks, but the size of the block mean square relative to the mean square for error does give an indication of how much precision was gained by blocking. If the block mean square is large (at least two or three times as large as the error means square) the test is more sensitive than it would have been with complete randomization. If the block mean square is about equal to or only slightly larger than the error mean square, the use of blocks has not improved the precision of the test. The block mean square should not be appreciably smaller than the error mean square. If it is, the method of conducting the study and the computations should be re-examined.

Assumptions—In addition to the assumption of homogeneous variance and normality, the randomized block design assumes that there is no interaction between treatments and blocks, i.e., that differences among treatments are about the same in all blocks. Because of this assumption, it is not advisable to have blocks that differ greatly—since they may cause an interaction with treatments.

DID YOU KNOW?

With only two treatments, the analysis of variance of a randomized block design is equivalent to the t test of paired replicates. The value of *F* will be equal to the value of t^2 and the inferences derived from the tests will be the same. The choice of tests is a matter of personal preference.

Latin Square Design

In the randomized block design, the purpose of blocking is to isolate a recognizable extraneous source of variation. If successful, blocking reduces the error mean square and hence gives a more sensitive test than could be obtained by complete randomization.

In some situations, however, we have a two-way source of variation that cannot be isolated by blocks alone. In an agricultural field, for example, fertility gradients may exist both parallel to and at right angles to plowed rows. Simple blocking isolates only one of these sources of variation, leaving the other to swell the error term and reduce the sensitivity of the test.

When such a two-way source of extraneous variation is recognized or suspected, the Latin square design may be helpful. In this design, the total number of plots or experimental units is made equal to the square of the number of treatments. In forestry and agricultural experiments, the plots are often (but not always) arranged in rows and columns with each row and column having a number of poles equal to the number of treatments being tested. The rows represent different levels of one source of extraneous variation while the columns represent different levels of the other source of extraneous variation. Thus, before the assignment of treatments, the field layout of a Latin square for testing five treatments might look like this:

		COLUMNS				
		1	2	3	4	5
ROWS	1					
	2					
	3					
	4					
	5					

Treatments are assigned to plots at random, but with the very important restriction that a given treatment cannot appear more than once in any row or any column.

The Latin square design can be used whenever there is a two-way heterogeneity that cannot be controlled simply by blocking. In greenhouse studies, distance from a window could be treated as a row effect while distance from the blower or heater might be regarded as a column effect. Though the plots are often physically arranged in rows or columns, this is not required. In testing the use of materials in a manufacturing process where different machines and machine operators will be involved, the variation between machines could be treated as a row effect and the variation due to operator could be treated as a column effect.

The Latin square should not be used if an interaction between rows and treatments or columns and treatments is suspected.

Factorial Experiments

In environmental practice, knowledge that interactions between elements of the environment occur and an understanding of what their influence or impact on the environment is, or can be, is important. Consider a comparison of corn yields following three rates or levels of nitrogen fertilization indicating that the yields depended on how much phosphorus was used along with the nitrogen. The differences in yield were smaller when no phosphorus was used than when the nitrogen applications were accompanied by 100 pounds per acre of phosphorus. In statistics this situation is referred to as an interaction between nitrogen and phosphorus. Another example: when leaf litter was removed from the forest floor, the catch of pine seedlings was much greater than when the litter was not removed; but for red oak the reverse was true—the seedling catch was lower where litter was removed. Thus, species and litter treatment were interacting.

Interactions are important in the interpretation of study results. In the presence of an interaction between species and litter treatment it obviously makes no sense to talk about the effects of litter removal without specifying the species. The nitrogen–phosphorus interaction means that it may be misleading to recommend a level of nitrogen without mentioning the associated level of phosphorus.

Factorial experiments are aimed at evaluating known or suspected interactions. In these experiments, each factor to be studied is tested at several levels and each level of a factor is tested at all possible combinations of the levels of the other factors. In a planting test involving three species of trees and four methods of preplanting site preparation, each method will be applied to each species, and the total number of treatment combinations will be 12. In a factorial test of the effects of two nursery treatments on the survival of 4 species of pine planted by three different methods, there would be 24 ($2 \times 4 \times 3 = 24$) treatment combinations.

The Split-Plot Design

When two or more types of treatment are applied in factorial combinations, it may be that one type can be applied on relatively small plots while the other type is best applied to larger plots. Rather than make all plots of the size needed for the second type, a split-plot design can be employed. In this design, the major (large-plot) treatments are applied to a number of plots with replication accomplished through any of the common designs (such as complete randomization, randomized blocks, Latin Square). Each major plot is then split into a number of subplots, equal to the number of minor (small-plot) treatments. Minor treatments are assigned at random to subplots within each major plot.

Subplots can also be split—If desired, the subplots can also be split for a third level of treatment, producing a split-split-plot design. The calculations follow the same general pattern but are more involved. A split-split-plot design has three separate error terms.

Comparisons among means in a split-plot design—For comparisons among major- or subplot treatments, F tests with a single degree of freedom may be made in the usual manner. Comparisons among major-plot treatments should be tested against the major-plot error mean square, while subplot treatment comparisons are tested against the subplot error. In addition, it is sometimes desirable to compare the means of two treatment combinations. This can get tricky, for the variation among such means may contain more than one source of error. A few of the more common cases are discussed below.

In general, the t test for comparing two equally replicated treatment means is

$$t = \frac{\text{Mean difference}}{\text{Standard error or the mean difference}} = \frac{\bar{D}}{s\bar{D}}$$

1. For the difference between two major-treatment means:

   $$s\bar{D} = \frac{\sqrt{2\ (\text{Major-plot error MS})}}{(m)(R)}$$; t has df equal to the df for the major-plot error.

   where

   R = number of replications of major treatments.

   m = number of subplots per major plot.

2. For the difference between two minor-treatment means:

   $$s\bar{D} = \frac{\sqrt{2\ (\text{Subplot error MS})}}{(R)(M)}$$; t has df equal to the df for subplot error

   where M = number of major-plot treatments.

3. For the difference between two minor treatments within a single major treatment:

   $$s\bar{D} = \frac{\sqrt{2\ (\text{Subplot error MS})}}{R}$$; df for t = df for the subplot error

4. For the difference between the means of two major treatments at a single level of a minor treatment, or between the means of two major treatments at different levels of a minor treatment:

   $$s\bar{D} = \sqrt{2}\,\frac{(m-1)(\text{Subplot error MS}) + \text{Major-plot error MS}}{(m)(R)}$$

   In this case, t will not follow the t distribution. A close approximation to the value of t required for significance at a level is given by

   $$t = \frac{(m-1)(\text{Subplot error MS})t_m + (\text{Major-plot error MS})t_m}{(m-1)(\text{Subplot error MS}) + (\text{Major-plot error MS})}$$

where

t_m = tabular value of t at a level for df equal to the df for the subplot error.

t_M = tabular value of t at a level for df equal to the df for the major-plot error.

Other symbols are as previously defined.

Missing Plots

A mathematician who had developed a complex electronic computer program for analyzing a wide variety of experimental designs was asked how he handled missing plots. His disdainful reply was, "We tell our research workers not to have missing plots."

This is good advice. But it is sometimes hard to follow, and particularly so in forest, environmental, and ecological research, where close control over experimental material is difficult, and studies may run for several years.

The likelihood of plots being lost during the course of a study should be considered when selecting an experimental design. Lost plots are least troublesome in the simple designs. For this reason, complete randomization and randomized blocks may be preferable to the more intricate designs when missing data can be expected.

In the complete randomization design, loss of one or more plots causes no computational difficulties. The analysis is made as though the missing plots never existed. Of course, a degree of freedom will be lost from the total and error terms for each missing plot and the sensitivity of the test will be reduced. If missing plots are likely, the number of replications should be increased accordingly.

In the randomized block design, completion of the analysis will usually require an estimate of the values for the missing plots. A single missing value can be estimated by

$$X = \frac{bB + tT - G}{(b-1)(t-1)}$$

where

b = Number of blocks

t = Number of treatments

B = Total of all other units in the block with a missing plot

T = Total of all other units that received the same treatment as the missing plot

G = Total of all observed units

If more than one plot is missing, the customary procedure is to insert guessed values for all but one of the missing units, which is then estimated by the above formula. This estimate is used in obtaining an estimated value for one of the guessed plots, and so on through each missing unit. Then the process is repeated with the first estimates replacing the guessed values. The cycle should be repeated until the new approximations differ little from the previous estimates.

The estimated values are now applied in the usual analysis-of-variance calculations. For each missing unit, one degree of freedom is deducted from the total and from the error term.

A similar procedure is used with the Latin square design, but the formula for a missing plot is

$$X = \frac{R(R+C+T)-2G}{(r-1)(r-2)}$$

where

r = Number of rows
R = Total of all observed units in the row with the missing plot
C = Total of all observed units in the column with the missing plot
T = Total of all observed unit in the missing plot treatment
G = Grand total of all observed units

With the split-plot design, missing plots can cause trouble. A single missing subplot value can be estimated by the equation

$$X = \frac{rP + m(T_{ij}) - (T_{i})}{(r-1)(m-1)}$$

where

r = Number of replications of major-plot treatments
P = Total of all observed subplots in the major plot having a missing subplot
m = Number of subplot treatments
T_{ij} = Total of all subplots having the same treatment combination as the missing unit
$T_{i.}$ = Total of all subplots having the same major-plot treatment as the missing unit

For more than one missing subplot the iterative process described for randomized blocks must be used. In the analysis, one df will be deducted from the total and subplot error terms for each missing subplot.

When data for missing plots are estimated, the treatment mean square for all designs is biased upward. If the proportion of missing plots is small, the bias can usually be ignored. Where the proportion is large, adjustments can be made as described in the standard references on experimental designs.

REGRESSION

SIMPLE LINEAR REGRESSION

An environmental researcher had an idea that she could tell how well a loblolly pine was growing from the volume of the crown. Very simple: big crown—good growth, small crown—poor growth. But she couldn't say how big and how good, or how small and how poor. What she needed was regression analysis: it would enable her to express a relationship between tree growth and crown volume in an equation. Given a certain crown volume, she could use the equation to predict what the tree growth was.

To gather data, she ran parallel survey lines across a large tract that was representative of the area in which she was interested. The lines were 5 chains apart. At each 2-chain mark along the lines, she measured the nearest loblolly pine of at least 5.6 inches d.b.h. (diameter at breast height; i.e., 4.5 ft above the forest floor on the uphill side of the tree) for crown volume and basal area growth over the past 10 years.

A portion of the data is printed below to illustrate the methods of calculation. Crown volume in hundreds of cubic feet is labeled *X* and basal area growth in square feet is labeled *Y*. Now, what can we tell the environmental researcher about the relationship?

X Crown volume	*Y* Growth	*X* Crown volume	*Y* Growth	*X* Crown volume	*Y* Growth
22	0.36	53	0.47	51	0.41
6	0.09	70	0.55	75	0.66
93	0.67	5	0.07	6	0.18
62	0.44	90	0.69	20	0.21
84	0.72	46	0.42	36	0.29
14	0.24	36	0.39	50	0.56
52	0.33	14	0.09	9	0.13
69	0.61	60	0.54	2	0.10
104	0.66	103	0.74	21	0.18
100	0.80	43	0.64	17	0.17
41	0.47	22	0.50	87	0.63
85	0.60	75	0.39	97	0.66
90	0.51	29	0.30	33	0.18
27	0.14	76	0.61	20	0.06
18	0.32	20	0.29	96	0.58
48	0.21	29	0.38	61	0.42
37	0.54	30	0.53		
67	0.70	59	0.58		
56	0.67	70	0.62		
31	0.42	81	0.66		
17	0.39	93	0.69		
7	0.25	99	0.71		
2	0.06	14	0.14		
Totals				3,050	26.62
Means ($n = 62$)				49.1935	0.42935

Often, the first step is to plot the field data on coordinate paper (see Figure 6.1). This is done to provide some visual evidence of whether the two variables are related. If there is a simple relationship, the plotted points will tend to form a pattern (a straight line or curve). If the relationship is very strong, the pattern will generally be distinct. If the relationship is weak, the points will be more spread out and the

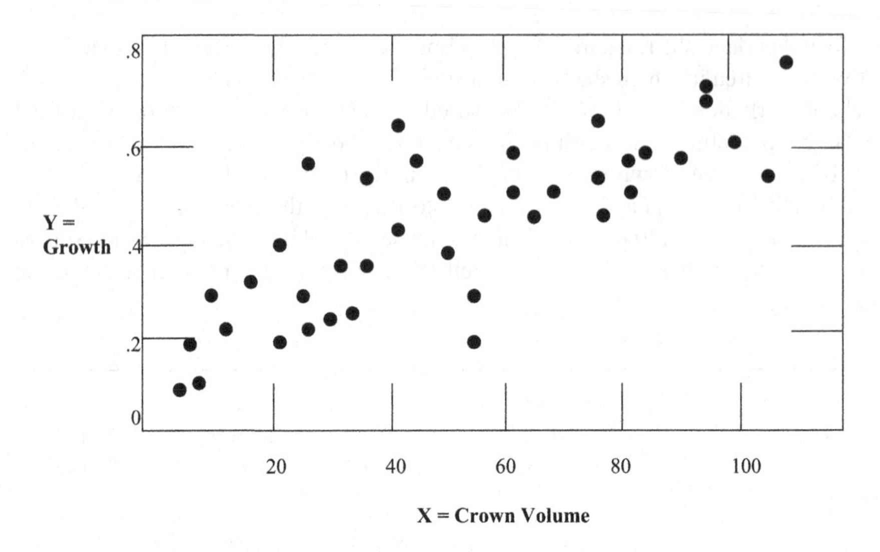

FIGURE 6.1 Plotting of growth *(F)* over crown volume *(X)*.

pattern less definite. If the points appear to fall pretty much at random, there may be no simple relationship or one that is so very poor as to make it a waste of time to fit any regression.

The type of pattern (straight line, parabolic curve, exponential curve, etc.) will influence the regression model to be fitted. In this particular case, we will assume a simple straight-line relationship.

After selecting the model to be fitted, the next step will be to calculate the corrected sums of squares and products. In the following equations, capital letters indicate uncorrected values of the variables; lower-case letters will be used for the corrected values ($y = Y - Y$).

$$\text{The corrected sum of squares for Y}: \Sigma y^2 = \sum^{n} Y^2 - \frac{\left(\sum^{n} y\right)^2}{n}$$

$$= \left(0.36^2 + 0.09^2\right) + \cdots + 0.42^2$$

$$- \frac{26.62^2}{62}$$

$$= 2.7826$$

$$\text{The corrected sum of squares for } X: \Sigma x^2 = \Sigma X^2 \frac{(\Sigma X)^2}{n}$$

$$= (22^2 + 6^2 + \cdots + 6^2) - \frac{3,050^2}{62}$$

$$= 59,397.6778$$

$$\text{The corrected sum of products}: \Sigma xy = \sum^{n} (XY) - \frac{\left(\sum^{n} X\right)\left(\sum^{n} Y\right)}{n}$$

$$= [(220)(.36) + (6)(.09) + \cdots + (61)(.42)] - \frac{(3,050)(26.62)}{62}$$

$$= 354.1477$$

The general form of equation for a straight line is $Y = a + bX$

In this equation, a and b are constants or regression coefficients that must be estimated. According to the principle of least squares, the best estimates of these coefficients are:

$$b = \frac{\Sigma xy}{\Sigma x^2} = \frac{354.1477}{59,397.6775} = 0.005962$$

$$a = \bar{Y} - b\,\bar{X} = 0.42935 \quad (0.005962)(49.1935) = 0.13606$$

Substituting these estimates in the general equation gives

$$\bar{Y} = 0.13606 + 0.005962X$$

Where $\hat{Y}$ is used to indicated that we are dealing with an estimated value of Y.

With this equation we can estimate the basal area growth for the past 10 years ($\hat{Y}$) from the measurements of the crown volume X.

Because Y is estimated from a known value of X, it is called the dependent variable and X the independent variable. In plotting on graph paper, the values of Y are usually (purely by convention) plotted along the vertical axis (ordinate) and the values of x along the horizontal axis (abscissa).

How Well Does the Regression Line Fit the Data?

A regression line can be thought of as a moving average. It gives an average value of Y associated with a particular value of X. Of course, some values of Y will be above the regression line (moving average) and some below, just as some values of Y are above or below the general average of Y.

The corrected sum of squares for Y (i.e., Σy^2) estimates the amount of variation of individual values of Y about the mean value of Y. A regression equation is a statement that part of the observed variation in Y (estimated by Σy^2) is associated with the relationship of Y to X. The amount of variation in Y that is associated with the regression on X is called the reduction or regression sum of squares.

$$\text{Reduction SS} = \frac{(\Sigma xy)^2}{\Sigma x^2} = \frac{(34.1477)^2}{(59{,}397.6775)} = 2.1115$$

As noted above, the total variation in Y is estimated by $\Sigma y^2 = 2.7826$ (as previously calculated).

The part of the total variation in Y that is not associated with the regression is called the residual sum of squares. It is calculated by

$$\text{Residual SS} = \Sigma y^2 - \text{Reduction SS} = 2.782 - 2.1115 = 0.6711$$

In analysis of variance we used the unexplained variation as a standard for testing the amount of variation attributable to treatments. We can do the same in regression. What's more, the familiar F test will serve.

Source of variation	df[1]	SS	MS[2]
Due to regression $\left[= \frac{(\Sigma xy)^2}{\Sigma x^2} \right]$	1	2.1115	2.1115
Residual (i.e., unexplained	60	0.6711	0.01118
Total (= Σy^2)	61	2.7826	

[1] As there are 62 values of Y, the total sum of squares has 61 df. The regression of Y on X has on df. The residual df's are obtained by subtraction.

[2] MS is, as always = SS/df

The regression is tested by

$$F = \frac{\text{Regression MS}}{\text{Residual MS}} = \frac{2.1115}{0.01118} = 188.86$$

As calculated F is much greater than tabular $F_{0.01}$ with 1/60 df, the regression is deemed significant at the 0.01 level.

Before we fitted a regression line to the data, Y had a certain amount of variation about its mean ($\bar{Y}$). Fitting the regression was, in effect, an attempt to explain part of this variation by the linear association of Y with X. But even after the line had been fitted, some variation was unexplained—that of Y about the regression line. When we tested the recession line above, we merely showed that the part of the variation in Y that is explained by the fitted line is significantly greater than the part that the line left unexplained. The test did not show that the line we fitted gives the best possible description of the data (a curved line might be even better). Nor does it mean that we have found the true mathematical relationship between the two variables. There is a dangerous tendency to ascribe more meaning to a fitted regression than is warranted.

It might be noted that the residual sum of squares is equal to the sum of the squared deviations of the observed values of Y from the regression line. That is,

$$\text{Residual SS} = \Sigma(Y - Y)^2 = \Sigma\left(Y \quad a \quad bX\right)^2$$

The principle of least squares says that the best estimates of the regression coefficients (a and b) are those that make this sum of squares a minimum.

Coefficient of Determination

The coefficient of determination, denoted R^2, is used in the context of statistical models whose main purpose is the prediction of future outcomes on the basis of other related information. Stated differently, the coefficient of determination is a ratio that measures how well a regression fits the sample data.

$$\text{Coefficient of determination } (R^2) = \frac{\text{Reduction SS}}{\text{Total SS}}$$

$$= \frac{2.1115}{2.7826} = 0.758823$$

When someone says, "76 percent of variation in Y was associated with X," she means that the coefficient of determination was 0.76. Note that R^2 is most often seen as a number between 0 and 1.0, used to describe how well a regression line fits a set of data. An R^2 near 1.0 indicates that the regression line fits the data well, while an R^2 closer to 0 indicates a regression line does not fit the data very well.

The coefficient of determination is equal to the square of the correlation coefficient.

$$\frac{\text{Reduction SS}}{\text{Total SS}} = \frac{(\Sigma xy)^2 / \Sigma x^2}{\Sigma y^2} = \frac{(\Sigma xy)^2}{(\Sigma x^2)(\Sigma y^2)} = R^2$$

In fact, most present-day users of regression refer to R^2 values rather than to coefficients of determination.

Confidence Intervals

Because it is based on sample data, a regression equation is subject to sample variation. Confidence limits (i.e., a pair of numbers used to estimate characteristics of a population) on the regression line can be obtained by specifying several values over the range of X and computed (refer to Figure 6.2)

where: X_0 = a selected value of X,

Degrees of freedom for t = df for residue MS

In the example we had:

$$\hat{Y} = 0.13606 + 0.005962X$$

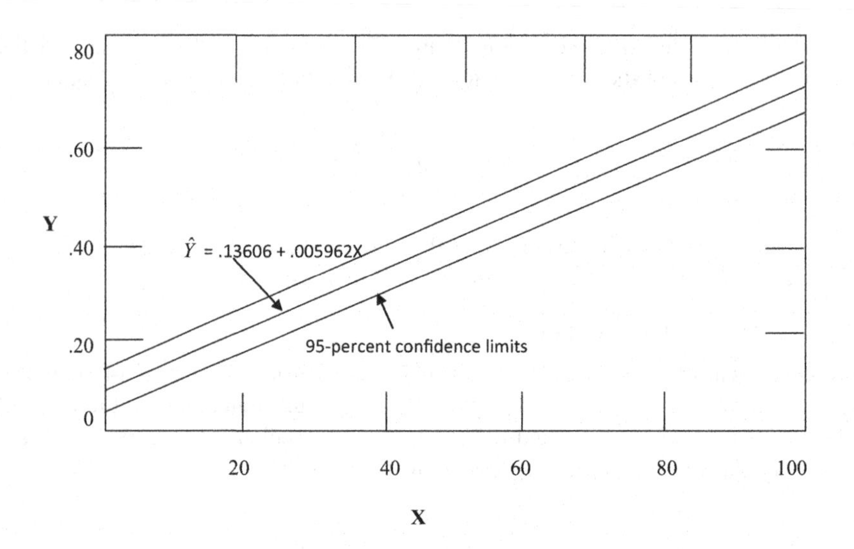

FIGURE 6.2 Confidence limits for the regress of *Y* on *X*.

Residual MS = 0.01118 with 60 df's

$$\Sigma x^2 = 59{,}397.6775$$

So, if we pick $X_0 = 28$ we have $\bar{Y} = 0.303$, and 95-percent confidence limits

$$= 0.303 \pm 2.000\sqrt{0.01118}\left(\frac{1}{62} + \frac{(28 - 49.1935)^2}{59{,}397.6775}\right)$$

$$= 0.270 \text{ to } 0.336$$

For other values of X_0 we would get:

		95-percent limits	
X_0	$\hat{Y}$	**Lower**	**Upper**
8	0.184	0.139	0.229
49.1935	0.429	0.402	0.456
70	0.553	0.521	0.585
90	0.673	0.629	0.717

Note that these are confidence limits on the regression of Y on X. They indicate the limits within which the true mean of Y for a given X will lie unless a one-in-twenty chance has occurred. The limits do not apply to a single predicted value of Y. The limits within which a single Y might lie are given by

$$\hat{Y} \pm t\sqrt{(\text{Residual MS})\left(+1/n + \frac{(X_0 - \bar{X})^2}{\Sigma x^2}\right)}$$

Assumptions—In addition to assuming that the relationship of Y to X is linear, the above method of fitting assumes that the variance of Y about the regression line is the same at all levels of X (the assumption of homogeneous variance or homoscedasticity—that is, the property of having equal variances). The fitting does not assume, nor does it require that the variation of Y about the regression line follows the normal distribution. However, the F test does assume normality, and so does the use of t for the computation of confidence limits.

There is also an assumption of independence of the errors (departures from regression) of the sample observations. The validity of this assumption is best insured by selecting the sample units at random. The requirements of independence may not be met if successive observations are made on a single unit or if the units are observed in clusters. For example, a series of observations of tree diameter made by means of a growth band would probably lack independence.

Selecting the sample units so as to get a particular distribution of the X values does not violate any of the regression assumptions, provided the Y values are a random sample of all Y's associated with the selected values of X. Spreading the sample over a wide range of X values will usually increase the precision with which the regression coefficients are estimated. This device must be used with caution, however, for if the Y values are not random, the regression coefficients and residual means squares may be improperly estimated.

Multiple Regression

It frequently happens that a variable (Y) in which we are interested is related to more than one independent variable. If this relationship can be estimated, it may enable us to make more precise predictions of the dependent variable than would be possible by a simple linear regression. This brings us up against multiple regression (describes the changes in a dependent variable associated with changes in one or more independent variables), which is a little more work but no more complicated than a simple linear regression.

The calculation methods can be illustrated with the following set of hypothetical data from an environmental study relating the growth of even-aged loblolly-shortleaf pine stands to the total basal area (X_1), the percentage of the basal area in loblolly pine (X_2), and loblolly pine site index (X_3).

Y	X_1	X_2	X_3
65	41	79	75
78	90	48	83
85	53	67	74
50	42	52	61
55	57	52	59
59	32	82	73
82	71	80	72
66	60	65	66
113	98	96	99
86	80	81	90
104	101	78	86
92	100	59	88
96	84	84	93
65	72	48	70
81	55	93	85
77	77	68	71
83	98	51	84
97	95	82	81
90	90	70	78
87	93	61	89
74	45	96	81
70	50	80	77
75	60	76	70
75	68	74	76
93	75	96	85
76	82	58	80
71	72	58	68
61	46	69	65
Sums 2,206	1,987	2,003	2,179
Means 78.7857	70.9643	71.5387	77.8214
($n = 28$)			

With this data we would like to fit an equation of the form

$$Y = a + b_1X_1 + b_2X_2 + b_3X_3$$

According to the principle of least squares, the best estimates of the X coefficients can be obtained by solving the set of least squares normal equations.

$$b_1 \text{ equation}: \left(\Sigma x_1^2\right)b_1 + \left(\Sigma x_1x_2\right)b_2 + \left(\Sigma x_1x_3\right)b_3 = \Sigma x_1y$$

$$b_2 \text{ equation}: \left(\Sigma x^1x_2\right)b_1 + \left(\Sigma x_2^2\right)b_2 + \left(\Sigma x_2x_3\right)b_2 = \Sigma x_2y$$

$$b_3 \text{ equation}: \left(\Sigma x_1 x_3\right) b_1 + \left(\Sigma x_2 x_3\right) b_2 + \left(\Sigma x_3^2\right) b_3 = \Sigma x_3 y$$

where $\Sigma x_i x_j = \Sigma X_i X_j - \dfrac{\left(\Sigma X_i\right)\left(\Sigma X_j\right)}{n}$

Having solved for the X coefficients (b_1, b_2, and b_3), we obtain the constant term by solving

$$a = \bar{Y} \text{--} b_1 \bar{X}_1 - b_2 \bar{X}_2 - b_3 \bar{X}_3$$

Derivation of the least squares normal equations requires a knowledge of differential calculus. However, for the general linear mode with a constant term

$$Y = a + b_1 X_1 + b_2 + \cdots + b_k X_k$$

the normal equations can be written quite mechanically once their pattern has been recognized. Every term in the first row contains an x_1, every term in the second row an x_2, and so forth down to the kth row, every term of which will have an x_k. Similarly, every term in the first column has an x_1 and a b_1 every term in the second column has an x_2 and a b_2, and so through the kth column, every term of which has an x_k and a b_k. On the right side of the equations, each term has a y times the x that is appropriate for a particular row. So, for the general linear model given above, the normal equations are:

$$b_1 \text{ equation}: \left(\Sigma x_1^2\right) b_1 + \left(\Sigma x_1 x_2\right) b_2 + \left(\Sigma x_1 x_3\right) b_3 + \cdots + (\Sigma x_1 x_k) b_k = \Sigma x_1 y$$

$$b_2 \text{ equation}: \left(\Sigma x^1 x_2\right) b_1 + \left(\Sigma x_2^2\right) b_2 + \left(\Sigma x_2 x_3\right) b_2 + \cdots + \left(\Sigma x_2 x_k\right) b_k = \Sigma x_2 y$$

$$b_3 \text{ equation}: \left(\Sigma x_1 x_3\right) b_1 + \left(\Sigma x_2 x_3\right) b_2 + \left(\Sigma x_3^2\right) b_3 + \cdots + \left(\Sigma x_3 x_k\right) b_k = \Sigma x_3 y$$

$$b_k \text{ equation}: \left(\Sigma x_1 x_k\right) b_1 + \left(\Sigma x_2 x_k\right) b_2 + \left(\Sigma x_3 x_k\right) b_3 + \cdots + \left(\Sigma x_k^2\right) b_k = \Sigma x_k y$$

Given the X coefficients, the constant term can be computed as

$$a = \bar{Y} - b_1 \bar{X}_1 - b_2 \bar{X}_2 - \cdots - b_k \bar{X}_k$$

Note that the normal equations for the general linear model include the solution for the simple linear regression

$$\left(\Sigma x_1^2\right) b_1 = \Sigma x_1 y$$

Hence,

$$b_1 = \left(\Sigma x_1 y\right) / \Sigma x_1^2 \text{ as previously given.}$$

In fact, all of this section on multiple regression can be applied to the simple linear regression as a special case.

The corrected sums of squares and products are computed in the familiar manner:

$$\Sigma y^2 = \Sigma Y^2 - \frac{(\Sigma Y)^2}{n} = (65^2 + \cdots + 61^2) - \frac{(2206)^2}{28} = 5{,}974.7143$$

$$\Sigma x_1{}^2 = \Sigma X_1{}^2 - \frac{(\Sigma X_1)^2}{n} = (41^2 + \cdots + 46^2) - \frac{(1987)^2}{28} = 11{,}436.9643$$

$$\Sigma x_1 y = \Sigma X_1 Y - \frac{(\Sigma X_1)(\Sigma Y)}{n} = (41)(65) + \cdots + (46)(61) - \frac{(1{,}987)(2{,}206)}{28}$$

$$= 6{,}428.7858$$

Similarly,

$\Sigma x_1 x_2 = -1{,}171.4642$ $\Sigma x_2 y = 2{,}632.2143$
$\Sigma x_1 x_3 = 3{,}458.8215$ $\Sigma x_3{}^2 = 2{,}606.1072$
$\Sigma x_2{}^2 = 5{,}998.9643$ $\Sigma x_3 y = 3{,}327.9286$
$\Sigma x_2 x_3 = 1{,}789.6786$

Putting these values in the normal equations gives:

$$11{,}436.9643\,b_1 - 1{,}171.4642\,b_2 + 3{,}458.8215\,b_3 = 6{,}428.7858$$

$$-1{,}171.4642\,b_1 + 5{,}998.9643\,b_2 + 1{,}789.6786\,b_3 = 2{,}632.2143$$

$$3{,}458.8215\,b_1 + 1{,}789.6786\,b_2 + 2{,}606.1072\,b_3 = 3{,}327.9286$$

These equations can be solved by any of the standard procedures for simultaneous equations. One approach (applied the above equations) is as follows:

1. Divide through each equation by the numerical coefficient of b_1.

$$b_1 - 0.102{,}427{,}897\,b_2 + 0.302{,}424{,}788\,b_3 = 0.562{,}105{,}960$$

$$b_1 - 5.120{,}911{,}334\,b_2 + 1.527{,}727{,}949\,b_3 = -2.246{,}943{,}867$$

$$b_1 + 0.517{,}424{,}389\,b_2 + 0.753{,}466{,}809\,b_3 = 0.962{,}156{,}792$$

2. Subtract the second equation from the first and the third from the first so as to leave two equations in b_2 and b_3.

$$5.018{,}483{,}437\,b_2 + 1.830{,}152{,}737\,b_3 = 2.809{,}049{,}827$$

$$-0.619,852,286\,b_2 \quad 0.451,042,021\,b_3 = -0.400,050,832$$

3. Divide through each equation by the numerical coefficient of b_2.

$$b^2 + 0.364,682,430\,b_3 = 0.559,740,779$$

$$b^2 + 0.727,660,494\,b_2 = 0.645,397,042$$

4. Subtract the second of these equations from the first, leaving one equation in b_3.

$$-0.362,978,064\,b_3 = -0.085,656,263$$

5. Solve for b_3

$$b_3 = \frac{-\ 0.085,656,263}{-\ 0.362,978,064} = 0.235,981,927$$

6. Substitute this value of b_3 in one of the equations (the first one, for example) of step 3 and solve for b_2.

$$b_2 + (0.364,682,43)(0.381,927) = 0.59,740,779$$

$$b_2 = 0.473,682,316$$

7. Substitute the solutions for b_2 and b_3 in one of the equations (the first one, for example) of step 1, and solve for b_1.

$$b_1 - (0.102,427,897)(0.473,682,316) + (0.302,424,788)(0.235,981,927)$$

$$= 0.562,105,960$$

$$b_1 = 0.539,257,459$$

8. As a check, add up the original normal equations and substitute the solutions for b_1, b_2, and b_3.

$$13,724.3216\,b_1 + 6,617.1787\,b_2 + 7,854.6073\,b_3 = 12,388.9287$$

$$12,388.92869 = 12,388.9287, \text{ check}$$

Given the values of b_1, b_2, and b_3 we can now compute

$$a = \bar{Y} - b_1\bar{X}_1 - b_2\bar{X} - b_3\bar{X}_3 = -11.7320$$

Thus, after rounding off the coefficients, the regression equation is

$$\hat{Y} = -11.732 + 0.539X_1 + 0.474X_2 + 0.236X_3$$

It should be noted that in solving the normal equations, more digits have been carried than would be justified by the rules for number of significant digits. Unless this is done, the rounding errors may make it difficult to check the computations.

Tests of Significance

Tests of significance refer to the methods of inference used to support or reject claims based on sample data. To test the significance of the fitted regression, the outline for the analysis of variance is

Source	Df
Reduction due to regression on X_1, X_2, and X_3	3
Residuals	24
Totals	27

The degrees of freedom for the total are equal to the number of observations minus 1. The total sum of squares is

$$\text{Total SS} = \Sigma y^2 = 5{,}974.7143$$

The degrees of freedom for the reduction are equal to the number of independent variables fitted, in this case 3. The reduction sum of squares for any least squares regression is

$$\text{Reduction SS} = \Sigma(\text{estimated coefficients})(\text{right side of their normal equations})$$

In this example there are three coefficients estimated by the normal equations, and so

$$\text{Reduction } \underset{3\,\text{df}}{\text{SS}} = b_2\left(\Sigma x_1 y\right) + b_2\left(\Sigma x_2 y\right) + b_3\left(\Sigma x_3 y\right)$$

$$= (0.53926)(6{,}428.7858) + (0.47368)(2{,}632.2143) + (0.23598)(3{,}327.9286)$$

$$= 5{,}498.9389$$

The residual df and sum of squares are obtained by subtraction. Thus, the analysis becomes

Source	df	SS	MS
Reduction due to X_1, X_2, and X_3	3	5,498.9389	1,832.9796
Residuals	24	475.7754	19.8240
Total	27	5,974.7143	

To test the regression, we compute

$$F_{3/24\ \mathrm{df}} = \frac{1{,}832.9796}{19.8240} = 92.46$$

which is significant at the 0.01 level.

Often, we will want to test individual terms of the regression. In the previous example we might want to test the hypothesis that the true value of b_3 is zero. This would be equivalent to testing whether the viable X_3 makes any contribution to the prediction of Y. If we decide a b_3 may be equal to zero, we might rewrite the equation in terms of X_1 and X_2. Similarly, we could test the hypothesis that b_1 and b_3 are both equal to zero.

To test the contribution of any set of the independent variables in the presence of the remaining variables:

1. Fit all independent variables and compute the reduction and residual sums of squares.
2. Fit a new regression that includes only the variables not being tested. Compute the reduction due to this regression.
3. The reduction obtained in the first step minus the reduction in the second step is the gain due to the variables being tested.
4. The mean square for the gain (step 3) is tested against the residual mean square from the first step.

Coefficient of Multiple Determination

As a measure of how well the regression fits the data, it is customary to compute the ratio of the reduction sum of squares to the total sum of squares. As mentioned earlier, this ratio is symbolized by R^2 and is sometimes called the coefficient of determination:

$$R^2 = \frac{\text{Reduction SS}}{\text{Total SS}}$$

For the regression of Y on X_1, X_2, and X_3,

$$R^2 = \frac{5{,}498.9389}{5{,}974.7143} = 0.92$$

The R^2 value is usually referred to saying that a certain percentage (92 in this case) of the variation in Y is associated with regression. The square root (R) of the ratio is called the multiple correlation coefficient.

The C-multipliers

Putting confidence limits on a multiple regression requires computation of the Gauss or c-multipliers. The c-multipliers are the elements of the inverse of the matrix of corrected sums of squares and products as they appear in the normal equations.

CURVILINEAR REGRESSIONS AND INTERACTIONS

CURVES

Many forms of curvilinear relationships can be fitted by the regression methods that have been described in the previous sections.

If the relationship between height and age is assumed to be hyperbolic so that

$$\text{Height} = \text{a} + \text{b} / \text{Age}$$

then we could let Y = Height and X_1 = 1/Age and fit

$$Y = a + b_1 X_1$$

Similarly, if the relationship between Y and X is quadratic

$$Y = a + bX + cX^2$$

We can let $X = X_1$ and $X^2 = X_1$ and fit

$$Y = a + b_1 X_1 + b_2 X_2$$

Functions such as

$$Y = aX^b$$

$$Y = a(b^x)$$

$$10^Y = aX^b$$

Which are nonlinear in the coefficients can sometimes be made linear by a logarithmic transformation. The equation

$$Y = aX^b$$

would become

$$\log Y = \log a + b(\log X)$$

which could be fitted by

$$Y' = a' + b_1 X_1$$

where $Y' = \log Y$, and
$X_1 = \log X$

The second equation transforms to

$$\log Y = \log a + (\log b)X$$

The third becomes

$$Y = \log a + b(\log X)$$

Both can be fitted by the linear model.

In making these transformations the effect on the assumption of homogeneous variance must be considered. If Y has homogeneous variance, log Y probably will not have—and vice versa.

Some curvilinear models cannot be fitted by the methods that have been described. Some examples are

$$Y = a + b^x$$

$$Y = a(X - b)^2$$

$$Y = a(X_1 - b)(X_2 - c)$$

Fitting these models requires more cumbersome procedures.

Interactions

Suppose that there is a simple linear relationship between Y and X_1. If the slope (b) of this relationship varies, depending on the level of some other independent variable (X_2), then X_1 and X_2 are said to interact. Such interactions can sometimes be handled by introducing interaction variables.

To illustrate, suppose that we know that there is a linear relationship between Y and X_1

$$Y = a + bX_1$$

Suppose further that we know or suspect that the slope (b) varies linearly with Z

$$b = a' + b'Z$$

This implies the relationship

$$Y = a + (a' + b'Z)X_1$$

or

$$Y = a + a'X_1 + b'X_1Z$$

where $X_2 = X_1Z$, and interaction variable.

If the Y-intercept is also a linear function of Z, then

$$a = a'' + b''Z$$

and the form of relationship is

$$Y = a'' + b''Z + a'X_1 + b'X_1Z$$

Group Regressions

Linear regressions of Y on X were fitted for each of the two groups. The basic data and fitted regressions were:

Group A										Sum	Mean
Y	3	7	9	6	8	13	10	12	14	82	9.111
X	1	4	7	7	2	9	10	6	12	58	6.444

$$n = 9 \quad \Sigma Y^2 = 848, \quad \Sigma XY = 609 \quad \Sigma X^2 = 480$$

$$\Sigma y^2 = 100.8889, \quad \Sigma xy = 80.5556, \quad \Sigma x^2 = 106.2222,$$

$$\hat{Y} = 4.224 + 0.7584X$$

$$\text{Residual SS} = 39.7980, \text{ with 7 df.}$$

Group A														Sum	Mean
Y	4	6	12	2	8	7	0	5	9	2	11	3	10	79	6.077
X	4	9	14	6	9	12	2	7	5	5	11	2	13	99	7.616

$$n = 13 \quad \Sigma Y^2 = 653, \quad \Sigma XY = 753, \quad \Sigma X^2 = 951$$

$$\Sigma y^2 = 172.9231, \quad \Sigma xy = 151.3846, \quad \Sigma x^2 = 197.0769$$

$$\hat{Y} = 0.228 + 0.7681X$$

$$\text{Residual SS} = 56.6370, \text{ with 11 df.}$$

Now, we might ask, are these really different regressions? Or could the data be combined to produce a single regression that would be applicable to both groups? If there is no significant difference between the residual mean squares for the two groups (this matter may be determined by Bartlett's test), the test described below helps to answer the question.

Testing for the Common Regressions

Simple linear regressions may differ either in their slope or in their level. In testing for common regressions, the procedure is to test first for common slopes. If the slopes differ significantly, the regressions are different, and no further testing is needed. If the slopes are not significantly different, the difference in level is tested. The analysis table is:

						Residuals		
Line	**Group**	**Df**	Σy^2	Σxy	Σx^2	**df**	**SS**	**MS**
1	A	8	100.8889	80.5556	106.2222	7	39.7980	
2	B	12	172.9231	151.3846	97.0769	11	56.6370	
3	Pooled Residuals					18	96.4350	5.3575
4	Difference for testing common slopes					1	0.0067	0.0067
5	Common slope	20	273.8120	231.9402	303.2991	19	96.4417	5.0759
6	Difference for testing trends					1	80.1954	80.1954
7	Single regression	21	322.7727	213.0455	310.5909	20	176.6371	

The first two lines in this table contain the basic data for the two groups. To the left are the total df for the groups (8 for *A* and 12 for *B*). In the center are the corrected sums of squares and products. The right side of the table gives the residual sum of squares and df. Since only simple linear regressions have been fitted, the residual df's of each group are one less than the total df. The residual sum of squares is obtained by first computing the reduction sum of squares for each group.

$$\text{Reduction SS} = \frac{(\Sigma xy)^2}{\Sigma x^2}$$

This reduction is then subtracted from the total sum of squares (Σy^2) to give to residuals.

Line 3 is obtained by pooling the residual df and residual sums of squares for the groups. Dividing the pooled sum of squares by the pooled df gives the pooled mean square.

The left side and center of line (we will skip line 4 for the moment) is obtained by pooling the total df and the corrected sums of square and products for the groups. These are the values that are obtained under the assumption of no difference in the slopes of the group regressions. If the assumption is wrong, the residuals about this common slope regression will be considerably larger than the mean square residual about the separate regressions. The residual df and sum of squares are obtained by fitting a straight line to this pooled data. The residual df's are, of course, one less than the total df. The residual sum of squares is, as usual,

$$\text{Residual SS} = 273.8120 - \frac{(231.9402)^2}{303.2991} = 96.4417$$

Now, the difference between these residuals (line 4 = 5 – line 3) provides a test of the hypothesis of common slopes. The error term for this test is the pooled mean square from line 3.

$$\text{Test of common slopes}: F_{1/18\text{ df}} = \frac{0.0067}{5.3575}$$

The difference is not significant.

If the slopes differed significantly, the groups would have different regressions, and we would stop here. Since the slopes did not differ, we now go on to test for a difference in the levels of the regression.

Line 7 is what we would have if we ignored the groups entirely, lumped all the original observations together, and fitted a single linear regression. The combined data are as follows:

$$n = (9 + 13) = 22 \text{ (so the df for total} = 21)$$

$$\Sigma Y = (82 + 79) = 161, \quad \Sigma Y^2 = (848 + 653) = 1{,}501$$

$$\Sigma Y^2 = 1{,}501 - \frac{(161)^2}{22} = 322.7727$$

$$\Sigma X = (585. + 99) = 157, \quad \Sigma X^2 = (480 + 951) = 1{,}431$$

$$\Sigma x^2 = 1{,}431 - \frac{(157)^2}{22} = 310.5909$$

$$\Sigma XY = (609 + 753) = 1{,}362, \Sigma xy = 1{,}362 - \frac{(157)(161)}{22} = 213.0455$$

From this we obtain the residual values on the right side of line 7.

$$\text{Residual SS} = 322.7727 - \frac{(213.0455)^2}{310.5909} = 176.6371$$

If there is a real difference among the levels of the groups, the residuals about this single regression will be considerably larger than the mean square residual about the regression that assumed the same slopes but different levels. This difference (line 6 = line 7 – line 5) is tested against the residual mean square from line 5.

$$\text{Test of levels}: F_{1/18\,\text{df}} = \frac{80.1954}{5.0759} = 15.80$$

As the levels differ significantly, the groups do not have the same regressions.

The test is easily extended to cover several groups, though there may be a problem in finding which groups are likely to have separate regressions and which can be combined. The test can also be extended to multiple regressions

Analysis of Covariance in a Randomized Block Design

A test was made of the effect of three soil treatments on the height growth of two-year-old seedlings. Treatments were assigned at random to the three plots within each of 11 blocks. Each plot was made up of 50 seedlings. Average five-year height growth was the criterion for evaluating treatments. Initial heights and five-year growths, all in feet, were:

Block	Treatment *A*		Treatment *B*		Treatment *C*		Block totals	
	Height	Growth	Height	Growth	Height	Growth	Height	Growth
1	3.6	8.9	3.1	10.7	4.7	12.4	11.4	32.0
2	4.7	10.1	4.9	14.2	2.6	9.0	12.2	33.3
3	2.6	6.3	0.8	5.9	1.5	7.4	4.9	19.6
4	5.3	14.0	4.6	12.6	4.3	10.1	14.2	36.7
5	3.1	9.6	3.9	12.5	3.3	6.8	10.3	28.9
6	1.8	6.4	1.7	9.6	3.6	10.0	7.1	26.0
7	5.8	12.3	5.5	12.8	5.8	11.9	17.1	37.0
8	3.8	10.8	2.6	8.0	2.0	7.5	8.4	26.3
9	2.4	8.0	1.1	7.5	1.6	5.2	5.1	20.7
10	5.3	12.6	4.4	11.4	5.8	13.4	15.5	37.4
11	3.6	7.4	1.4	8.4	4.8	10.7	9.8	26.5
Sums	42.0	106.4	34.0	113.6	40.0	104.4	116.0	324.4
	3.82	9.67	3.09	10.33	3.64	9.49	3.52	9.83

The analysis of variance of growth is:

Source	df	SS	MS
Blocks	10	132.83	-----
Treatment	2	4.26	2.130
Error	20	68.88	3.444
Total	32	205.97	

$$F\text{ (for testing treatments)}_{2/20\text{ df}} = \frac{2.130}{3.444}$$

Not significant at 0.05 level.

There is no evidence of a real difference in growth due to treatments. This is, however, the reason to believe that for young seedlings, growth is affected by initial height. A glance at the block totals seems to suggest that plots with the greatest initial height had the greatest five-year growth. The possibility that effects of treatment are being obscured by differences in initial heights raises the question of how the treatments would compare if adjusted for differences in initial heights.

If the relationship between height growth and initial height is linear and if the slope of the regression is the same for all treatments, the test of adjusted treatment means can be made by an analysis of covariance as described below. In this analysis, the growth will be labeled Y and initial height X.

Computationally, the first step is to obtain total, block, treatment, and error sums of squares of X (SS_x) and sums of products of x and Y (SP_{xy}), just as has already been done for Y.

$$\text{For } X: \text{C.T.}_{\cdot x} = \frac{116.0^2}{33} = 407.76$$

$$\text{Total SS}_x = \left(3.6^2 + \cdots + 4.8^2\right) \quad \text{C.T.}_{\cdot x} = 73.2$$

$$\text{Block SS}_x = \left(\frac{11.4^2 + \cdots + 9.8^2}{3}\right) \quad \text{C.T.}_{\cdot x} = 4.31$$

$$\text{Treatment SS}_x = \left(\frac{42.0^2 + 34.0^2 + 40.0^2}{11}\right) \quad \text{C.T.}_{\cdot x} = 3.15$$

$$\text{Error SS}_x = \text{Total} - \text{Block} - \text{Treatment} = 15.80$$

$$\text{For } XY: \text{C.T.}_{\cdot xy} = \frac{(116.0)(324.4)}{33} = 1{,}140.32$$

$$\text{Total SP}_{xy} = (3.6)(8.9) + \cdots + (4.8)(10.7) \quad \text{C.T.}_{\cdot xy} = 103.99$$

$$\text{Treatment SP}_{xy} = \left(\frac{(11.4)(32.0) + \cdots + (9.8)(26.5)}{3}\right) \quad \text{C.T.}_{\cdot xy} = 82.71$$

$$\text{Treatment SP}_{xy} = \left(\frac{(42.0)(106.4) + (34.0)(113.6) + (40.0)(104.4)}{11}\right) - \text{C.T.}_{\cdot xy} = -3.30$$

$$\text{Error SP}_{xy} = \text{Total} - \text{Block} - \text{Treatment} = 24.58$$

These computed terms are arranged in a manner similar to the test of group regressions (which is exactly what the covariance analysis is). One departure is that the total line is put at the top.

					Residuals		
Source	**df**	**SS_y**	**SP_{xy}**	**SS_x**	**df**	**SS**	**MS**
Total	32	205.97	103.99	73.26			
Blocks	10	132.83	82.71	54.31			
Treatment	2	4.26	−3.30	3.15			
Error	20	68.88	24.58	15.80	19	30.641	1.613

On the error line, the residual sum of squares after adjusting for a linear regression is

$$\text{Residual SS} = SS_y - \frac{(SP_{xy})^2}{SS_x} = 68.88 - \frac{24.58^2}{15.80} = 30.641$$

This sum of squares has 1 df less than the unadjusted sum or squares.

To test treatments, we first pool the unadjusted df and sums of squares and products for treatment and error. The residual terms for this pooled line are then computed just as they were for the error line

					Residuals	
	df	**SS_y**	**SP_{xy}**	**SS_x**	**df**	**ss**
Treatment +error--	22	73.14	21.28	18.95	21	49.244

Then to test for a difference among treatments after adjustment for the regression of growth on initial height, we compute the difference in residuals between the error and the treatment + error lines

	Df	**SS**	**MS**
Difference for testing adjusted treatments	2	18.603	9.302

The mean square for the difference in residual is now tested against the residual mean square for error.

$$F_{2/19\ \text{df}} = \frac{9.302}{1.613} = 5.77$$

Thus, after adjustment, the difference in treatment means is found to be significant at the 0.05 level. It may also happen that differences that were significant before adjustment are not significant afterward.

If the independent variable has been affected by treatments, interpretation of a covariance analysis requires careful thinking. The covariance adjustment may have the effect of removing the treatment differences that are being tested. On the other hand, it may be informative to know that treatments are or are not significantly different in spite of the covariance adjustment. The beginner who is uncertain of the interpretations would do well to select as covariates only those that have not been affected by treatments.

The covariance test may be made in a similar manner for any experimental design and, if desired (and justified), adjustment may be made for multiple or curvilinear regressions.

The entire analysis is usually presented in the following form:

						Adjusted	
Source	**df**	**SS_y**	**SP_y**	**SS_x**	**df**	**SS**	**MS**
Total	32	205.97	103.99	73.26			
Blocks	10	132.83	82.71	54.31			
Treatment	2	4.26	–3.30	3.15			
Error	20	68.88	24.58	15.80	19	30.641	1.613
Treatment +Error --	22	73/14	21.28	18.95	21	49.244	-----
Difference for testing adjusted treatment means					2	18.603	9.302

$$\text{Unadjusted treatments}: F_{2/20\text{ df}} = \frac{2.130}{3.444}. \text{ Not significant.}$$

$$\text{Adjusted treatments}: F_{2/19\text{ df}} = \frac{9.302}{1.613} = 5.77. \text{ Significant at the 0.05 level.}$$

Adjusted Means

If we wish to know what the treatment means are after adjustment for regression, the equation is

$$\text{Adjusted } \bar{Y}_i = \bar{Y}_i - b(\bar{X} - X)$$

where

$\bar{Y}$ = Unadjusted mean for treatment i

$$b = \text{Coefficient of the linear regression} = \frac{\text{Error SP}_{xy}}{\text{Error SS}_x}$$

$\bar{X}_i$ = Mean of the independent variable for treatment i

$\bar{X}$ = Mean of × for all treatments

In the example we had $\bar{X}_A = 3.82$, $\bar{X}_B = 3.09$, $\bar{X}_C = 3.64$, $\bar{X} = 3.52$, and

$$b = \frac{24.58}{15.80} = 1.56$$

So, the unadjusted and adjusted mean growths are

	Mean growths	
Treatment	**Unadjusted**	**Adjusted**
A	9.67	9.20
B	10.33	11.00
C	9.49	9.30

Tests among Adjusted Means

In an earlier section we encountered methods of making further tests among the means. Ignoring the covariance adjustment, we could for example make an F test for prespecified comparisons such as $A + C$ vs. B, or A vs. C. Similar tests can also be made after adjustment for covariance, though they involve more labor. The F test will be illustrated for the comparison B vs. $A + C$ after adjustment.

As might be suspected, to make the F test we must first compute sums of squares and products of X and Y for the specified comparison:

$$\text{SS}_y = \frac{\left[2(\Sigma Y_B) - (\Sigma Y_A + \Sigma Y_c)\right]^2}{\left[2^2 + 1^2 + 1^2\right]\left[11\right]} = \frac{\left[2(113.6) - (106.4 + 104.4)\right]^2}{66} = 4.08$$

$$\text{SS}_x = \frac{\left[2(\Sigma X_B) - (\Sigma X_A + \Sigma X_c)\right]^2}{\left[2^2 + 1^2 + 1^2\right]\left[11\right]} = \frac{\left[2(34.0) - (42.0 + 40.0)\right]^2}{66} = 2.97$$

$$\text{SP}_{xy} = \frac{\left[2(\Sigma Y_B) - (\Sigma Y_A + \Sigma Y_c)\right]\left[2(\Sigma X_B) - (\Sigma X_A + \Sigma X_c)\right]}{\left[2^2 + 1^2 + 1^2\right]\left[11\right]} = -3.48$$

From this point on, the F test of $A + B$ vs. C is made in exactly the same manner as the test of treatments in the covariance analysis.

					Residuals		
Source	**df**	**SS_y**	**SP_{xy}**	**SS_x**	**Df**	**SS**	**MS**
2*B*-(*A*+*C*)	1	4.08	−3.48	2.97	---	---	----
Error	20	68.88	24.58	15.80	19	30.641	1.613
Sum	21	72.96	21.10	18.77	20	49.241	-----
Difference for testing adjusted comparison					1	18.600	18.600

$F_{1/19\ df} = 11.531$. Significant at the 0.01 level.

REFERENCES

Addelman, S., 1969. The generalized randomized block design. *The American Statistician*, 23(4): 35–36.

Addelman, S., 1970. Variability of treatments and experimental units in the design and analysis of experiments. *Journal of the American Statistical Association*, 65(331): 1095–1108.

Freese, F., 1969. *Elementary Statistical Methods for Foresters*. Washington, DC: USDA.

Hamburg, M., 1987. *Statistical Analysis for Decision Making*, 4th ed. New York: Harcourt Brace Jovanovich Publishers.

Mayr, E., 1970. *Populations, Species, and Evolution: An Abridgement of Animal Species and Evolution*. Cambridge, MA: Belknap Press.

Mayr, E., 2002. The biology of race and the concept of equality. *Daedalus*, 131(Winter): 89–94. http://www.goodurmj.com/Mayr.html.

Spellman, F.R., 2020. *Environmental Engineers Mathematics Handbook*. Boca Raton, FL: CRC Press.

7 Frequency Measures
Incidence, Prevalence, and Mortality Rates

FREQUENCY MEASURES

There are two types of measures commonly used in epidemiology: a measure of central location and a frequency measure. A measure of central location provides a single value that summarizes an entire distribution of data. In contrast, a frequency measure characterizes only part of the distribution. Frequency measures compare one part of the distribution to another part of the distribution, or the entire distribution. Common frequency measures are ratios, proportions, and rates. All three frequency measures have the same basic form:

$$\frac{\text{numerator}}{\text{denominator}} \times 10^n$$

RECALL THAT

$10^0 = 1$ (anything raised to the 0 power equals 1)
$10^1 = 10$ (anything raised to the 1st power is the value itself)
$10^2 = 10 \times 10 = 100$
$10^3 = 10 \times 10 \times 10 = 1{,}000$

The fraction of (numerator/denominator) can be multiplied by 1, 10, 100, 1000, and so on. This multiplier varies by measure.

In public health practice, ratios and proportions are used to characterize populations by age, sex, race, exposures, and other variables. Ratios, proportions, and rates are also used to describe three aspects of the human condition: morbidity (disease), mortality (death), and natality (birth).

RATIO

A ratio is a comparison of any two values or is the relative magnitude of two quantities. In simpler terms, we can say that a *ratio* is the established relationship between two numbers. For example, if someone says, I'll give you four to one the Redskins over the Cowboys in the Super Bowl, what does that person mean?

Four to one, or 4:1, is a ratio. If someone gives you 4 to 1, it's his/her $4 to your $1.

As another more pertinent example, for an average of 3 ft^3 of screenings are removed from each million gallons of wastewater treated, the ratio of screenings removed (cubic feet) to treated wastewater (MG) is 3:1. Ratios are normally written using a colon (such as 2:1), or written as a fraction (such as 2/1).

Note that ratio is calculated by dividing one interval- or ratio-scale variable by the other and the numerator and denominator need not be related. Therefore, one could compare apples with bananas or oranges with number of physician visits.

The method used to calculate ratio is

$$\frac{\text{Number or rate of events, items, persons, etc. in one group}}{\text{Number or rate of events items, persons, etc. in another group}}$$

After the numerator is divided by the denominator the result is often expressed as the result "to one" or written as the result: "1."

Note that in certain ratios, the numerator and denominator are different categories of the same variable, such as dogs and cats, or persons 11–19 years and 20–29 years of age. In other ratios, the numerator and denominator are completely different variables, such as the number of hospitals in a county and the size of the population living in that county.

Example 7.1

PROBLEM:

In a hypothetical survey, between 1984 and 1988, as part of a National Health Survey, 5,532 people ages 42–75 years were enrolled in a follow-up study. At the time of enrollment, each study participant was classified as having or not having diabetes. During 1992–1994, enrollees were documented either to have died or were still alive. The results are summarized as follows.

	Original enrollment (1984–1988)	Dead at follow-up (1992–1994)
Diabetic men	111	77
Nondiabetic men	2,876	767
Diabetic women	222	67
Nondiabetic women	2,323	312

Of the men enrolled in the follow-up study, 2,876 were nondiabetic and 111 were diabetic. Calculate the ratio of nondiabetic to diabetic men.

SOLUTION:

$$\text{Ratio} = 2{,}876/111 \times 1 = 25.9:1$$

In epidemiology, ratios are used as both descriptive measures and as analytic tools. As a descriptive measure, ratios can describe the male-to-female ratio of participants in a study, or the ratio of controls to cases (e.g., two controls per case). As an analytic tool, ratios can be calculated for occurrence of illness, injury, or death between two groups.

Again, it is important to note that the numerators and denominators of a ratio can be related or unrelated. Simply stated, you are free to use a ratio to compare the number of males in a population with the number of females, or to compare the number of residents in a population with the number of pharmacies or dollars spent on prescribed medicines.

Example 7.2

PROBLEM:

A city of 3,000,000 people has 422 clinics. Calculate the ratio of clinics per person.

SOLUTION:

$$422/3{,}000{,}000 \times 10^n = 0.0001406 \text{ clinics per person}$$

To obtain a more easily understood result, you could set $10^n = 10^4$. Then the ratio becomes:

$$0.0001406 \times 10{,}000 = 1.406 \text{ clinics per } 10{,}000 \text{ persons}$$

Example 7.3

PROBLEM:

State 1's infant mortality rate in 2001 was 10.5 per 1,000 live births. State 2's infant mortality rate in 2001 was 3.6 per 1,000 live births. Calculate the ratio of infant mortality rate in State 1 to that in State 2.

SOLUTION:

$$10.5/3.6 \times 1 = 2.9:1$$

Thus, State 1's infant mortality rate was 2.9 times as high as State 2's infant mortality rate in 2001.

Death-to-Case Ratio

A commonly used epidemiologic ratio is the death-to-case ratio, which is the number of deaths attributed to a particular disease during a specified period divided by the number of new cases of that disease identified during the same period. It is used as a measure of the severity of illness: the death-to-case ratio for rabies is close to 1 (i.e., almost everyone who develops rabies dies from it), whereas the death-to-case ratio for the common cold is close to 0.

Example 7.4

PROBLEM:

In the United States in 2002, a total of 15,075 new cases of tuberculosis were reported (CDC, 2004). During the same year, 802 deaths were attributed to tuberculosis. Calculate the tuberculosis death-to-case ratio for 2002.

SOLUTION:

The tuberculosis death-to-case ratio for 2002 can be calculated as 802/15,075. Dividing both numerator and denominator by the numerator yields 1 death per 18.8 new cases. Dividing both numerator and denominator by the denominator (and multiplying by $10^n = 100$) yields 5.3 deaths per 100 new cases. Both expressions are correct.

DID YOU KNOW?

Ratios, proportions, and rates are not three distinctly different kinds of frequency measures. They are all ratios: proportions are a particular type of ratio, and some rates are a particular type of proportion. In epidemiology, however, we often shorten the terms for these measures in a way that makes it sound as though they are completely different. When we call a measure a ratio, we usually mean a nonproportional ratio; when we call a measure a proportion, we usually mean a proportional ratio that doesn't measure an event over time, and when we use the term "rate," we frequently refer to a proportional ratio that does measure an event in a population over time (CDC, 2012).

PROPORTION

A *proportion* is a statement that two ratios are equal. Stated differently, a proportion is the comparison of a part to the whole. It is a type of ratio in which the numerator is included in the denominator. For example, 1 is to 2 as 3 is to 6, so 1:2 = 3:6. In this case, 1 has the same relation to 2 that 3 has to 6. And what exactly is that relation? What do you think? You're right: 1 is half the size of 2 and 3 is half the size of 6. Or, alternately, 2 is twice the size of 1, and 6 is twice the size of 3. In another example, proportion can be used to describe what fraction of clinic patients tested positive for HIV, or what percentage of the population is younger than 25 years of age. A proportion may be expressed as a decimal, a fraction, or a percentage.

Working with Ratio and Proportion

When working with ratio and/or proportion, the following key points are important to remember.

(1) One place where fractions are used in calculations is when *ratios* and *proportions* are used, such as calculating solutions.

(2) A *ratio* is usually stated in the form A is to B as C is to D, and we can write it as two fractions that are equal to each other:

$$\frac{A}{B} = \frac{C}{D}$$

(3) Cross-multiplying solves ratio problems; that is, we multiply the left numerator (A) by the right denominator (D) and say that is equal to the left denominator (B) times the right numerator (C):

$$A \times D = B \times C$$

$$AD = BC$$

(4) If one of the four items is unknown, dividing the two known items that are multiplied together by the known item that is multiplied by the unknown solves the ratio. For example, if 2 pounds of alum are needed to treat 500 gallons of water, how many pounds of alum will we need to treat 10,000 gallons? We can state this as a ratio: 2 pounds of alum is to 500 gallons of water as "pounds of alum are to 10,000 gallons."

This is set up in this manner:

$$\frac{1 \text{ lb.alum}}{500\text{-gal water}} = \frac{x \text{ lb.alum}}{10{,}000\text{-gal water}}$$

Cross-multiplying:

$$(500)\,(x) = (1) \times (10{,}000)$$

Transposing

$$x = \frac{1 \times 10{,}000}{500}$$

$$x = 20 \text{ lb.alum}$$

For calculating proportion, for example, five gallons of fuel costs $5.40. How much do 15 gallons cost?

$$\frac{5\text{-gal}}{\$5.40} = \frac{15 \text{ gal}}{\$y}$$

$$5 \times y = 15 \times 5.40 = 81$$

$$y = \frac{81}{5} = \$16.20$$

Example 7.5

PROBLEM:

If a pump will fill a tank in 20 hours at 4gpm (gallons per minute), how long will it take a 10 gpm pump to fill the same tank?

SOLUTION:

First, analyze the problem. Here, the unknown is some number of hours. But should the answer be larger or smaller than 20 hours? If a 4 gpm pump can fill the tank in 20 hours, a larger pump (10 gpm) should be able to complete the filling in less than 20 hours. Therefore, the answer should be less than 20 hours.

Now set up the proportion:

$$\frac{x\,\text{h}}{20\,\text{h}} = \frac{4\text{ gpm}}{10\text{ gpm}}$$

$$x = \frac{(4)(20)}{(10)}$$

$$x = 8\,\text{h}$$

Example 7.6

PROBLEM:

Solve for x in the proportion problem given below.

SOLUTION:

$$\frac{36}{180} = \frac{x}{4450}$$

$$\frac{(4450)(36)}{180} = x$$

$$= 890$$

Example 7.7

PROBLEM:

Solve for the unknown value x in the problem given below.

SOLUTION:

$$\frac{3.4}{2} = \frac{6}{x}$$

$$(3.4)(x) = (2)(6)$$

$$x = \frac{(2)(6)}{3.4}$$

$$x = 3.53$$

Example 7.8

PROBLEM:

1 lb. of chlorine is dissolved in 65 gallons of water. To maintain the same concentration, how many pounds of chlorine would have to be dissolved in 150 gallons of water?

SOLUTION:

$$\frac{1\text{ lb.}}{65\text{-gal}} = \frac{x\text{ lbs.}}{150\text{ gal}}$$

$$(65)(x) = (1)(150)$$

$$x = \frac{(1)(150)}{65}$$

$$= 2.3\text{ lb.}$$

Example 7.9

PROBLEM:

It takes 5 workers 50 hours to complete a job. At the same rate, how many hours would it take 8 workers to complete the job?

SOLUTION:

$$\frac{5\text{ workers}}{8\text{ workers}} = \frac{x\text{ hours}}{50\text{ hours}}$$

$$x = \frac{(5)(50)}{8}$$

$$x = 31.3\text{ hours}$$

Example 7.10

PROBLEM:

Calculate the proportion of men in a public health follow-up study who were diabetics.

Numerator = 191 diabetic men
Denominator = Total number of men = 191 + 3,345 = 3,536
Proportion = (191/3,536) × 100 = 5.40 percent

Example 7.11

PROBLEM:

Calculate the proportion of deaths among men. Numerator = deaths in men.

$$= 78 \text{ deaths in diabetic men} + 777 \text{ deaths in nondiabetic men}$$

$$= 855 \text{ deaths in men}$$

Notice that the numerator (855 deaths in mend) is a subset of the denominator.

$$\text{Denominator} = \text{all deaths}$$

$$= 855 \text{ deaths in men} + 78 \text{ deaths in diabetic women}$$

$$+ 505 \text{ deaths in nondiabetic women}$$

$$= 1438 \text{ deaths}$$

$$\text{Proportion} = 855/1,438 = 35\%$$

DID YOU KNOW?

Proportions can easily be converted to ratios. If the numerator is the number of women (192) who attended a clinic and the denominator is all the clinic attendees (314), the proportion of clinic attendees who are women is 192/314, or 61 percent (more than half). To convert to a ratio, subtract the numerator from the denominator to get the number of clinic patients who are not women, i.e., the number of men (314–192 = 122 men.) Thus, ratio of women to men could be calculated from the proportion as:

$$\text{Ratio} = 193/\left(314 - 193\right) \times 1$$

$$= 193/122$$

$$= 1.6 \text{ to } 1 \text{ female-to-male ratio}$$

RATE

In epidemiology, a *rate* is a measure of the frequency with which an event occurs in a defined population over a specified period of time (CDC, 2012). A rate is a measure of risk because rate puts disease frequency in the perspective of the size of the population. It is particularly useful for comparing disease frequency in different locations, at different times, or among different groups of persons with potentially different sized populations.

On the other hand, rate has a different connotation or meaning to non-epidemiologists. For example, the speedometer of a ship indicates the ship's speed or rate of travel in miles or kilometers per hour. This rate is always reported per some unit of time. It is not uncommon for some epidemiologists to interpret rate as a description of how quickly disease occurs in a population, for example, 75 new cases of colon cancer per 1,000 women per year. What this measure conveys is a sense of the speed with which disease occurs in a population and seems to imply that this pattern has occurred and will continue to occur for the foreseeable future. This rate is an *incidence rate*, described later in the text.

Some epidemiologists use the term rate more loosely, referring to proportions with case counts in the numerator and size of population in the denominator as rates. Thus, an *attack rate* is the proportion of the population that develops illness during an outbreak. For example, 22 of 240 persons develop diarrhea after attending a picnic. (An alternative and more accurate phrase for attack rate is *incidence proportion or attack rate*.) A *prevalence rate* is the proportion of the population that has a health condition at a point in time. These terms are defined in detail in the next section.

MORBIDITY FREQUENCY

DID YOU KNOW?

The presence of disease in a population, or the probability (risk) of its occurrence is described by using morbidity frequency measures. In public health terms, disease includes illness, injury, or disability. Morbidity frequency measures include incidence proportion (or attack rate or risk), secondary attack rate, incidence rate (or person-time rate), point prevalence, and period prevalence.

INCIDENCE PROPORTION OR RISK

Incidence rates are the most common way of measuring and comparing the frequency of disease in populations. *Incident proportion* is the proportion of an initially disease-free population that develops disease, becomes injured, or dies during a specified (usually limited) period of time. Synonyms commonly used to describe incidence proportion are attack rate, risks, probability of getting disease, and cumulative incidence. Incidence proportion is a proportion because the persons in the numerator, those who develop disease, are all included in the denominator (the entire population). The method for calculating incidence proportion (risk) is

$$\frac{\text{Number of new cases of disease or injury during specified period}}{\text{Size of population at start of period}}$$

Example 7.12

PROBLEM:

In a study of diabetics, 110 of the 195 diabetic men died during the 14-year follow-up period. Calculate the risk of death for these men.

Numerator = 110 deaths among the diabetic men
Denominator = 195 diabetic men
$10^n = 10^2 = 100$

SOLUTION:

$$\text{Risk} = (110/195 \times 100) = 56.4\%$$

Example 7.13

PROBLEM:

In an outbreak of coronavirus-19 among attendees of a corporate party, 97 persons attended the party, 30 of whom contracted the disease. Calculate the risk of illness among persons who attended the party.

Numerator = 30 persons who contracted the disease
Developed = 97 persons who attended the party
$10^n = 10^2 = 100$

$$\text{Risk} = \text{"Party-specific attack rate"} = (30/97) \times 100 = 30.9\%$$

DID YOU KNOW?

Because all of the persons with new cases of disease (numerator) are also represented in the denominator, a risk is also a proportion.

DID YOU KNOW?

At the start of the observation period the denominator of an incidence proportion is the number of persons. The denominator should be limited to the "population at risk" for developing disease, i.e., persons who have the potential to get the disease and be included in the numerator. For example, if the numerator represents new cases of cancer of the ovaries, the denominator should be restricted to women, because men do not have ovaries. This is easily accomplished because census data by sex are readily available (CDC, 2012).

Attack Rate

The term *attack rate*, in the outbreak setting, is often used as a synonym for risk. It is the risk of getting the disease during a specified period, such as the duration of an outbreak. A variety of attack rates can be calculated. For example, *overall attack rate* is the total number of new cases divided by the total population. A *food-specific attack rate* is the number of persons who ate a specified food and became ill, divided by the total number of persons who ate that food. A secondary attack rate is sometimes calculated to document the difference between community transmission of illness versus transmission of illness in a household, barracks, or other closed population. It is calculated as:

$$\frac{\text{Number of cases among contacts of primary cases} \times 10^n}{\text{Total number of contacts}}$$

Keep in mind that the total number of contacts in the denominator is calculated as the total population in the households of the primary cases, minus the number of primary cases. For a secondary attack rate, 10^n usually is 100 percent.

Example 7.14

PROBLEM:

Consider an outbreak of coronavirus in which 16 persons in 16 different households all became ill. If the population of the community was 900, then the overall attack rate was 16/900 × 100 percent = 1.78 percent. One week later, 15 persons in the same households as these "primary" cases developed coronavirus. If the 16 households included 92 persons, calculate the secondary attack rate.

SOLUTION:

$$\text{Secondary attack rate} = (15/(92-16)) \times 100\% = 19.7\%$$

Incidence Rate

Incidence rate (sometimes referred to as *person-time rate*) is a measure of incidence that incorporates time directly into the denominator. A person-time rate is generally calculated from a long-term cohort follow-up study, wherein enrollees are followed over time and the occurrence of new cases of disease is documented. In general, each person is observed from an established starting time until one of four "end points" is reached: onset of disease, death, migration out of the study ("lost to follow-up"), or the end of the study. Comparable to the incidence proportion, the numerator of the incidence rate is the number of new cases identified during the period of observation. However, the denominator differs. The denominator is the sum of the time each person was observed, totaled for all persons. This denominator represents the total time the population at risk was being watched for disease. Thus, the incidence rate is the ratio of the number of cases to the total time the population is at risk of disease (CDC, 2012). The method for calculating incidence rate is

$$\frac{\text{Number of new cases of disease or injury during specified period}}{\text{Time each person was observed, totaled for all persons}}$$

In a continuing follow-up study of morbidity, each study participant may be followed or observed for several years. For example, one person followed for five years without developing disease is said to contribute five person-years of follow-up.

Simply stated, an incidence rate describes how quickly disease occurs in a population. It is based on person-time, so it has some advantages over an incidence proportion. Because person-time is calculated for each subject, it can accommodate persons coming into and leaving the study. Note that person-time has one important drawback. Person-time assumes that the probability of disease during the study period is constant, so that ten persons followed for one year equals one person followed for ten years. Because the risk of many chronic diseases increases with age, this assumption is often invalid.

Prevalence

Prevalence or *prevalence rate* is the proportion of persons in a population who have a particular disease or attribute at a specified point in time or over a specified period of time. Prevalence differs from incidence in that prevalence includes all cases, both new and preexisting, in the population at the specified time, whereas incidence is limited to new cases only.

Point prevalence refers to the prevalence measured at a particular point in time. It is the proportion of person with a particular disease or attribute on a particular date. On the other hand, *period prevalence* refers to prevalence measured over an interval of time. It is the proportion of persons with a particular disease or attribute at any time during the interval.

The methods used for calculating prevalence of disease and prevalence of an attribute are

$$\begin{matrix}\text{Prevalence}\\ \text{of disease}\end{matrix} = \frac{\text{All new and pre-existing cases during a given time period}}{\text{Population during the same time period}} \times 10^n$$

$$\begin{matrix}\text{Prevalence of}\\ \text{an attribute}\end{matrix} = \frac{\text{Persons having a particular attribute during a given time period}}{\text{Population during the same time period}} \times 10^n$$

Note: The value of 10^n is usually 1 or 100 for common attributes. The value of 10^n might be 1,000, 100,000, or even 1,000,000 for rate attributes and for most diseases.

Example 7.15

PROBLEM:

Suppose a survey of 1,200 women who gave birth in Vermont in 2002 was conducted and a total of 450 reported taking a multivitamin at least 4 times a week during the month before becoming pregnant. Calculate the prevalence of frequent multivitamin use in this group.

SOLUTION:

Numerator = 450 multivitamin users
Denominator = 1,200 women
Prevalence = (450/1,200) × 100 = 37.5 percent

DID YOU KNOW?

Incidence and prevalence are often confused. Incidence refers to the proportion, or rate, of persons who develop a condition during a particular time period, whereas prevalence refers to the proportion of persons who have a condition at or during a particular time period. So, incidence and prevalence are similar, but incidence includes new cases only, whereas prevalence includes new and preexisting cases.

Proportionate Mortality

Proportionate mortality is the proportion of deaths in a specified population during a period of time that are attributable to different causes. Each cause is expressed as a percentage of all deaths, and the sum of the causes adds up to 100 percent. These proportions are not rates because the denomination is all deaths, not the size of the population in which the deaths occurred.

MORTALITY FREQUENCY MEASURES

Mortality Rate

Morbidity and mortality measures are often the same mathematically; it's just a matter of what needs to be measured, illness or death. *Mortality rate* is a measure of the frequency of occurrence of death in a defined population during a specified interval. The formula for the mortality of a defined population, over a specified period of time, is:

$$\frac{\text{Deaths occurring during a given time period}}{\text{Size of the population among which the deaths occurred}}$$

Measures of Mortality

Frequently used measures of mortality include:

- **Crude death rate (crude mortality rate)**—The crude mortality rate is the mortality rate from all causes of death for a population. In the United States in 2003, a total of 2,419,921 deaths occurred. The estimated population was 290,809,777. The crude mortality rate in 2003 was, therefore, (2,419/290,809,777) × 100,000, or 832.1 deaths per 100,000 population (10^n = 1,000 or 10,000; CDC, 2006)
- **Cause-specific mortality rate**—It is the mortality rate from a specified cause for a population. The numerator is the number of deaths attributed to a specific cause. The denominator remains the size of the population at the midpoint of the time period. The fraction is usually expressed per 100,000 population. In the United States in 2003, a total of 108,256 deaths were attributed to accidents (unintentional injuries), yielding a cause-specific mortality rate of 37.2 per 100,000 population (10^n = 100,000; CDC, 2006)
- **Age-specific mortality rate**—It is a mortality rate limited to a particular age group. The numerator is the number of deaths in that age group; the denominator is the number of persons in that age group in the population. In the United States in 2003, a total of 130,761 deaths occurred among persons aged 25–44 years, or an age-specific mortality rate of 153.0 per 10,000 25–44-year-olds (CDC, 2006). Some specific types of age-specific mortality rates are neonatal, postneonatal, and infant mortality rates, as described in the following sections (10^n = 100 or 1,000).
- **Infant mortality rate**—It is perhaps the most commonly used measure for comparing health status among nations. It is calculated as follows:

$$\frac{\text{Number of deaths among children} < 1 \text{ year of age reported during a given time period}}{\text{Number of live births reported during the same time period}} \times 1{,}000$$

  The infant mortality rate is generally calculated on an annual basis. The question is: is the infant mortality rate truly a rate? No, because the denominator is not the size of the mid-year population of children <1 year of age in 2003. In fact, the age-specific death rate for child <1 years of age of 2003 was 694.7 per 100,000 (CDC, 2006; 10^n = 1,000).
- **Neonatal mortality rate**—It covers birth up to but not including 28 days. The numerator of the neonatal mortality rate therefore is the number of deaths among children under 28 days of age during a given time period. The denominator of the neonatal mortality rate, like that of the infant mortality rate, is the number of live births reported during the same time period. The neonatal mortality rate is usually expressed per 1,000 live births. In 2003, the neonatal mortality rate in the United States was 4.7 per 1,000 live births (CDC, 2006; 10^n = 1,000).

- **Postneonatal mortality rate—It** is defined as the period from 28 days of age up to but not including 1 year of age. The numerator of the postneonatal mortality rate therefore is the number of deaths among children from 28 days up to but not including 1 year of age during a given time period. The denominator is the number of live births reported during the same time period. The postneonatal mortality rate is usually expressed per 1,000 live births. In 2003, the postneonatal mortality rate in the United States was 2.3 per 1,000 live births (CDC, 2006; 10^n = 1,000).
- **Maternal mortality rate**—It is really a ratio used to measure mortality associated with pregnancy. The numerator is the number of deaths during a given time period among women while pregnant or within 42 days of termination of pregnancy, irrespective of the duration and the site of the pregnancy, from any cause related to or aggravated by the pregnancy or its management, but not from accidental or incidental causes. The denominator is the number of live births reported during the same time period. Maternal mortality rate is usually expressed per 100,000 live births. In 2003, the U.S. maternal mortality rate was 8.9 per 100,000 live births (CDC, 2006; 10^n = 1,000).
- **Sex-specific mortality rate**—It is a mortality rate among either males or females. Both numerator and denominator are limited to one sex.
- **Race-specific mortality rate**—It is a mortality rate related to a specified racial group. Both numerator and denominator are limited to the specified race.
- **Combinations of specific mortality rates**—It can be further stratified by combinations of cause, age, sex, and/or race (CDC, 2012).
- **Age-adjusted mortality rates**—It can be used to compare the rates in one area with the rates in another area, or to compare rates over time. Keep in mind, however, because mortality rates obviously increase with age, a higher mortality rate among one population than among another might simply reflect the fact that the first population is older than the second.

REFERENCES

CDC, 2004. *Reported Tuberculosis in the United States, 2003.* Atlanta, GA: U.S. Department of Health and Human Services, CDC.

CDC, 2006. *National Center for Injury Prevention and Control.* Atlanta, GA. Accessed 09/25/2020 @ https://www.cdc.gov/injury/wisqars.

CDC, 2012. *Principles of Epidemiology in Public Heath Practice*, 3rd ed. Atlanta, GA: Centers for Disease Control and Prevention.

8 Measures of Association and Public Health Impact

Comparison is the key to epidemiologic analysis. Occasionally, an observance of an incidence rate among a population may seem high, and the question may arise as to whether it is actually higher than what should be expected based on, let's say, the incidence rates in other communities. Or it may be observed that, among a group of case-patients in an outbreak, several reports having eaten at a particular restaurant. The next question is whether the restaurant is just a popular one, or have more case-patients eaten there than would be expected? This is where comparison comes into the picture; it is the way to address concerns by comparing the observed group with another group that represents the expected level.

A *measure of association* quantifies the relationship between exposure and disease among the two groups. Exposure is used loosely to mean not only exposure to foods, mosquitoes, a partner with a sexually transmissible disease, or a toxic waste dump, but also inborn characteristics of a person (for example age, race, sex), biologic characteristics (immune status), acquired characteristics (marital status) activities (occupation, leisure activities), or conditions under which they live (socioeconomic status or access to medical care). Examples of measures of association include risk ratio (relative risk), rate ratio, odds ratio, and proportionate mortality ratio.

RISK RATIO

A *risk ratio* (RR), also called relative risk, compares the risk of a health event (disease, injury, risk factor, or death) among one group with the risk among another group. It does so by dividing the risk (incidence proportion, attack rate) in group 1 by the risk (incidence proportion, attack rate) in group 2. The two groups are differentiated by such demographic factors as sex (e.g., males versus females) or by exposure to a suspected risk factor (e.g., did or did not eat macaroni salad). Regularly, the group of primary interest is identified as the exposed group, and the comparison group is identified as the unexposed group. The formula for risk ratio (rr) is:

$$\text{Risk Ratio} = \frac{\text{Risk of disease (incidence proportion, attack rate) in group of primary interest}}{\text{Risk of disease (incidence proportion, attack rate) in comparison group}}$$

DID YOU KNOW?

A risk ratio of 1.0 indicates identical risk among the two groups. A risk ratio greater than 1.0 indicates an increased risk for the group in the numerator, usually the exposed group. A risk ratio less than 1.0 indicates a decreased risk for the exposed group, indicating that perhaps exposure actually protects against disease occurrence (CDC, 2012).

RATE RATIO

A rate ratio compares or measures the incidence rates, person-time rates, or mortality rates of two groups. As with risk ratio, the two groups are typically singled out or differentiated by demographic factors or by exposure to a suspected causative agent. The rate for the group of primary interest is divided by the rate for the comparison group.

$$\text{Risk ratio} = \frac{\text{Rate for group of primary interest}}{\text{Rate for comparison group}}$$

DID YOU KNOW?

The interpretation of the value of a rate ratio is similar to that of the risk ratio. That is, a rate ratio of 1.0 indicates equal rates in the two groups, a ratio greater than 1.0 indicates an increased risk for the group in the numerator, and a rate ratio less than 1.0 indicates a decreased risk for the group in the numerator (CDC, 2012).

ODDS RATIO

Another measure of association is an *odds ratio*; it quantifies the relationship between an exposure with two categories and health outcome. Moreover, the odds ratio is the measure of choice in a case-control study. A case-control study is based on enrolling a group of persons with disease ("case-patients") and a comparable group without disease ("controls"). The number of persons in the control group is usually decided by the investigation. Often, the size of the population from which the case-patients came is not known. As a result, risks, rates, risk ratios, or rater ratios can't be calculated from the typical case-control study. However, you can calculate an odds ratio and interpret it as an approximation of the risk ratio, particularly when the disease is uncommon in the population.

MEASURES OF PUBLIC HEALTH IMPACT

To place the association between an exposure and an outcome into a meaningful public health context, a measure of public health impact is used. Whereas a measure of association quantifies the relationship between exposure and disease, and thus begins to provide insight into causal relationships, measures of public health impact reflect the burden that an exposure contributes to the frequency of disease in the population. Two measures of public health impact often used are the attributable proportion and efficacy or effectiveness.

Attributable Proportion

The measure of the public health impact of a causative factor is known as the *attributable proportion*, or the *attributable risk percent*. The calculation of this measure (remember, epidemiology is all about measuring) assumes that the occurrence of disease in the unexposed group represents the baseline or expected risk for that disease. Moreover, it assumes that if the risk of disease in the exposed group is higher than the risk in the unexposed group, the difference can be attributed to the exposure. It represents the expected reduction in disease if the exposure could be removed (or never actually existed).

DID YOU KNOW?

Appropriate use of attributable proportion depends on a single risk factor being responsible for a condition. When multiple risk factors may interact (e.g., physical activity and age or health status), this measure may not be appropriate.

Attributable proportion is calculated as follows:

$$\frac{\text{Risk for exposed group} - \text{risk for unexposed group}}{\text{Risk for exposed group}}$$

Note: Attributable proportion can be calculated for rates in the same way.

Vaccine Efficacy

In measuring the proportionate reduction in cases among vaccinated persons, epidemiologists use *vaccine efficacy* and *vaccine effectiveness*. Vaccine efficacy is used when a study is carried out under ideal conditions, for example, during a clinical trial. Vaccine effectiveness is used when a study is carried out under typical field (i.e., less than perfectly controlled) conditions.

In calculating the risk of disease among vaccinated and unvaccinated persons and determining the percentage reduction in the risk of disease among vaccinated persons relative to unvaccinated persons the *vaccine efficacy/effectiveness* (VE) is measured. The greater the percentage reduction of illness in the vaccinated group, the greater the vaccine efficacy/effectiveness. The basic formula is written as:

$$\frac{\text{Risk among unvaccinated group} - \text{risk among vaccinated group}}{\text{Risk among unvaccinated group}}$$

$$\text{OR} : 1 - \text{risk ratio}$$

SUMMARY OF PART 2

Frequency measures are used quite commonly in epidemiology because many of the variables encountered in field epidemiology are nominal-scale variables. (A nominal-scale variable is divided into two or more categories, for example yes or no, agree/disagree, etc.) Frequency measures include ratios, proportions, and rates. Ratios and proportions are useful for describing the characteristics of populations. Proportions and rates are used for quantifying morbidity and mortality. These measures allow epidemiologists to infer risk among different groups, detect groups at high risk, and develop hypotheses about causes—that is, why these groups might be at increased risk. Keep in mind that the hallmark of epidemiologic analysis is comparison, such as comparison of observed amount of disease in a population with the expected amount of disease.

REFERENCE

CDC, 2012. *Principles of Epidemiology in Public Heath Practice*, 3rd ed. Atlanta, GA: Centers for Disease Control and Prevention.

Part 3

Wastewater

9 The Science of Wastewater

***Note*:** Well, it has been called sewage, waste discharge, sewerage, slop, excreta, feces, droppings, manure, ordure, poop, doo-doo, scat, stools, cow-patties, guano, dropping, and titles not suitable for publication herein, or simply and totally, including all of the above, wastewater or the inglorious ingredients that make up domestic/municipal wastewater. Whatever you wish to call it, most people simply say it is filth or foul matter. Is that a fair and true assessment of wastewater? Like every human view on every conceivable subject, humans have their own opinions and/or descriptions of just about anything and everything. So, is wastewater really all of the despicable descriptors listed above?

Depends on your point of view.

I have always had a different view of wastewater. I view it as a resource.

A resource?

Absolutely. Consider, for example, that when wastewater is properly and extensively treated to ultimate drinking water quality and is in compliance with USEPA and other agencies' requirements it can be reused—instead of wasted. Moreover, treated wastewater can be used in industry for cooling industrial equipment. In this text, I point out that wastewater has an even more valuable use especially for environmental scientists, namely, for those concerned or involved in wastewater-based epidemiology (WBE).

Parts 1 and 2 of this text have presented a very basic foundation for what follows in this part, and WBE aspects and procedures are covered in Part 4 of the text. Before discussing WBE it is important to have an understanding of wastewater; that is, what domestic/municipal wastewater is, its characteristics and its constituents

INTRODUCTION

The Code of Federal Regulations (CFR) 40 CFR Part 403, regulations were established in the late 1970s and early 1980s to help Publicly Owned Treatment Works (POTW) control industrial discharges to sewers. These regulations were designed to prevent pass-through and interference at the treatment plants and interference in the collection and transmission systems.

Pass-through occurs when pollutants literally "pass through" a POTW without being properly treated, and cause the POTW to have an effluent violation or increase the magnitude or duration of a violation.

Interference occurs when a pollutant discharge causes a POTW to violate its permit by inhibiting or disrupting treatment processes, treatment operations, or processes related to sludge use or disposal.

CONSTITUENTS AND CHARACTERISTICS OF WASTEWATER*

To get close and maybe even personal with wastewater (personal in the sense for those who work with wastewater by sampling, testing, and analyzing whatever it is he or she is looking for within the wastewater) a knowledge of what wastewater is, what is in the wastewater (constituents within), and the characteristics of wastewater is important. Knowledge of the constituents/characteristics most commonly associated with municipal wastewater is essential to the environmental professional working with wastewater-based epidemiology. Note that in discussing the constituents and characteristics of domestic/municipal wastewater we must start with original substance, water, and then move on to how this water is changed (contaminated) to wastewater. Thus, important characteristics and constituents in both so-called "clean" water and contaminated wastewater are discussed.

PHYSICAL CHARACTERISTICS OF WATER/WASTEWATER

The physical characteristics of water/wastewater we are interested in are more germane to the discussion at hand—namely, a category of parameters/characteristics that can be used to describe water quality. One such category is the physical characteristics of water, those that are apparent to the senses of smell, taste, sight, and touch. Solids content, turbidity, color, taste and odor, and temperature also fall into this category.

Solids

Other than gases, all contaminants of water contribute to the *solids content*. Classified by their size and state, by their chemical characteristics, and by their size distribution, solids can be dispersed in water in both suspended and dissolved forms. With regard to size, solids in water and wastewater can be classified as suspended, settleable, colloidal, or dissolved. Solids are also characterized as being *volatile* or *nonvolatile*. The distribution of solids is determined by computing the percentage of filterable solids by size range. Solids typically include inorganic solids such as silt, sand, gravel, and clay from riverbanks and organic matter such as plant fibers and microorganisms from natural or human-made sources. We use the term "siltation" to describe the suspension and deposition of small sediment particles in water bodies. In flowing water, many of these contaminants result from the erosive action of water flowing over surfaces.

Sedimentation and siltation can severely alter aquatic communities. Sedimentation may clog and abrade fish gills, suffocate eggs and aquatic insect larvae on the bottom, and fill in the pore space between bottom cobbles where fish lay eggs. Suspended silt and sediment interfere with recreational activities and aesthetic enjoyment at streams and lakes by reducing water clarity and filling in lakes. Sediment may also carry other pollutants into surface waters. Nutrients and toxic chemicals may attach

* Based on F.R. Spellman's *Handbook of Water and Wastewater Treatment Plant Operations*, 4th ed. (2020). Boca Raton, FL: CRC Press.

to sediment particles on land and ride the particles into surface waters where the pollutants may settle with the sediment or detach and become soluble in the water column.

Suspended solids are a measure of the weight of relatively insoluble materials in the ambient water. These materials enter the water column as soil particles from land surfaces or sand, silt, and clay from stream bank erosion of channel scour. Suspended solids can include both organic (detritus and biosolids) and inorganic (sand or finer colloids) constituents.

In water, suspended material is objectionable because it provides adsorption sites for biological and chemical agents. These adsorption sites provide attached microorganisms a protective barrier against the chemical action of chlorine. In addition, suspended solids in water may be degraded biologically resulting in objectionable by-products. Thus, the removal of these solids is of great concern in the production of clean, safe drinking water and wastewater effluent.

In water treatment, the most effective means of removing solids from water is by filtration. It should be pointed out, however, that not all solids, such as colloids and other dissolved solids, can be removed by filtration. In wastewater treatment, suspended solids are an important water-quality parameter and are used to measure the quality of the wastewater influent, to monitor performance of several processes, and to measure the quality of effluent. Wastewater is normally 99.9 percent water and 0.1 percent solids. If a wastewater sample is evaporated, the solids remaining are called *total solids*. The USEPA has set a maximum suspended-solids standard of 30 mg/L for most treated wastewater discharges.

Turbidity

One of the first things that are noticed about water is its clarity. The clarity of water is usually measured by its *turbidity*. Turbidity is a measure of the extent to which light is either absorbed or scattered by suspended material in water. Both the size and surface characteristics of the suspended material in the water influence absorption and scattering. Although algal blooms can make waters turbid, in surface water, most turbidity is related to the smaller inorganic components of the suspended solids burden, primarily the clay particles. Microorganisms and vegetable material may also contribute to turbidity. Wastewaters from industry and households usually contain a wide variety of turbidity-producing materials. Personal care products such as detergents, soaps, cosmetics, and various emulsifying agents contribute to turbidity.

In water treatment, turbidity is useful in defining drinking water quality. In wastewater treatment, turbidity measurements are particularly important whenever ultraviolet radiation (UV) is used in the disinfection process. For UV to be effective in disinfecting wastewater effluent, UV light must be able to penetrate the stream flow. Obviously, stream flow that is turbid works to reduce the effectiveness of irradiation (penetration of light).

The colloidal material associated with turbidity provides absorption sites for microorganisms and chemicals that may be harmful or cause undesirable tastes and odors. Moreover, the adsorptive characteristics of many colloids work to provide protection sites for microorganisms from disinfection processes. Turbidity in running waters interferes with light penetration and photosynthetic reactions.

Color

Color is another physical characteristic by which the quality of water can be judged. Pure water is colorless. Water takes on color when foreign substances such as organic matter from soils, vegetation, minerals, and aquatic organisms are present. Color can also be contributed to water by municipal and industrial wastes. Color in water is classified as either *true color* or *apparent color.* Water whose color is partly due to dissolved solids that remain after removal of suspended matter is known as true color. Color contributed by suspended matter is said to have apparent color. In water treatment, true color is the most difficult to remove.

***Note*:** Water has an intrinsic color, and this color has a unique origin. Intrinsic color is easy to discern, as can be seen in Crater Lake, Oregon, which is known for its intense blue color. The appearance of the lake varies from turquoise to deep navy blue, depending on whether the sky is hazy or clear. Pure water and ice have a pale blue color.

The obvious problem with colored water is that it is not acceptable to the public. That is, given a choice, the public prefers clear, uncolored water. Another problem with colored water is the affect it has on laundering, papermaking, manufacturing, textiles, and food processing. The color of water has a profound impact on its marketability for both domestic and industrial use.

In water treatment, color is not usually considered unsafe or unsanitary, but is a treatment problem in regard to exerting a chlorine demand, which reduces the effectiveness of chlorine as a disinfectant. In wastewater treatment, color is not necessarily a problem but instead is an indicator of the *condition* of the wastewater. Condition refers to the age of the wastewater, which, along with odor, provides a qualitative indication of its age. Early in the flow, wastewater is a light brownish-gray color. The color of wastewater containing dissolved oxygen (DO) is normally gray. Black-colored wastewater usually accompanied by foul odors, containing little or no DO, is said to be septic. Table 9.1 provides wastewater color information. As the travel time in the collection system increases (flow becomes increasingly more septic), and more anaerobic conditions develop, the color of the wastewater changes from gray to dark gray and ultimately to black.

TABLE 9.1
Significance of Color in Wastewater Influent

Color	Problem indicated
Gray	None
Red	Blood or other industrial wastes
Green, Yellow	Industrial wastes (e.g., paints) not pretreated
Red or other soil color	Surface runoff into influent; industrial flows
Black	Septic conditions or industrial flows

Taste and Odor

Taste and odor are used jointly in the vernacular of water science. The term "odor" is used in wastewater; taste, obviously, is not a consideration. Domestic sewage should have a musty odor. Bubbling gas and/or foul odor may indicate industrial wastes, anaerobic (septic) conditions, and operational problems. Refer to Table 9.2 for typical wastewater odors, possible problems, and solutions.

In wastewater, odors are of major concern, especially to those who reside in close proximity to a wastewater treatment plant. These odors are generated by gases produced by decomposition of organic matter or by substances added to the wastewater. Because these substances are volatile, they are readily released to the atmosphere at any point where the waste stream is exposed, particularly if there is turbulence at the surface.

Most people would argue that all wastewater is the same—that is, it has a disagreeable odor. It is hard to argue against the disagreeable odor. However, one wastewater operator told me that wastewater "smelled great—smells just like money to me—money in the bank," she said.

This was an operator's view. We also received another opinion of odor problems resulting from wastewater operations. This particular opinion, given by an odor control manager, was quite different. His statement was "that odor control is a never-ending problem." She also pointed out that to combat this difficult problem, odors must be contained. In most urban plants, it has become necessary to physically cover all source areas such as treatment basins, clarifiers, aeration basins, and contact tanks to prevent odors from leaving the processes. These contained spaces

TABLE 9.2
Odors in Wastewater Treatment Plant

Odor	Location	Problem	Possible solution
Earthy, musty	Primary and secondary units	No problem (Normal)	None required
Hydrogen sulfide	Influent	Septic (rotten egg odor)	Aerate, chlorinate, oxonizate
	Primary clarifier	Septic sludge	Remove sludge
	Activated sludge	Septic sludge	Remove sludge
	Trickling filters	Septic conditions	More air/less BOD
	Secondary clarifier Clarifiers	Septic conditions	Remove sludge
	Chlorine contact	Septic conditions	Remove sludge
	General plant	Septic conditions	Good housekeeping
Chlorine	Chlorine contact tank	Improper chlorine dosage	Adjust chlorine dosage
Industrial odors	General plant	Inadequate pretreatment	Enforce sewer use regulation

must then be positively vented to wet-chemical scrubbers to prevent the buildup of a toxic concentration of gas.

As mentioned, in drinking water, taste and odor are not normally a problem until the consumer complains. The problem is, of course, that most consumers find taste and odor in water aesthetically displeasing. As mentioned, taste and odor do not directly present a health hazard, but they can cause the customer to seek water that tastes and smells good, but may not be safe to drink. Most consumers consider water tasteless and odorless. Thus, when consumers find that their drinking water has a taste or odor, or both, they automatically associate the drinking water with contamination.

Water contaminants are attributable to contact with nature or human use. Taste and odor in water are caused by a variety of substances such as minerals, metals, and salts from the soil, constituents of wastewater, and end products produced in biological reactions. When water has a taste but no accompanying odor, the cause is usually inorganic contamination. Water that tastes bitter is usually alkaline, while salty water is commonly the result of metallic salts. However, when water has both taste and odor, the likely cause is organic materials. The list of possible organic contaminants is too long to record here; however, petroleum-based products lead the list of offenders. Taste- and odor-producing liquids and gases in water are produced by biological decomposition of organics. A prime example of one of these is hydrogen sulfide, known best for its characteristic "rotten egg" taste and odor. Certain species of algae also secrete an oily substance that may produce both taste and odor. When certain substances combine (such as organics and chlorine), the synergistic effect produces taste and odor.

In water treatment, one of the common methods used to remove taste and odor is to oxidize the materials that cause the problem. Oxidants, such as potassium permanganate and chlorine, are used. Another common treatment method is to feed powdered activated carbon before the filter. The activated carbon has numerous small openings that absorb the components that cause the odor and taste. These contained spaces must then be positively vented to wet-chemical scrubbers to prevent the buildup of toxic concentrations of gas.

Temperature

Heat is added to surface and groundwater in many ways. Some of these are natural, some artificial. For example, heat is added by natural means to Yellowstone Lake, Wyoming. The lake, one of the world's largest freshwater lakes, resides in a caldera, situated at more than 7,700 feet (the largest high altitude lake in North America). When one attempts to swim in Yellowstone Lake (without a wetsuit), the bitter cold of the water literally takes one's breath away. However, if it were not for the hydrothermal discharges that occur in Yellowstone, the water would be even colder. In regard to human-heated water, this most commonly occurs whenever a raw water source is used for cooling water in industrial operations. The influent to industrial facilities is at normal ambient temperature. When it is used to cool machinery and industrial processes, however, and then discharged back to the receiving body, it is often heated. The problem with heat or temperature increases in surface waters is that it affects the solubility of oxygen in water, the rate of bacterial activity, and the rate at which gases are transferred to and from the water.

Note: It is important to point out that in the examination of water or wastewater, temperature is not normally used to evaluate either. However, temperature is one of the most important parameters in natural surface-water systems. Surface waters are subject to great temperature variations.

Water temperature does determine, in part, how efficiently certain water treatment processes operate. For example, temperature has an effect on the rate at which chemicals dissolve and react. When water is cold, more chemicals are required for efficient coagulation and flocculation to take place. When water temperature is high, the result may be a higher chlorine demand because of the increased reactivity and there is often an increased level of algae and other organic matter in raw water. Temperature also has a pronounced effect on the solubility of gases in water.

Ambient temperature (temperature of the surrounding atmosphere) has the most profound and universal effect on temperature of shallow natural water systems. When water is used by industry to dissipate process waste heat, the discharge locations into surface waters may experience localized temperature changes that are quite dramatic. Other sources of increased temperatures in running water systems result because of clear-cutting practices in forests (where protective canopies are removed) and from irrigation flows returned to a body of running water.

In wastewater treatment, the temperature of wastewater varies greatly, depending upon the type of operations being conducted at a particular installation. However, wastewater is generally warmer than that of the water supply, because of the addition of warm water from industrial activities and households. Wide variation in the wastewater temperature indicates heated or cooled discharges, often of substantial volume. They have any number of sources. For example, decreased temperatures after a snowmelt or rain event may indicate serious infiltration. In the treatment process itself, temperature not only influences the metabolic activities of the microbial population but also has a profound effect on such factors as gas-transfer rates and the settling characteristics of the biological solids.

Chemical Characteristics of Water

Another category used to define or describe water quality is its chemical characteristics. The most important chemical characteristics are total dissolved solids (TDS), alkalinity, hardness, fluoride, metals, organics, and nutrients. Chemical impurities can be either natural, human-made (industrial), or be deployed in raw water sources by enemy forces. Some chemical impurities cause water to behave as either an acid or a base. Because either condition has an important bearing on the water treatment process, the pH value must be determined. Generally, the pH influences the corrosiveness of the water, chemical dosages necessary for proper disinfection, and the ability to detect contaminants. The principal contaminants found in water are shown in Table 9.3. These chemical constituents are important because each one affects water use in some manner; each one either restricts or enhances specific uses. The pH of water is important. As pH rises, for example, the equilibrium (between

TABLE 9.3
Chemical Constituents Commonly Found in Water

Constituents	
Calcium	Fluorine
Magnesium	Nitrate
Sodium	Silica
Potassium	Total Dissolved Solids
Iron	Hardness
Manganese	Color
Bicarbonate	pH
Carbonate	Turbidity
Sulfate	Temperature
Chloride	

bicarbonate and carbonate) increasingly favors the formation of carbonate, which often results in the precipitation of carbonate salts. If you have ever had flow in a pipe system interrupted or a heat-transfer problem in your water heater system, then carbonate salts that formed a hard-to-dissolve scale within the system most likely the cause. It should be pointed out that not all carbonate salts have a negative effect on their surroundings. Consider, for example, the case of blue marl lakes; they owe their unusually clear, attractive appearance to carbonate salts.

Water has been called the *universal solvent*. This is, of course, a fitting description. The solvent capabilities of water are directly related to its chemical characteristics or parameters. As mentioned, in water-quality management, total dissolved solids, alkalinity, hardness, fluorides, metals, organics, and nutrients are the major chemical parameters of concern.

Total Dissolved Solids

Because of water's solvent properties, minerals dissolved from rocks and soil as water passes over and through it produce total dissolved solids (comprised of any minerals, salts, metals, cations, or anions dissolved in water). TDS constitutes a part of total solids (TS) in water; it is the material remaining in water after filtration. Dissolved solids may be organic or inorganic. Water may be exposed to these substances within the soil, on surfaces, and in the atmosphere. The organic dissolved constituents of water come from the decay products of vegetation, from organic chemicals, and from organic gases.

Dissolved solids can be removed from water by distillation, electro-dialysis, reverse osmosis, or ion exchange. It is desirable to remove these dissolved minerals, gases, and organic constituents because they may cause psychological effects and produce aesthetically displeasing color, taste, and odors. While it is desirable to remove many of these dissolved substances from water, it is not prudent to remove them all. This is the case, for example, because pure, distilled water has a flat taste.

Further, water has an equilibrium state with respect to dissolved constituents. Thus, if water is out of equilibrium or undersaturated, it will aggressively dissolve materials it comes into contact with. Because of this problem, substances that are readily dissolvable are sometimes added to pure water to reduce its tendency to dissolve plumbing.

Alkalinity

Another important characteristic of water is its *alkalinity*, which is a measure of water's ability to neutralize acid or really an expression of buffering capacity. The major chemical constituents of alkalinity in natural water supplies are the bicarbonate, carbonate, and hydroxyl ions. These compounds are mostly the carbonates and bicarbonates of sodium, potassium, magnesium, and calcium. These constituents originate from carbon dioxide (from the atmosphere and as a by-product of microbial decomposition of organic material) and from their mineral origin (primarily from chemical compounds dissolved from rocks and soil). Highly alkaline waters are unpalatable; however, this condition has little known significance for human health. The principal problem with alkaline water is the reactions that occur between alkalinity and certain substances in the water. Alkalinity is important for fish and aquatic life because it protects or buffers against rapid pH changes. Moreover, the resultant precipitate can foul water system appurtenances. In addition, alkalinity levels affect the efficiency of certain water treatment processes, especially the coagulation process.

Hardness

Hardness is due to the presence of multivalent metal ions, which come from minerals dissolved in water. Hardness is based on the ability of these ions to react with soap to form a precipitate or soap scum. In freshwater, the primary ions are calcium and magnesium; however, iron and manganese may also contribute. Hardness is classified as *carbonate* hardness or *noncarbonate* hardness. Carbonate hardness is equal to alkalinity but a noncarbonate fraction may include nitrates and chlorides. Hardness is either temporary or permanent. Carbonate hardness (temporary hardness) can be removed by boiling. Noncarbonate hardness cannot be removed by boiling and is classified as permanent.

Hardness values are expressed as an equivalent amount or equivalent weight of calcium carbonate (*equivalent weight* of a substance is its atomic or molecular weight divided by *n*). Water with a hardness of less than 50 parts per million (ppm) is soft. Above 200 ppm, domestic supplies are usually blended to reduce the hardness value. The classification used by the U.S. Geological Survey is shown in Table 9.4.

The impact of hardness can be measured in economic terms. Soap consumption points this out; that is, soap consumption represents an economic loss to the water user. When washing with a bar of soap, there is a need to use more soap to "get a lather," whenever washing in hard water. There is another problem with soap and hardness. When using a bar of soap in hard water, when lather is finally built up, the water has been "softened" by the soap. The precipitate formed by the hardness and soap (soap curd) adheres to just about anything (tubs, sinks, dishwashers) and

TABLE 9.4
Classification of Hardness

Range of hardness (mg/liter (ppm) as $CaCO_3$)	Descriptive classification
1–50	Soft
51–150	Moderately hard
151–300	Hard
Above 300	Very hard

may stain clothing, dishes, and other items. There also is a personal problem: the residues of the hardness-soap precipitate may precipitate into the pores, causing skin to feel rough and uncomfortable. Today these problems have been largely reduced by the development of synthetic soaps and detergents that do not react with hardness. However, hardness still leads to other problems, scaling and laxative effect. Scaling occurs when carbonate hard water is heated and calcium carbonate and magnesium hydroxide are precipitated out of solution, forming a rock-hard scale that clogs hot water pipes and reduces the efficiency of boilers, water heaters, and heat exchangers. Hardness, especially with the presence of magnesium sulfates, can lead to the development of a laxative effect on new consumers.

The use of hard water does offer some advantages, though, in that (1) hard water aids in growth of teeth and bones; (2) hard water reduces toxicity to many by poisoning with lead oxide from lead pipelines; and (3) soft waters are suspected to be associated with cardiovascular diseases (Rowe & Abdel-Magid, 1995).

Fluoride

We purposely fluoridate a range of everyday products, notably toothpaste and drinking water, because for decades we have believed that fluoride in small doses has no adverse effects on health to offset its proven benefits in preventing dental decay. The jury is still out, however, on the real benefits of fluoride, even in small amounts. Fluoride is seldom found in appreciable quantities in surface waters and appears in groundwater in only a few geographical regions. However, fluoride is sometimes found in a few types of igneous or sedimentary rocks. Fluoride is toxic to humans in large quantities. Fluoride is also toxic to some animals. For example, certain plants used for fodder have the ability to store and concentrate fluoride. When animals consume this forage, they ingest an enormous overdose of fluoride. Animals' teeth become mottled, they lose weight, give less milk, grow spurs on their bones, and become so crippled they must be destroyed (Koren, 1991).

Fluoride used in small concentrations (about 1.0 mg/L in drinking water) can be beneficial. Experience has shown that drinking water containing a proper amount of fluoride can reduce tooth decay by 65 percent in children of ages 12–15. When large concentrations are used (>2.0 mg/L), discoloration of teeth may result. Adult teeth are not affected by fluoride. EPA sets the upper limits for fluoride based on ambient temperatures, because people drink more water in warmer climates; therefore, fluoride concentrations should be lower in these areas.

Note: How does fluoridization of a drinking water supply actually work to reduce tooth decay? Fluoride combines chemically with tooth enamel when permanent teeth are forming. The result is teeth that are harder, stronger, and more resistant to decay.

Metals

Although iron and manganese are most commonly found in groundwaters, surface waters may also contain significant amounts at times. Metal ions are dissolved in groundwater and surface water when the water is exposed to rock or soil containing the metals, usually in the form of metal salts. Metals can also enter with discharges from sewage treatment plants, industrial plants, and other sources. The metals most often found in the highest concentrations in natural waters are calcium and magnesium. These are usually associated with a carbonate anion and come from the dissolution of limestone rock. As mentioned under the discussion of hardness, the higher the concentration of these metal ions, the harder the water; however, in some waters other metals can contribute to hardness. Calcium and magnesium are nontoxic and normally absorbed by living organisms more readily than the other metals. Therefore, if the water is hard, the toxicity of a given concentration of a toxic metal is reduced. Conversely, in soft, acidic water the same concentrations of metals may be more toxic. In natural water systems, other nontoxic metals are generally found in very small quantities. Most of these metals cause taste problems well before they reach toxic levels. Fortunately, toxic metals are present in only minute quantities in most natural water systems. However, even in small quantities, toxic metals in drinking water are harmful to humans and other organisms. Arsenic, barium, cadmium, chromium, lead, mercury, and silver are toxic metals that may be dissolved in water. Arsenic, cadmium, lead, and mercury, all cumulative toxins, are particularly hazardous. These particular metals are concentrated by the food chain, thereby posing the greatest danger to organisms near the top of the chain.

Organics

Organic chemicals in water primarily emanate from synthetic compounds that contain carbon, such as PCBs, dioxin, and DDT (all toxic organic chemicals). These synthesized compounds often persist and accumulate in the environment because they do not readily breakdown in natural ecosystems. Many of these compounds can cause cancer in people and birth defects in other predators near the top of the food chain, such as birds and fish. The presence of organic matter in water is troublesome for the following reasons: "(1) color formation, (2) taste and odor problems, (3) oxygen depletion in streams, (4) interference with water treatment processes, and (5) the formation of halogenated compounds when chlorine is added to disinfect water" (Tchobanglous & Schroeder, 1985).

Generally, the source of organic matter in water is from decaying leaves, weeds, and trees; the amount of these materials present in natural waters is usually low. The general category of "organics" in natural waters includes organic matter whose origins could be from both natural sources and from human activities. It is important to distinguish natural organic compounds from organic compounds that are solely man-made (anthropogenic), such as pesticides and other synthetic organic compounds.

Many organic compounds are soluble in water, and surface waters are more prone to contamination by natural organic compounds that are groundwaters. In water, dissolved organics are usually divided into two categories: *biodegradable* and *nonbiodegradeable*. Biodegradable (break down) material consists of organics that can be utilized for nutrients (food) by naturally occurring microorganisms within a reasonable length of time. These materials usually consist of alcohols, acids, starches, fats, proteins, esters, and aldehydes. They may result from domestic or industrial wastewater discharges, or they may be end products of the initial microbial decomposition of plant or animal tissue. The principal problem associated with biodegradable organics is the effect resulting from the action of microorganisms. Moreover, some biodegradable organics can cause color, taste, and odor problems.

Oxidation and *reduction* play an important accompanying role in microbial utilization of dissolved organics. In oxidation, oxygen is added, or hydrogen is deleted from elements of the organic molecule. Reduction occurs when hydrogen is added to or oxygen is deleted from elements of the organic molecule. The oxidation process is by far more efficient and is predominant when oxygen is available. In *oxygen-present (aerobic)* environments, the end products of microbial decomposition of organics are stable and acceptable compounds. On the other hand, *oxygen-absent (anaerobic)* decomposition results in unstable and objectionable end products.

The quantity of oxygen-consuming organics in water is usually determined by measuring the *biochemical oxygen demand (BOD)*: the amount of dissolved oxygen needed by aerobic decomposers to break down the organic materials in a given volume of water over a five-day incubation period at 20°C (68°F).

Nonbiodegradeable organics are resistant to biological degradation. For example, constituents of woody plants such as tannin and lignic acids, phenols, and cellulose are found in natural water systems and are considered refractory (resistant to biodegradation). In addition, some polysaccharides with exceptionally strong bonds and benzene with its ringed structure are essentially nonbiodegradeable. An example is benzene associated with the refining of petroleum. Some organics are toxic to organisms and thus are nonbiodegradeable. These include the organic pesticides and compounds that combine with chlorine. Pesticides and herbicides have found widespread use in agriculture, forestry (silviculture), and mosquito control. Surface streams are contaminated via runoff and wash off by rainfall. These toxic substances are harmful to some fish, shellfish, predatory birds, and mammals. Some compounds are toxic to humans.

Nutrients

Nutrients (Biostimulants) are essential building blocks for healthy aquatic communities, but excess nutrients (especially nitrogen and phosphorous compounds) over stimulate the growth of aquatic weeds and algae. Excessive growth of these organisms, in turn, can clog navigable waters, interfere with swimming and boating, outcompete native submerged aquatic vegetation, and, with excessive decomposition, lead to oxygen depletion. Oxygen concentrations can fluctuate daily during algae blooms, rising during the day as algae perform photosynthesis and falling at night as algae continue to respire, which consumes oxygen. Beneficial bacteria also consume oxygen as they decompose the abundant organic food supply in dying algae cells.

Plants require large amounts of the nutrients carbon, nitrogen, and phosphorus, otherwise growth will be *limited*.

Carbon is readily available from a number of natural sources including alkalinity, decaying products of organic matter, and from dissolved carbon dioxide from the atmosphere. Since carbon is readily available, it is seldom the *limiting nutrient*. This is an important point because it suggests that identifying and reducing the supply of a particular nutrient can control algal growth. In most cases, nitrogen and phosphorous are essential growth factors and are the limiting factors in aquatic plant growth. Freshwater systems are most often limited by phosphorus.

Nitrogen gas (N_2), which is extremely stable, is the primary component of the earth's atmosphere. Major sources of nitrogen include runoff from animal feedlots, fertilizer runoff from agricultural fields, from municipal wastewater discharges, and from certain bacteria and blue-green algae that can obtain nitrogen directly from the atmosphere. In addition, certain forms of acid rain can also contribute nitrogen to surface waters.

Nitrogen in water is commonly found in the form of nitrate (NO_3). Nitrate in drinking water can lead to a serious problem. Specifically, nitrate poisoning in infant humans, including animals, can cause serious problems and even death. Bacteria commonly found in the intestinal tract of infants can convert nitrate to highly toxic nitrites (NO_2). Nitrite can replace oxygen in the bloodstream and result in oxygen starvation that causes a bluish discoloration of the infant ("blue baby" syndrome).

In aquatic environments, phosphorus is found in the form of phosphate. Major sources of phosphorus include phosphates in detergents, fertilizer and feedlot runoff, and municipal wastewater discharges.

Chemical Characteristics of Wastewater

The chemical characteristics of wastewater consist of three parts: (1) organic matter, (2) inorganic matter, and (3) gases. Metcalf and Eddy (2003) point out that in "wastewater of medium strength, about 75% of the suspended solids and 40% of the filterable solids are organic in nature." The organic substances of interest in this discussion include proteins, oil and grease, carbohydrates, and detergents (surfactants).

Organic Substances

Proteins are nitrogenous organic substances of high molecular weight found in the animal kingdom and to a lesser extent in the plant kingdom. The amount present varies from a small percentage found in tomatoes and other watery fruits and in the fatty tissues of meat, to a high percentage in lean meats and beans. All raw foodstuffs, plant and animal, contain proteins. Proteins consist wholly or partially of very large numbers of amino acids. They also contain carbon, hydrogen, oxygen, sulfur, phosphorous, and a fairly high and constant proportion of nitrogen. The molecular weight of proteins is quite high.

Proteinaceous materials constitute a large part of the wastewater biosolids; biosolids particles that do not consist of pure protein will be covered with a layer of protein that will govern their chemical and physical behavior (Coakley, 1975). Moreover, the protein content ranges from 15 percent to 30 percent of the organic matter present for digested biosolids, and from 28 percent to 50 percent in the case of activated biosolids. Proteins and urea are the chief sources of nitrogen in wastewater. When proteins are present in large quantities, microorganisms decompose them, producing end products that have objectionable foul odors. During this decomposition process, proteins are hydrolyzed to amino acids and then further degraded to ammonia, hydrogen sulfide, and to simple organic compounds.

Oils and *grease* are other major components of foodstuffs. They are also usually related to spills or other releases of petroleum products. Minor oil and grease problems can result from wet weather runoff from highways or the improper disposal in storm drains of motor oil. They are insoluble in water but dissolve in organic solvents such as petroleum, chloroform, and ether. Fats, oils, waxes, and other related constituents found in wastewater are commonly grouped under the term "grease." Fats and oils are contributed in domestic wastewater in butter, lard, margarine, and vegetable fats and oils. Fats, which are compounds of alcohol and glycerol, are among the more stable organic compounds and are not easily decomposed by bacteria. However, they can be broken down by mineral acids resulting in the formation of fatty acid and glycerin. When these glycerides of fatty acids are liquid at ordinary temperature they are called oils, and those that are solids are called fats.

The grease content of wastewater can cause many problems in wastewater treatment unit processes. For example, high grease content can cause clogging of filters, nozzles, and sand beds (Gilcreas et al., 1975). Moreover, grease can coat the walls of sedimentation tanks and decompose and increase the amount of scum. Additionally, if grease is not removed before discharge of the effluent, it can interfere with the biological processes in the surface waters and create unsightly floating matter and films (Rowe & Abdel-Magid, 1995). In the treatment process, grease can coat trickling filters and interfere with the activated sludge process, which, in turn, interferes with the transfer of oxygen from the liquid to the interior of living cells (Sawyer et al., 1994).

Carbohydrates, which are widely distributed in nature and found in wastewater, are organic substances that include starch, cellulose, sugars, and wood fibers; they contain carbon, hydrogen, and oxygen. Sugars are soluble while starches are insoluble in water. The primary function of carbohydrates in higher animals is to serve as a source of energy. In lower organisms, e.g., bacteria, carbohydrates are utilized to synthesize fats and proteins as well as energy. In the absence of oxygen, the end products of decomposition of carbohydrates are organic acids, alcohols, as well as gases such as carbon dioxide and hydrogen sulfide. The formation of large quantities of organic acids can affect the treatment process by overtaxing the buffering capacity of the wastewater resulting in a drop in pH and a cessation of biological activity.

Detergents (*surfactants*) are large organic molecules that are slightly soluble in water and cause foaming in wastewater treatment plants and in the surface waters into which the effluent is discharged. Probably the most serious effect detergents can have on wastewater treatment processes is in their tendency to reduce the oxygen uptake in biological processes. According to Rowe and Abdel-Magid (1995),

> Detergents affect wastewater treatment processes by (1) lowering the surface, or interfacial, tension of water and increase its ability to wet surfaces with which they come in contact; (2) emulsify grease and oil, deflocculate colloids; (3) induce flotation of solids and give rise to foams; and (4) may kill useful bacteria and other living organisms.

Since the development and increasing use of synthetic detergents, many of these problems have been reduced or eliminated.

Inorganic Substances

Several inorganic components are common to both wastewater and natural waters and are important in establishing and controlling water quality. Inorganic load in water is the result of discharges of treated and untreated wastewater, various geologic formations, and inorganic substances left in the water after evaporation. Natural waters dissolve rocks and minerals with which they come in contact. As mentioned, many of the inorganic constituents found in natural waters are also found in wastewater. Many of these constituents are added via human use. These inorganic constituents include pH, chlorides, alkalinity, nitrogen, phosphorus, sulfur, toxic inorganic compounds, and heavy metals.

When the *pH* of a water or wastewater is considered, we are simply referring to the hydrogen ion concentration. Acidity, the concentration of hydrogen ions, drives many chemical reactions in living organisms. A pH value of 7 represents a neutral condition. A low pH value (less than 5) indicates acidic conditions; a high pH (greater than 9) indicates alkaline conditions. Many biological processes, such as reproduction, cannot function in acidic or alkaline waters. Acidic conditions also aggravate toxic contamination problems because sediments release toxicants in acidic waters.

Many of the important properties of wastewater are due to the presence of weak acids and bases and their salts. The wastewater treatment process is made up of several different unit processes (these are discussed later). It can be safely stated that one of the most important unit processes in the overall wastewater treatment process is disinfection. pH has an effect on disinfection. This is particularly the case in regard to disinfection using chlorine. For example, with increases in pH, the amount of contact time needed for disinfection using chlorine increases. Common sources of acidity include mine drainage, runoff from mine tailings, and atmospheric deposition.

In the form of the Cl^- ion, *chloride* is one of the major inorganic constituents in water and wastewater. Sources of chlorides in natural waters are (1) leaching of chloride form rocks and soils; (2) in coastal areas, salt-water intrusion; (3) from agricultural, industrial, domestic, and human wastewater; and (4) from infiltration of groundwater into sewers adjacent to salt water. The salty taste produced by chloride concentration in potable water is variable and depends on the chemical composition of the water. In wastewater, the chloride concentration is higher than in raw water because sodium chloride (salt) is a common part of the diet and passes unchanged through the digestive system. Because conventional methods of waste treatment do not remove chloride to any significant extent, higher than usual chloride concentrations can be taken as an indication that the body of water is being used for waste disposal (Metcalf & Eddy, 2003).

As mentioned earlier, *alkalinity* is a measure of the buffering capacity of water and in wastewater helps to resist changes in pH caused by the addition of acids. Alkalinity is caused by chemical compounds dissolved from soil and geologic formations and is mainly due to the presence of hydroxyl and bicarbonate ions. These compounds are mostly the carbonates and bicarbonates of calcium, potassium, magnesium, and sodium. Wastewater is usually alkaline. Alkalinity is important in wastewater treatment because anaerobic digestion requires sufficient alkalinity to ensure that the pH will not drop below 6.2; if alkalinity does drop below this level, the methane bacteria cannot function. For the digestion process to operate successfully the alkalinity must range from about 1,000 mg/L to 5,000 mg/L as calcium carbonate. Alkalinity in wastewater is also important when chemical treatment is used, in biological nutrient removal, and whenever ammonia is removed by air stripping.

In domestic wastewater, "nitrogen compounds result from the biological decomposition of proteins and form urea discharged in body waste" (Peavy et al., 1987). In wastewater treatment, biological treatment cannot proceed unless *nitrogen*, in some form, is present. Nitrogen must be present in the form of organic nitrogen (N), ammonia (NH_3), nitrite (NO_2), or nitrate (NO_3). Organic nitrogen includes such natural constituents as peptides, proteins, urea, nucleic acids, and numerous synthetic organic materials. Ammonia is present naturally in wastewaters. It is produced primarily by de-aeration of organic nitrogen-containing compounds and by hydrolysis of area. Nitrite, an intermediate oxidation state of nitrogen, can enter a water system through use as a corrosion inhibitor in industrial applications. Nitrate is derived from the oxidation of ammonia.

Nitrogen data is essential in evaluating the treatability of wastewater by biological processes. If nitrogen is not present in sufficient amounts, it may be necessary to add it to the waste to make it treatable. When the treatment process is complete, it is important to determine how much nitrogen is in the effluent. This is important because the discharge of nitrogen into receiving waters may stimulate algal and aquatic plant growth. These, of course, exert a high oxygen demand at nighttime, which adversely affects aquatic life and has a negative impact on the beneficial use of water resources.

Phosphorus (P) is a macronutrient that is necessary for all living cells and is a ubiquitous constituent of wastewater. It is primarily present in the form of phosphates—the salts of phosphoric acid. Municipal wastewaters may contain 10–20 mg/L phosphorus as P, much of which comes from phosphate builders in detergents. Because of noxious algal blooms that occur in surface waters, there is much interest in controlling the amount of phosphorus compounds that enter surface waters in domestic and industrial waste discharges and natural runoff. This is particularly the case in the United States because approximately 15 percent of the population contributes wastewater effluents to lakes, resulting in *eutrophication* (or cultural enrichment) of these water bodies. Eutrophication occurs when the nutrient levels exceed the affected waterbody's ability to assimilate them. Eutrophication leads to significant changes in water quality. Reducing phosphorus inputs to receiving waters can control this problem.

Sulfur (S) is required for the synthesis of proteins and is released in their degradation. The sulfate ion occurs naturally in most water supplies and is present in wastewater as well. Sulfate is reduced biologically to sulfide, which in turn can combine with hydrogen to form hydrogen sulfide (H_2S). H_2S is toxic to animals and plants. H_2S in interceptor systems can cause severe corrosion to pipes and appurtenances. Moreover, in certain concentrations, H_2S is a deadly toxin.

Toxic inorganic compounds such as copper, lead, silver, arsenic, boron, and chromium are classified as priority pollutants and are toxic to microorganisms. Thus, they must be taken into consideration in the design and operation of a biological treatment process. When introduced into a treatment process, these contaminants can kill off the microorganisms needed for treatment and thus stop the treatment process.

Heavy metals are major toxicants found in industrial wastewaters; they may adversely affect the biological treatment of wastewater. Mercury, lead, cadmium, zinc, chromium, and plutonium are among the so-called heavy metals—those with a high atomic mass. (It should be noted that the term "heavy metals" is rather loose and is taken by some to include arsenic, beryllium, and selenium, which are not really metals and are better termed toxic metals). The presence of any of these metals in excessive quantities will interfere with many beneficial uses of water because of their toxicity. Urban runoff is a major source of lead and zinc in many water bodies.

***Note*:** Lead is a toxic metal that is harmful to human health; there is *no* safe level for lead exposure. It is estimated that up to 20 percent of the total lead exposure in children can be attributed to a waterborne route, i.e., consuming contaminated water). The lead comes from the exhaust of automobiles using leaded gasoline, while zinc comes from tire wear.

Biological Characteristics of Water and Wastewater

Specialists or practitioners who work in the water or wastewater treatment field must not only have a general understanding of the microbiological principles, but also must have some knowledge of the biological characteristics of water and wastewater. This knowledge is especially important for those involved with evaluating wastewater-based epidemiology for biological microbes. This knowledge begins with an understanding that water may serve as a medium in which thousands of biological species spend part, if not all, of their life cycles. It is important to understand that, to some extent, all members of the biological community are water-quality parameters, because their presence or absence may indicate in general terms the characteristics of a given body of water.

The presence or absence of certain biological organisms is of primary importance to the water/wastewater specialist. These are, of course, the *pathogens*. Pathogens are organisms that are capable of infecting or transmitting diseases in humans and animals. It should be pointed out that these organisms are not native to aquatic systems and usually require an animal host for growth and reproduction. They can, however, be transported by natural water systems. These waterborne pathogens include species of bacteria, viruses, protozoa, and parasitic worms (helminthes). In the following sections a brief review of each of these species is provided.

Bacteria

The word *bacteria* (singular: bacterium) comes from the Greek word meaning "rod" or" staff," a shape characteristic of many bacteria. Recall that bacteria are single-celled microscopic organisms that multiply by splitting in two (binary fission). In order to multiply they need carbon dioxide if they are autographs, from organic compounds (dead vegetation, meat, sewage) if they are heterotrophs. Their energy comes either from sunlight if they are photosynthetic or from chemical reactions if they are chemosynthetic. Bacteria are present in air, water, earth, rotting vegetation, and the intestines of animals. Human and animal wastes are the primary source of bacteria in water. These sources of bacterial contamination include runoff from feedlots, pastures, dog runs, and other land areas where animal wastes are deposited. Additional sources include seepage or discharge from septic tanks and sewage treatment facilities. Bacteria from these sources can enter wells that are either open at the land surface, or do not have watertight casings or caps. Gastrointestinal disorders are common symptoms of most diseases transmitted by waterborne pathogenic bacteria. In wastewater treatment processes, bacteria are fundamental, especially in the degradation of organic matter, which takes place in trickling filters, activated biosolids processes, and biosolids digestion.

Viruses

A *virus* is an entity that carries the information needed for its replication but does not possess the machinery for such replication (Sterritt & Lester, 1988). Thus, they are obligate parasites that require a host in which to live. They are the smallest biological structures known, so they can only be seen with the aid of an electron microscope. Waterborne viral infections are usually indicated by disorders with the nervous system rather than of the gastrointestinal tract. Viruses that are excreted by human beings may become a major health hazard to public health. Waterborne viral pathogens are known to cause poliomyelitis and infectious hepatitis. Testing for viruses in water is difficult because (1) they are small; (2) they are of low concentrations in natural waters; (3) there are numerous varieties; (4) they are unstable; and (5) there are limited identification methods available. Because of these testing problems and the uncertainty of viral disinfection, direct recycling of wastewater and the practice of land application of wastewater is a cause of concern (Peavy et al., 1987).

Protozoa

Protozoa (singular: protozoan) are mobile, single-celled, complete, self-contained organisms that can be free-living or parasitic, pathogenic or nonpathogenic, microscopic or macroscopic. Protozoa range in size from two to several hundred microns in length. They are highly adaptable and widely distributed in natural waters, although only a few are parasitic. Most protozoa are harmless, only a few cause illness in humans—*Entamoeba histolytica* (amebiasis) being one of the exceptions.

Because aquatic protozoa form cysts during adverse environmental conditions, they are difficult to deactivate by disinfection and must undergo filtration to be removed.

Worms (Helminthes)

Worms are the normal inhabitants in organic mud and organic slime. They have aerobic requirements but can metabolize solid organic matter not readily degraded by other microorganisms. Water contamination may result from human and animal waste that contains worms. Worms pose hazards primarily to those persons who come into direct contact with untreated water. Thus, swimmers in surface water polluted by sewage or stormwater runoff from cattle feedlots and sewage plant operators are at particular risk.

REFERENCES

Coakley, P., 1975. Developments in our knowledge of sludge dewatering behavior, *8th Public Health Engineering Conference held in the Department of Civil Engineering.* Loughborough: University of Technology.

Gilcreas, F.W., Sanderson, W.W., & Elmer, R.P., 1975. Method for the Gravimetric determination of oil and grease. *Sewage and Industrial wastes*, 32(8): 1210–1221.

Koren, H., 1991. *Handbook of Environmental Health and Safety: Principles and Practices.* Chelsea, MI: Lewis Publishers.

Metcalf & Eddy, Inc., 2003. *Wastewater Engineering: Treatment, Disposal, Reuse*, 4th ed. New York: McGraw-Hill.

Peavy, H.S., Rowe, D.R., & Tchobanglous, G., 1987. *Environmental Engineering.* New York: McGraw-Hill, Inc.

Rowe, D.R., & Abdel-Magid, I.M., 1995. *Handbook of Wastewater Reclamation and Reuse.* Boca Raton, FL: Lewis Publishers.

Sawyer, C.N., McCarty, A.L., & Parking, G.F., 1994. *Chemistry for Environmental Engineering.* New York: McGraw-Hill.

Sterritt, R.M., & Lester, J.M., 1988. *Microbiology for Environmental and Public Health Engineers.* London: E. and F.N. Spoon.

Tchobanglous, G. & Schoreder, E.D., 1985. *Water Quality.* Reading, MA: Addison-Wesley.

10 Wastewater Spawned Infectious Disease

The infectious disease agents associated with or spawned in municipal wastewater are those found in the domestic/municipal sanitary waste of the population and from businesses that handle or process meats, fish, and other food products. These microbial pathogens include a large number of bacteria, viruses, and protozoa (parasites, including helminthes—worms). Important examples are members of the bacterial genera *Salmonella* and *Shigella*; the infection hepatitis, Rota and Norwalk virus (and in this text special focus is placed on coronavirus, the COVID-19 virus); and the parasites (protozoan's and worms) associated with giardiasis, cryptosporidiosis, taeniasis, and ascariasis (see Table 10.1 for a more complete list). It is reasonable to assume that any or all of these infectious agents might be present in raw sewage, including coronavirus.

DID YOU KNOW?

Taeniasis in humans is a parasitic infection caused by the tapeworm species *Taenia saginata* (beef tapeworm), *Taenia solium* (pork tapeworm), and *Taenia asiatica* (Asian tapeworm). Humans can become infected with these tapeworms by eating raw or undercooked beef or pork. People with taeniasis may not know they may have a tapeworm infection because symptoms are usually mild or nonexistent (CDC, 2013).

An estimated 807 million–1.2 billion people in the world are infected with *Ascaris lumbricoides* (sometimes call Ascaris or ascariasis). Ascaris, hookworm, and whipworm are parasitic worms known as soil-transmitted helminthes (STH; CDC, 2020).

BACTERIA, VIRUSES, PROTOZOA, AND HELMINTHS

In this section a basic, fundamental review of the microorganisms of concern, the microbial constituents of domestic/municipal wastewater, is provided. Again, the following is meant as a basic review for environmental professionals and/or a basic primer for the non-science professional. To set the foundation for the material presented in this section it begins with a fundamental discussion of the human cell.

TABLE 10.1
Examples of Pathogens Associated with Raw Domestic Sewage

Pathogen class	Examples	Disease
Bacteria	*Shigella sp.*	Bacillary dysentery
	Salmonella sp.	Salmonellosis (gastroenteritis)
	Salmonella typhi	Typhoid fever
	Vibrio cholera	Cholera
	Enteropathogenic *Escherichia coli*	A variety of gastroenteric diseases
	Yersinia sp.	Yersiniosis (gastroenteritis)
	Campylobacter jejuni	
Viruses	Hepatitis A virus	Infectious hepatitis
	COVID-19	Coronavirus
	Norwalk viruses	Acute gastroenteritis
	Rotaviruses	Acute gastroenteritis
	Polioviruses	Poliomyelitis
	Coxsackie viruses	"flu-like" symptoms
	Echoviruses	"flu-like" symptoms
Protozoa	*Entamoeba histolytica*	Amebiasis (amoebic dysentery)
	Giardia lamblia	Giardiasis (gastroenteritis)
	Cryptosporidium sp.	Cryptosporidiosis (gastroenteritis)
	Balantidium coli	Balantidiasis (gastroenteritis)
Helminthes	*Ascaris sp*	Ascariasis (roundworm infection)
	Taenia sp.	Taeniasis (tapeworm infection)
	Necator americanus	Ancylostomiasis (hookworm infection)
	Trichuris trichuria	Trichuriasis (whipworm infection)

Source: USEPA (2020). Public Health Concerns About Infectious Disease Agents. Accessed August 28 2020 @ https://www3.epa.gov/npdes/pubs/mstr-ch5.pdf.

The Cell

The structural and fundamental unit of both plants and animals, no matter how complex, is the cell. Since the nineteenth century, scientists have known that all living things, whether animal or plant, are made up of cells. A typical bacterial cell, for example, is an entity, isolated from other cells by a membrane or cell wall. The cell membrane contains protoplasm and the nucleus (see Figure 10.1). The protoplasm within the cell is a living mass of viscous, transparent material. Within the protoplasm is a dense spherical mass called the nucleus or nuclear material. In a typical mature plant cell, the cell wall is rigid and is composed of nonliving material, while in the typical animal cell, the wall is an elastic living membrane. Cells exist in a very great variety of sizes and shapes, as well as functions. Their average size ranges from bacteria too small to be seen with the light microscope to the largest known single

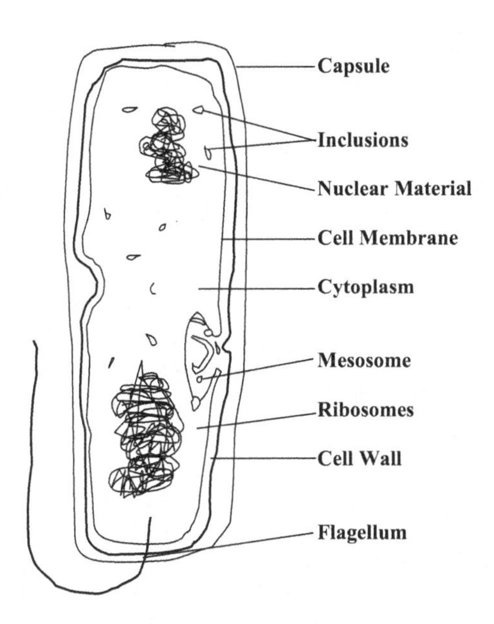

FIGURE 10.1 Bacterial cell. Illustration by F. Spellman (2001).

cell, the ostrich egg. Microbial cells also have an extensive size range, some being larger than human cells (Kordon, 1992).

Note: The nucleus cannot always be observed in bacteria.

Structure of the Bacterial Cell

The structural form and various components of the bacterial cell are probably best understood by referring to the simplified diagram of a rod-form bacterium shown in Figure 10.1. When studying Figure 10.1, keep in mind that cells of different species may differ greatly, both in structure and chemical composition; for this reason no typical bacterium exists. Figure 10.1 shows a generalized bacterium used for the discussion that follows. (Not all bacteria have all of the features shown in the figure, and some bacteria have structures not shown in the figure.)

Capsules

Bacterial *capsules* (see Figure 10.1) are organized accumulations of gelatinous materials on cell walls, in contrast to *slime layers* (a water secretion that adheres loosely to the cell wall and commonly diffuses into the cell), which are unorganized accumulations of similar material. The capsule is usually thick enough to be seen under the ordinary light microscope (macrocapsule), while thinner capsules (microcapsules) can be detected only by electron microscopy (Singleton & Sainsbury 1994). The production of capsules is determined largely by genetics as well as environmental conditions, and depends on the presence or absence of capsule-degrading enzymes and other growth factors. Varying in composition, capsules are mainly composed of water; the organic contents are made up of complex polysaccharides, nitrogen-containing substance, and polypeptides. Capsules confer several advantages when bacteria grow in their normal habitat. For example, they help to (1) prevent desiccation; (2) resist phagocytosis by host phagocytic cells; (3) prevent infection by bateriophages; and (4) aid bacterial attachment to tissue surfaces in plant and animal hosts or to surfaces of solids objects in aquatic environments. Capsule formation often correlates with pathogenicity.

Flagella

Many bacteria are motile, and this ability to move independently is usually attributed to a special structure, the *flagella* (singular: flagellum). Depending on species, a cell may have a single flagellum (see Figure 10.1) (*monotrichous* bacteria; *trichous* means "hair"); one flagellum at each end (*amphitrichous* bacteria; *amphi* means "on both sides"); a tuft of flagella at one or both ends (*lophotrichous* bacteria; *lopho* means "tuft"); or flagella that arise all over the cell surface (*peritrichous* bacteria; *peri* means "around").

A flagellum is a threadlike appendage extending outward from the plasma membrane and cell wall. Flagella are slender, rigid, locomotor structures, about 20 μm across and up to 15–20 μm long. Flagellation patterns are very useful in identifying bacteria and can be seen by light microscopy, but only after being stained with special techniques designed to increase their thickness. The detailed structure of flagella can be seen only in the electron microscope.

Bacterial cells benefit from flagella in several ways. They can increase the concentration of nutrients or decrease the concentration of toxic materials near the bacterial surfaces by causing a change in the flow rate of fluids. They can also disperse flagellated organisms to areas where colony formation can take place. The main benefit of flagella to organisms is their increased ability to flee from areas that might be harmful.

Cell Wall

The main structural component of most prokaryotes is the rigid *cell wall* (see Figure 10.1). Functions of the cell wall include (1) providing protection for the delicate protoplast from osmotic lysis (bursting); (2) determining a cell's shape; (3) acting as a permeability layer that excludes large molecules and various antibiotics and plays an active role in regulating the cell's intake of ions; and (4) providing a solid support for flagella. Cell walls of different species may differ greatly in structure, thickness, and composition. The cell wall accounts for about 20–40 percent of the dry weight of a bacterium.

Plasma Membrane (Cytoplasmic Membrane)

Surrounded externally by the cell wall and composed of a lipoprotein complex, the *plasma membrane* or cell membrane is the critical barrier, separating the inside from outside the cell (see Figure 10.1). About 7–8 μm thick and comprising 10–20 percent of a bacterium's dry weight, the plasma membrane controls the passage of all material into and out of the cell. The inner and outer faces of the plasma membrane are embedded with water-loving (hydrophilic) lips, whereas the interior is hydrophobic. Control of material into the cell is accomplished by screening, as well as by electric charge. The plasma membrane is the site of the surface charge of the bacteria.

In addition to serving as an osmotic barrier that passively regulates the passage of material into and out of the cell, the plasma membrane participates in the entire active transport of various substances into the bacterial cell. Inside the membrane, many highly reactive chemical groups guide the incoming material to the proper points for further reaction. This active transport system provides bacteria with certain advantages, including the ability to maintain a fairly constant intercellular ionic state in the presence of varying external ionic concentrations. In addition to participating in the uptake of nutrients, the cell membrane transport system participates in waste excretion and protein secretions.

Cytoplasm

Within a cell and bounded by the cell membrane is a complicated mixture of substances and structures called the *cytoplasm* (see figure 10.1). The cytoplasm is a water-based fluid containing ribosomes, ions, enzymes, nutrients, storage granules (under certain circumstances), waste products, and various molecules involved in synthesis, energy metabolism, and cell maintenance.

Mesosome

A common intracellular structure found in the bacterial cytoplasm is the Mesosome (see Figure 10.1). *Mesosomes* are invaginations of the plasma membrane in the shape of tubules, vesicles, or lamellae. Their exact function is unknown. Currently, many bacteriologists believe that Mesosomes are artifacts generated during the fixation of bacteria for electron microscopy.

Nucleoid (Nuclear Body or Region)

The *nuclear region* of the prokaryotic cell is primitive and a striking contrast to that of the eucaryotic cell (see Figure 10.1). Prokaryotic cells lack a distinct nucleus, the function of the nucleus being carried out by a single, long, double strand of DNA that is efficiently packaged to fit within the Nucleoid. The Nucleoid is attached to the plasma membrane. A cell can have more than one Nucleoid when cell division occurs after the genetic material has been duplicated.

Ribosomes

The bacterial cytoplasm is often packed with ribosomes (see Figure 10.1). *Ribosomes* are minute, rounded bodies made of RNA and are loosely attached to the plasma membrane. Ribosomes are estimated to account for about 40 percent of a bacterium's dry weight; a single cell may have as many as 10,000 ribosomes. Ribosomes are the site of protein synthesis and are part of the translation process.

Inclusions

Inclusions (or storage granules) are often seen within bacterial cells (see Figure 10.1). Some inclusion bodies are not bound by a membrane and lie free in the cytoplasm. A single-layered membrane about 2–4 μm thick encloses other inclusion bodies. Many bacteria produce polymers that are stored as granules in the cytoplasm.

BACTERIA

The simplest wholly contained life systems are *bacteria or prokaryotes*, which are the most diverse group of microorganisms. As mentioned, they are among the most common microorganisms in water, are primitive, unicellular (single celled) organisms, possessing no well-defined nucleus, that present a variety of shapes and nutritional needs. Bacteria contain about 85 percent water and 15 percent ash or mineral matter. The ash is largely composed of sulfur, potassium, sodium, calcium, and chlorides, with small amounts of iron, silicon, and magnesium. Bacteria reproduce by binary fission.

Note: *Binary fission* occurs when one organism splits or divides into two or more new organisms.

Bacteria, once called the smallest living organisms (now it is known that smaller forms of matter exhibit many of the characteristics of life), range in size from 0.5 µm to 2 µm in diameter and is about 1–10 µm long.

Note: A *micron* is a metric unit of measurement equal to 1 thousandth of a millimeter. To visualize the size of bacteria, consider that about 1,000 bacteria lying side by side would reach across the head of a straight pin.

Bacteria are categorized into three general groups based on their physical form or shape (though almost every variation has been found; see Table 10.2). The simplest form is the sphere. Spherical shaped bacteria are called *cocci* (meaning "berries"). They are not necessarily perfectly round but may be somewhat elongated, flattened on one side, or oval. Rod shaped bacteria is called *bacilli.* Spiral shaped bacteria (called Spirilla), which have one or more twists and are never straight, make up the third group (see Figure 10.2). Such formations are usually characteristic of a particular genus or species. Within these three groups are many different arrangements. Some exist as single cells; others as pairs, as packets of four or eight, as chains, and as clumps.

Most bacteria require organic food to survive and multiply. Plant and animal material that gets into the water provides the food source for bacteria. Bacteria convert food to energy and use the energy to make new cells. Some bacteria can use inorganics (e.g., minerals such as iron) as an energy source and exist and multiply even when organics (pollution) are not available.

TABLE 10.2
Forms of Bacteria

Form	Technical Name		Example
	Singular	Plural	
Sphere	Coccus	Cocci	Streptococcus
Rod	Bacillus	Bacilli	Bacillus typhosis
Curved or spiral	Spirillum	Spirilla	Spirillum cholera

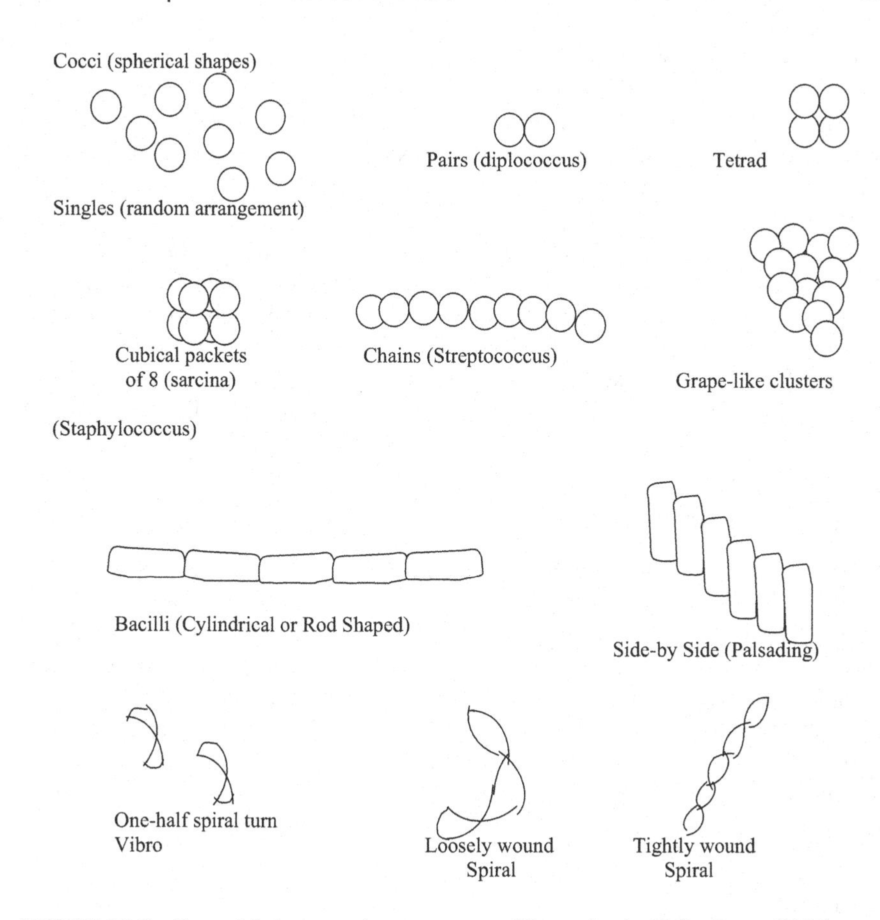

FIGURE 10.2 Bacterial shapes and arrangements. Illustration by F. Spellman (2001).

Bacteria Growth Factors

Several factors affect the rate at which bacteria grow, including temperature, pH, and oxygen levels. The warmer the environment, the faster the rate of growth. Generally, for each increase of 10°C, the growth rate doubles. Heat can also be used to kill bacteria. Most bacteria grow best at neutral pH. Extreme acidic or basic conditions generally inhibit growth, though some bacteria may require acidic and some require alkaline conditions for growth.

Bacteria are aerobic, anaerobic, or facultative. If *aerobic*, they require free oxygen in the aquatic environment. *Anaerobic* bacteria exist and multiply in environments that lack dissolved oxygen. *Facultative* bacteria (e.g., iron bacteria) can switch from an aerobic to anaerobic growth or grow in an anaerobic or aerobic environment.

Under optimum conditions, bacteria grow and reproduce very rapidly. As stated previously, bacteria reproduce by *binary fission*. An important point to consider in connection with bacterial reproduction is the rate at which the process can take place. The total time required for an organism to reproduce and the offspring to reach maturity is called *generation time*. Bacteria growing under optimal conditions can double their number about every 20–30 minutes. Obviously, this generation time is very short compared with that of higher plants and animals. Bacteria continue to grow at this rapid rate as long as nutrients hold out—even the smallest contamination can result in a sizable growth in a very short time.

Note: Even though wastewater can contain bacteria counts in the millions per mL, in wastewater treatment, under controlled conditions, bacteria can help to destroy and to identify pollutants. In such a process, bacteria stabilize organic matter (e.g., activated sludge processes), and thereby assist the treatment process in producing effluent that does not impose an excessive oxygen demand on the receiving body. Coliform bacteria can be used as an indicator of pollution by human or animal wastes.

Destruction of Bacteria

In water and wastewater treatment, the destruction of bacteria is usually called *disinfection*.

Disinfection does not mean that all microbial forms are killed. That would be *sterilization*. However, disinfection does reduce the number of disease-causing organisms to an acceptable number. Growing bacteria are easy to control by disinfection. Some bacteria, however, form spores—survival structures—which are much more difficult to destroy.

Note: Inhibiting the growth of microorganisms is termed antisepsis, while destroying them is called *disinfection*.

Waterborne Bacteria

All surface waters contain bacteria. Waterborne bacteria, as we have said, are responsible for infectious epidemic diseases. Bacterial numbers increase significantly during storm events when streams are high. Heavy rainstorms increase stream contamination by washing material from the ground surface into the stream. After the initial washing occurs, few impurities are left to be washed into the stream, which may then carry relatively "clean" water. A river of good quality shows its highest bacterial numbers during rainy periods; however, a much-polluted stream may show the highest numbers during low flows, because of the constant influx of pollutants. Water and wastewater operators are primarily concerned with bacterial pathogens responsible for disease. These pathogens enter potential drinking water supplies through fecal contamination and are ingested by humans if the water is not properly treated and disinfected.

Note: Regulations require that owners of all public water supplies collect water samples and deliver them to a certified laboratory for bacteriological examination at least monthly. The number of samples required is usually in accordance with Federal Standards, which generally require that one sample per month be collected for every 1,000 persons served by the waterworks.

PROTOZOA

Protozoans (or "first animals") are a large group of eucaryotic organisms of more than 50,000 known species belonging to the Kingdom Protista that have adapted a form of cell to serve as the entire body. In fact, protozoans are one-celled animal-like organisms with complex cellular structures. In the microbial world, protozoans are giants, many times larger than bacteria. They range in size from 4 μm to 500 μm. The largest ones can almost be seen by the naked eye. They can exist as solitary or independent organisms (or for example, the stalked ciliates (see Figure 10.3) such as *Vermicelli* sp.), or they can colonize like the sedentary *Carchesium* sp. Protozoa get their name because they employ the same type of feeding strategy as animals. That is, they are heterotrophic, meaning they obtain

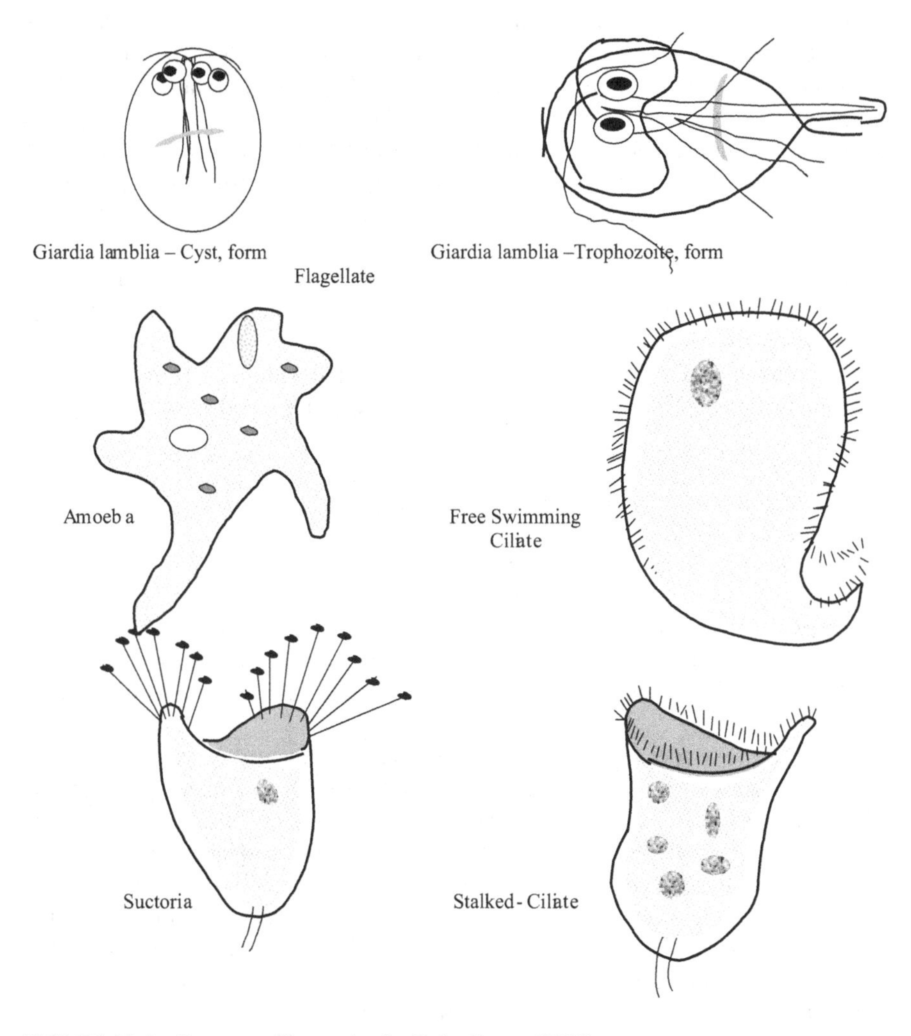

FIGURE 10.3 Protozoa. Illustration by F. Spellman (2001).

cellular energy from organic substances such as proteins. Most are harmless, but some are parasitic. Some forms have two life stages: *active trophozoites* (capable of feeding) and *dormant cysts*.

The major groups of protozoans are based on their method of locomotion (motility). For example, the *Mastigophora* are motile by means of one or more *flagella* (the whip-like projection that propels the free-swimming organisms—*Giardia lamblia* is a flagellated protozoan); the *Ciliophora* by means of shortened modified flagella called *cilia* (short hair-like structures that beat rapidly and propel them through the water); the *Sarcodina* by means of amoeboid movement (streaming or gliding action—the shape of amoebae change as they stretch, then contract, from place to place); and the *Sporozoa*, which are nonmotile; they are simply swept along, riding the current of the water.

Protozoa consume organics to survive; their favorite food is bacteria. Protozoa are mostly aerobic or facultative in regards to oxygen requirements. Toxic materials, pH, and temperature affect protozoan rates of growth in the same way as they affect bacteria.

Most protozoan life cycles alternate between an active growth phase (*trophozoites*) and a resting stage (*cysts*). Cysts are extremely resistant structures that protect the organism from destruction when it encounters harsh environmental conditions—including chlorination.

Note: Those protozoans not completely resistant to chlorination require higher disinfectant concentrations and longer contact time for disinfection than normally used in water treatment.

The protozoa and the waterborne diseases associated with them of most concern to the public health officials, wastewater-based epidemiology, and wastewater operators are:

- *Entamoeba histolytica*—Amoebic dysentery
- *Giardia lamblia*—Giardiasis
- *Cryptosporidium*—Cryptosporidiosis

In wastewater treatment, protozoa are a critical part of the purification process and can be used to indicate the condition of treatment processes. Protozoa normally associated with wastewater include amoeba, flagellates, free-swimming ciliates, and stalked ciliates.

Amoebae are associated with poor wastewater treatment of a young biosolids mass (see Figure 10.3). They move through wastewater by a streaming or gliding motion. Moving the liquids stored within the cell wall effects this movement. They are normally associated with an effluent high in BODs and suspended solids.

Flagellates (flagellated protozoa) have a single, long hair-like or whip-like projection (flagella) that is used to propel the free-swimming organisms through wastewater and to attract food (see Figure 10.3). Flagellated protozoans are normally associated with poor treatment and a young biosolids. When the free-swimming ciliated protozoan is the predominate organisms, the plant effluent will contain large amounts of BODs and suspended solids.

The *free-swimming ciliated protozoan* uses its tiny, hair-like projections (cilia) to move itself through the wastewater and to attract food (see figure 10.3). The free-swimming ciliated protozoan is normally associated with a moderate biosolids age and effluent quality. When the free-swimming ciliated protozoan is the predominate organisms, the plant effluent will normally be turbid and contain a high amount of suspended solids.

The *stalked ciliated protozoan* attaches itself to the wastewater solids and uses its cilia to attract food (see Figure 10.3). The stalked ciliated protozoan is normally associated with a plant effluent that is very clear and contains low amounts of both BODs and suspended solids.

Rotifers make up a well-defined group of the smallest, simplest multicellular microorganisms and are found in nearly all aquatic habitats (see Figure 10.4). Rotifers are a higher life form associated with cleaner waters. Normally found in well-operated wastewater treatment plants, they can be used to indicate the performance of certain types of treatment processes.

VIRUSES

Viruses are very different from other microorganisms. Consider their size relationship, for example. Relative to size, if protozoans are the Goliaths of microorganisms, then viruses are the Davids. Stated more specifically and accurately, viruses are intercellular parasitic particles that are the smallest living infectious materials known—the midgets of the microbial world. Viruses are very simple life forms consisting of a central molecule of genetic material surrounded by a protein shell called a *capsid* and sometimes by a second layer called an *envelope.* They contain no mechanisms by which to obtain energy or reproduce on their own; thus to live, viruses must have a host. After they invade the cells of their specific host (animal, plant, insect, fish, or

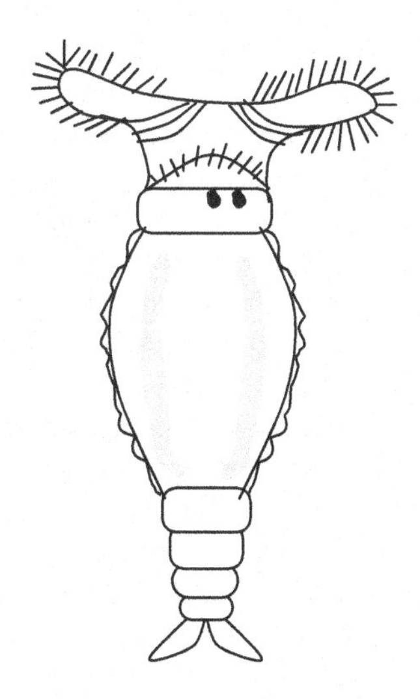

FIGURE 10.4 Philodina, a common rotifer. Illustration by F. Spellman (2001).

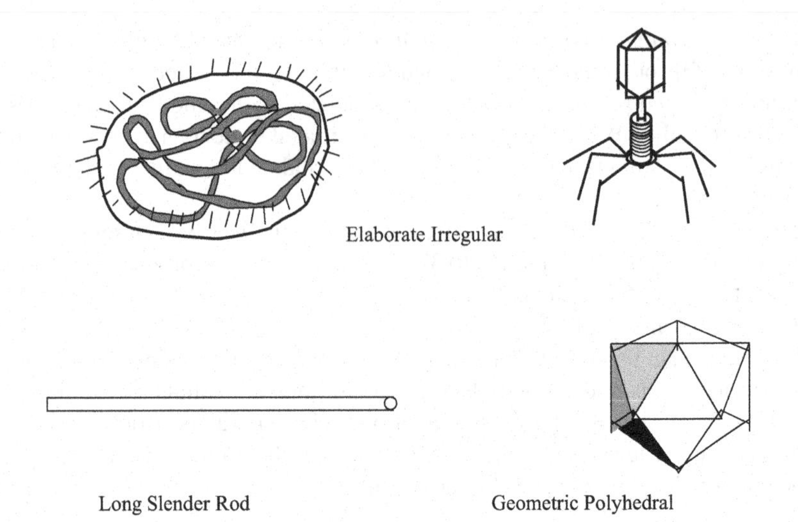

FIGURE 10.5 Virus shapes.

even bacteria), they take over the host's cellular machinery and force it to make more viruses. In the process, the host cell is destroyed and hundreds of new viruses are released into the environment. Viruses occur in many shapes, including long slender rods, elaborate irregular shapes, and geometric polyhedrals (see Figure 10.5).

Viruses contain no mechanisms by which to obtain energy or reproduce on their own, thus viruses must have a host to survive. After they invade the cells of their specific host (animal, plant, insect fish, or even bacteria), they take over the cellular machinery of the host and force it to make more viruses. In the process, the host cell is destroyed and hundreds of new viruses are released into the environment. The viruses of most concern to the waterworks operator are the pathogens that cause hepatitis, viral gastroenteritis, and poliomyelitis.

Smaller and different from bacteria, viruses are prevalent in water contaminated with sewage. Detecting viruses in water supplies is a major problem because of the complexity of non-routine procedures involved, although experience has shown that the normal coliform index can be used as a rough guide for viruses as for bacteria. More attention must be paid to viruses, however, whenever surface water supplies have been used for sewage disposal. Viruses are difficult to destroy by normal disinfection practices, requiring increased disinfectant concentration and contact time for effective destruction.

Because this text is focused on wastewater-based epidemiology and its present ability and usage to provide pertinent information related to the current viral pandemic, COVID-19 issue, a symbolic illustration of the coronavirus is shown in Figure 10.6. The rendition of the COVID-19 virus shown here has become quite familiar to the public because of its often shown appearance on social media and other media sources. Again, this text is focused on wastewater-based epidemiology and is dedicated to currency in regard to what is happening today, at the present time. Thus, Figure 10.6 is apropos to the material presented herein.

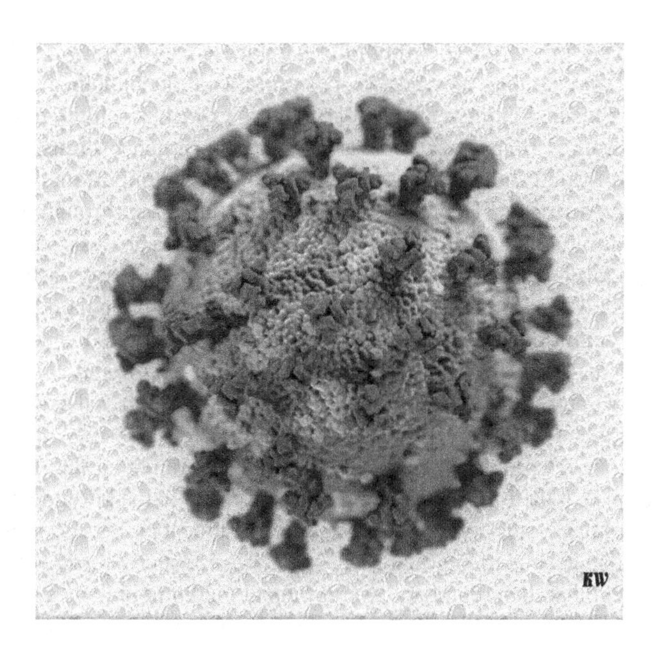

FIGURE 10.6 Coronavirus. Illustration by Kat Welsh-Ware and F. Spellman.

Note: Viruses that infect bacterial cells cannot infect and replicate within cells of other organisms. It is possible to utilize the specificity to identify bacteria, a procedure called *phage typing*.

HELMINTHS: WORMS

Along with inhabiting organic mud, worms also inhabit biological slimes; they have been found in activated sludge and in trickling filter slimes (wastewater treatment processes). Microscopic in size, they range in length from 0.5 mm to 3 mm and in diameter from 0.01 mm to 0.05 mm. Most species have a similar appearance. They have a body that is covered by cuticle, are cylindrical, nonsegmented, and taper at both ends.

These organisms continuously enter wastewater treatment systems, primarily through attachment to soils that reach the plant through inflow and infiltration (I & I). They are present in large, often highly variable numbers, but as strict aerobes, they are found only in aerobic treatment processes where they metabolize solid organic matter.

When nematodes are firmly established in the treatment process, they can promote microfloral activity and decomposition. They crop bacteria in both the activated sludge and trickling filter systems. Their activities in these systems enhance oxygen penetration by tunneling through floc particles and biofilm. In activated sludge processes, they are present in relatively small numbers because the liquefied environment is not a suitable habitat for crawling, which they prefer over the free-swimming mode. In trickling filters where the fine stationary substratum is suitable to permit crawling and mating, nematodes are quite abundant.

Along with preferring the trickling filter habitat, nematodes play a beneficial role in this habitat; for example, they break loose portions of the biological slime coating the filter bed. This action prevents excessive slime growth and filter clogging. They also aid in keeping slime porous and accessible to oxygen by tunneling through slime. In the activated sludge process, the nematodes play important roles as agents of better oxygen diffusion. They accomplish this by tunneling through floc particles. They also act as parameters of operational conditions in the process, such as low dissolved oxygen levels (anoxic conditions) and the presence of toxic wastes.

Environmental conditions have an impact on the growth of nematodes. For example, in anoxic conditions their swimming and growth is impaired. The most important condition they indicate is when the wastewater strength and composition has changed. Temperature fluctuations directly affect their growth and survival; population decreases when temperatures increase.

Aquatic flatworms (improperly named because they are not all flat) feed primarily on algae. Because of their aversion to light, they are found in the lower depths of pools. Two varieties of flatworms are seen in wastewater treatment processes: *microtubellarians* are more than flat and average about 0.5–5 mm in size, and *macrotubellarians* (planarians) are more flat than round and average about 5 20 mm in body size. Flatworms are very hardy and can survive in wide variations in humidity and temperature. As inhabitants of sewage sludge, they play an important part in sludge stabilization and as bioindicators or parameters of process problems. For example, their inactivity or sluggishness might indicate a low dissolved oxygen level or the presence of toxic wastes.

Surface waters grossly polluted with organic matter (especially domestic sewage) have a fauna that is capable of thriving in very low concentrations of oxygen. A few species of tubificid worms dominate this environment. Pennak reported (1989) that the bottoms of severely polluted streams can be literally covered with a "writhing" mass of these tubificids.

The *Tubifex* (commonly known as sludge worms) are small, slender, reddish worms that normally range in length from 25 mm to about 50 mm. They are burrowers; their posterior end protrudes to obtain nutrients (see Figure 10.7). When found in streams, *Tubifex* are indicators of pollution.

THE BASICS OF INFECTIOUS DISEASE TRANSMISSION

Now that the infectious disease agents associated with domestic/municipal wastewater have been introduced and described, the focus shifts to the conditions that are necessary to produce infectious disease in a population. The conditions necessary to produce infectious disease in a population include: the disease agent must be present; the agent must be present in sufficient strength to be infectious; and vulnerable individuals must come into contact with the agent in a way that causes infection and disease. With regard to raw wastewater it is safe to assume that pathogenic organisms are present; therefore, the first of the above criteria is always met. The actual concentration of pathogenic agents in wastewater is a function of the disease morbidity in the contributing population. Table 10.3 lists the number of pathogenic microorganisms found in raw, untreated wastewater.

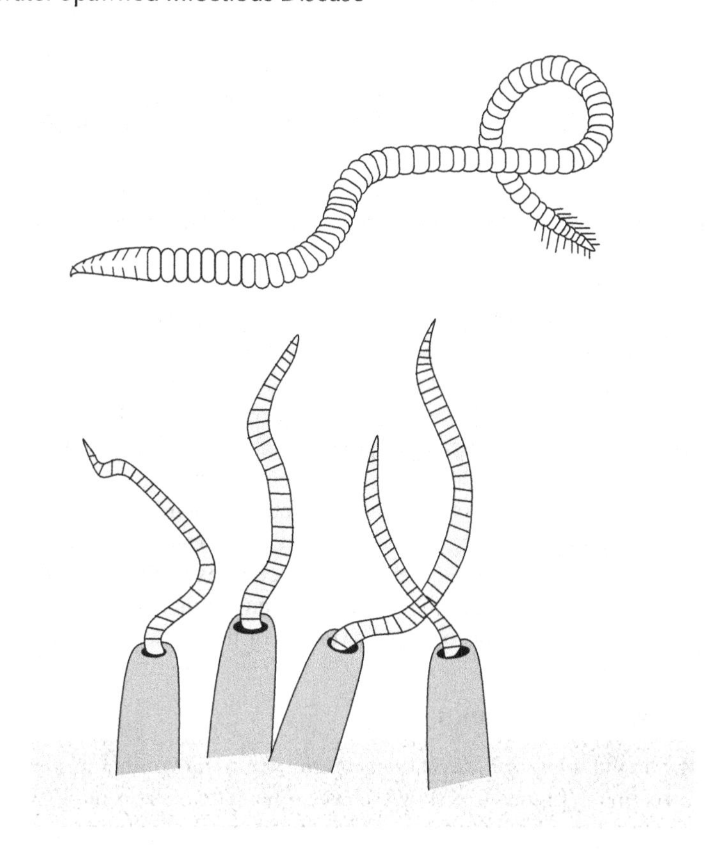

FIGURE 10.7 Tubcificid worm(s). Illustration by F. Spellman.

TABLE 10.3
Typical Numbers of Microorganisms Found in Raw, Untreated Wastewater

Microbe	Number per 100 mL of effluent in raw wastewater
Fecal coliform (most probable number)	1,000,000,000
Salmonella (most probable number)	8,000
Shigella (most prominent number)	1,000
Enter virus (Plaque-forming units)	50,000
Helminth ova	800
Giardia lamblia cysts	10,000

Source: EPA. 1991 and 1992a; Dean and Smith, 1973; Feachem et al., 19809; Engineering Science, 1987; Gerba, 1983 and Logsdon et al., 1985.

The second of the above criteria—that the infectious agent be present in sufficient concentration—is difficult to determine, at times. This is the case because limited human very limited dosage-response data that have been reported indicate much difference in the severity of sickness among those exposed to identified dosages of pathogens (Bryan, 1974). Moreover, another complication is that it usually takes more than a single organism to produce a detectable disease response in an individual in the exposed population.

The third and final link in the infectious disease transmission chain is the exposure of the susceptible human population to infectious agents. The primary route of exposure to wastewater-associated pathogen is by ingestion. For example and based on several personal investigation, wastewater workers/operators who chew tobacco and/or eat their bag lunches without paying attention to personal hygiene practices such as washing hands before inserting a plug of tobacco or lunch or other food items in their mouths have shown a prevalence to be contaminated, made ill, and some cases worse. Note that there are other routes of exposure such as respiratory and ocular. If reclaimed water is to be used in the production of human food crops, particularly those that are eaten raw, then there is a chance of exposure through ingestion. As a result, there is a greater need to reduce pathogen numbers to low levers prior to soil application, or at least prior to crop harvesting or livestock exposure.

Risk of Contracting Infectious Disease

The primary methodology to prevent wastewater from contaminating the public is wastewater treatment. Figure 10.8 shows a basic schematic of an example wastewater treatment process providing primary and secondary treatment using the *activated*

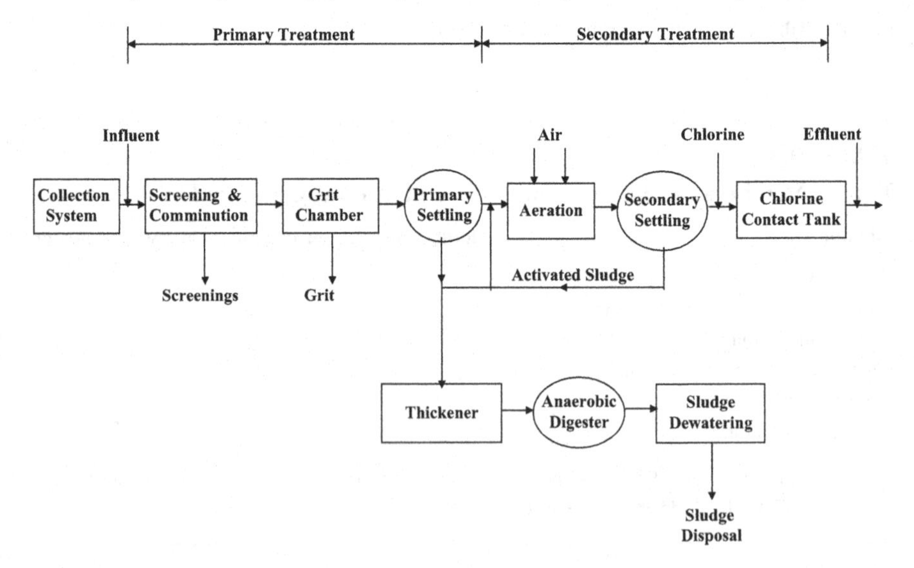

FIGURE 10.8 Schematic of an example wastewater treatment process providing primary and secondary treatment using the activated sludge process.

sludge process. This is the model, the prototype, the paradigm used in this book and around the globe. Though it is true that in secondary treatment (which provides BOD removal beyond that is achievable by simple sedimentation) there are actually three commonly used approaches—trickling filter, activated sludge, and oxidation ponds—we focus, for instructive and illustrative purposes, on the activated sludge process throughout this handbook. The purpose of Figure 10.8 is to allow the reader to follow the treatment process step-by-step as it is actually configured in the real world and to assist in the understanding of how all the various unit processes sequentially follow and tie into each other. Note that wastewater treatment is a series of individual steps (unit processes) that treat the waste stream as it makes its way through the entire process.

Note that the principal sources of domestic wastewater in a community are the residential areas and commercial districts. Other important sources include institutional and recreational facilities and storm water (runoff) and groundwater (infiltration). Each source produces wastewater with specific characteristics.

Wastewater is generated by five major sources: human and animal wastes, household wastes, industrial wastes, storm water runoff, and groundwater infiltration.

1. *Human and animal wastes*—they contain the solid and liquid discharges of humans and animals and are considered by many to be the most dangerous from a human health viewpoint. The primary health hazard is presented by the millions of bacteria, viruses, and other microorganisms (some of which may be pathogenic) present in the waste stream.
2. *Household wastes*—they are wastes, other than human and animal wastes, discharged from the home. Household wastes usually contain paper, household cleaners, detergents, trash, garbage, and other substances the homeowner discharges into the sewer system.
3. *Industrial wastes*—they include industry specific materials, which can be discharged from industrial processes into the collection system. They typically contain chemicals, dyes, acids, alkalis, grit, detergents, and highly toxic materials.
4. *Storm water runoff*—many collection systems are designed to carry both the wastes of the community and storm water runoff. In this type of system when a storm event occurs, the waste stream can contain large amounts of sand, gravel, and other grit as well as excessive amounts of water.
5. *Groundwater infiltration*—groundwater will enter older improperly sealed collection systems through cracks or unsealed pipe joints. Not only can this add large amounts of water to wastewater flows but also additional grit.

Wastewater can be classified according to the sources of flows: domestic, sanitary, industrial, combined, and stormwater:

1. *Domestic (sewage) wastewater*—it mainly contains human and animal wastes, household wastes, small amounts of groundwater infiltration, and small amounts of industrial wastes.

2. *Sanitary wastewater*—it consists of domestic wastes and significant amounts of industrial wastes. In many cases, the industrial wastes can be treated without special precautions. However, in some cases the industrial wastes will require special precautions or a pretreatment program to ensure the wastes do not cause compliance problems for the wastewater treatment plant.
3. *Industrial wastewater*—it consists of industrial wastes only. Often the industry will determine that it is safer and more economical to treat its waste independent of domestic waste.
4. *Combined wastewater*—it is the combination of sanitary wastewater and storm water runoff. All the wastewater and storm water of the community is transported through one system to the treatment plant.
5. *Storm water*—it is a separate collection system (no sanitary waste) that carries storm water runoff including street debris, road salt, and grit.

DID YOU KNOW?

Wastewater contains many different substances, which can be used to characterize it. The specific substances and amounts or concentrations of each will vary, depending on the source. Thus, it is difficult to "precisely" characterize wastewater. Instead, wastewater characterization is usually based on and applied to an average domestic wastewater. Wastewater is characterized in terms of its physical, chemical, and biological characteristics.

Note: Keep in mind that other sources and types of wastewater can dramatically change the characteristics.

Note: To gain understanding and knowledge about the primary means of removing, reducing, or deactivating pathogens in wastewater, the following section provides a short and basic discussion of the activated sludge process commonly used to treat wastewater.

Basic Wastewater Treatment

Preliminary Treatment

The initial stage in the wastewater treatment process (following collection and influent pumping) is *preliminary treatment.* Raw influent entering the treatment plant may contain many kinds of materials (trash). The purpose of preliminary treatment is to protect plant equipment by removing these materials which could cause clogs, jams, or excessive wear to plant machinery. In addition, the removal of various materials at the beginning of the treatment process saves valuable space within the treatment plant.

Preliminary treatment may include many different processes, each designed to remove a specific type of material which is a potential problem for the treatment process. Processes include wastewater collections—influent pumping, screening, shredding, grit removal, flow measurement, preaeration, chemical addition, and flow equalization—the major processes are shown in Figure 10.8. In this section, we describe and discuss each of these processes and their importance in the treatment process.

Note: As mentioned, not all treatment plants will include all of the processes shown in Figure 10.8. Specific processes have been included to facilitate discussion of major potential problems with each process and its operation; this is information that may be important to the wastewater operator.

Screening

The purpose of *screening* is to remove large solids such as rags, cans, rocks, branches, leaves, roots, etc., from the flow before the flow moves on to downstream processes.

Note: Typically, a treatment plant will remove anywhere from 0.5 ft^3 to 12 ft^3 of screenings for each million gallons of influent received.

A *bar screen* traps debris as wastewater influent passes through. Typically, a bar screen consists of a series of parallel, evenly spaced bars or a perforated screen placed in a channel (see Figure 10.9). The wastestream passes through the screen and the large solids *(screenings)* are trapped on the bars for removal.

The bar screen may be coarse (2–4-inch openings) or fine (0.75–2.0-inch openings). The bar screen may be manually cleaned (bars or screens are placed at an angle of 30° for easier solids removal—see Figure 10.9) or mechanically cleaned (bars are placed at 45° to 60° angle to improve mechanical cleaner operation).

The screening method employed depends on the design of the plant, the amount of solids expected and whether the screen is for constant or emergency use only.

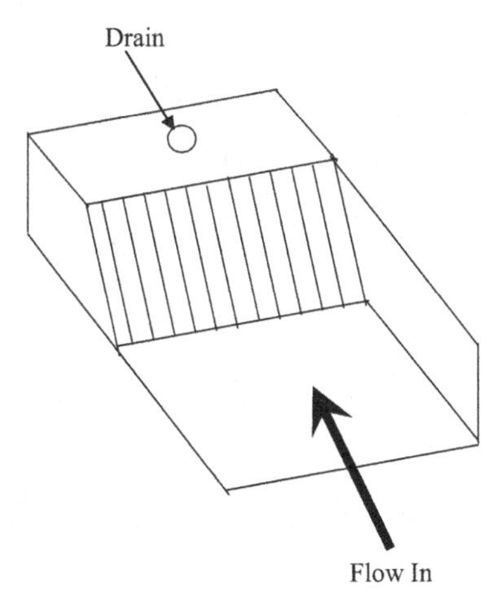

FIGURE 10.9 Basic bar screen. *Note*: The screenings must be removed frequently enough to prevent accumulation which will block the screen and cause the water level in front of the screen to build up.

Shredding

As an alternative to screening, *shredding* can be used to reduce solids to a size, which can enter the plant without causing mechanical problems or clogging. Shredding processes include comminution (comminute means to cut up) and barminution devices.

Comminution

The *comminutor* is the most common shredding device used in wastewater treatment. In this device all the wastewater flow passes through the grinder assembly. The grinder consists of a screen or slotted basket, a rotating or oscillating cutter, and a stationary cutter. Solids pass through the screen and are chopped or shredded between the two cutters. The comminutor will not remove solids, which are too large to fit through the slots, and it will not remove floating objects. These materials must be removed manually. Maintenance requirements for comminutors include aligning, sharpening and replacing cutters, and corrective and preventive maintenance performed in accordance with plant O & M Manual.

Preaeration

In the *preaeration process* (diffused or mechanical), we aerate wastewater to achieve and maintain an aerobic state (to freshen septic wastes), strip off hydrogen sulfide (to reduce odors and corrosion), agitate solids (to release trapped gases and improve solids separation and settling), and to reduce BOD_5. All of this can be accomplished by aerating the wastewater for 10–30 minutes. To reduce BOD_5, preaeration must be conducted from 45 to 60 minutes.

Chemical Addition

Chemical addition to the wastestream is done (either via dry chemical metering or solution feed metering) to improve settling, reduce odors, neutralize acids or bases, reduce corrosion, reduce BOD_5, improve solids and grease removal, reduce loading on the plant, add or remove nutrients, add organisms, and/or aid subsequent downstream processes. The particular chemical and amount used depends on the desired result. Chemicals must be added at a point where sufficient mixing will occur to obtain maximum benefit. Chemicals typically used in wastewater treatment include chlorine, peroxide, acids and bases, miner salts (ferric chloride, alum, etc.), and bioadditives and enzymes.

Equalization

The purpose of *flow equalization* (whether by surge, diurnal, or complete methods) is to reduce or remove the wide swings in flow rates normally associated with wastewater treatment plant loading; it minimizes the impact of storm flows. The process can be designed to prevent flows above maximum plant design hydraulic capacity; to reduce the magnitude of diurnal flow variations; and to eliminate flow variations. Flow equalization is accomplished using mixing or aeration equipment, pumps, and flow measurement. Normal operation depends on the purpose and requirements of the flow equalization system. Equalized flows allow the plant to perform at optimum levels by providing stable hydraulic and organic loading. The downside to flow equalization is in additional costs associated with construction and operation of the flow equalization facilities.

Aerated Systems

Aerated grit removal systems use aeration to keep the lighter organic solids in suspension while allowing the heavier grit particles to settle out. Aerated grit removal may be manually or mechanically cleaned; however, the majority of the systems are mechanically cleaned. In normal operation, the aeration rate is adjusted to produce the desired separation, which requires observation of mixing and aeration and sampling of fixed suspended solids. Actual grit removal is controlled by the rate of aeration. If the rate is too high, all of the solids remain in suspension. If the rate is too low, both the grit and the organics will settle out.

Cyclone Degritter

The *cyclone degritter* uses a rapid spinning motion (centrifugal force) to separate the heavy inorganic solids or grit from the light organic solids. This unit process is normally used on primary sludge rather than the entire wastewater flow. The critical control factor for the process is the inlet pressure. If the pressure exceeds the recommendations of the manufacturer, the unit will flood and grit will carry through with the flow. Grit is separated from the flow and discharged directly to a storage container. Grit removal performance is determined by calculating the percent removal for inorganic (fixed) suspended solids.

Primary Treatment (Sedimentation)

The purpose of primary treatment (primary sedimentation or primary clarification) is to remove settleable organic and flotable solids. Normally, each primary clarification unit can be expected to remove 90–95 percent settleable solids, 40–60 percent total suspended solids (TSS) and 25–35 percent BOD_5.

Note: Performance expectations for settling devices used in other areas of plant operation is normally expressed as overall unit performance rather than settling unit performance.

Sedimentation may be used throughout the plant to remove settleable and floatable solids. It is used in primary treatment, secondary treatment, and advanced wastewater treatment processes. In this section, we focus on primary treatment or primary clarification, which uses large basins in which primary settling is achieved under relatively quiescent conditions (see Figure 10.8). Within these basins, mechanical scrapers collect the primary settled solids into a hopper, from which they are pumped to a sludge-processing area. Oil, grease, and other floating materials (scum) are skimmed from the surface. The effluent is discharged over weirs into a collection trough.

In primary sedimentation, wastewater enters a settling tank or basin. The velocity is reduced to approximately 1 foot per minute.

Note: Notice that the velocity is based on minutes instead of seconds, as was the case in the grit channels. A grit channel velocity of 1 ft/sec would be 60 ft/min.

Solids, which are heavier than water, settle to the bottom while solids which are lighter than water float to the top. Settled solids are removed as sludge and floating solids are removed as scum. Wastewater leaves the sedimentation tank over an effluent weir and on to the next step in treatment. Detention time, temperature, tank design, and condition of the equipment control the efficiency of the process.

Types of Sedimentation Tanks

Sedimentation equipment includes septic tanks, two-story tanks, and plain settling tanks or clarifiers. All three devices may be used for primary treatment while plain settling tanks are normally used for secondary or advanced wastewater treatment processes.

Septic Tanks

Septic tanks are prefabricated tanks that serve as a combined settling and skimming tank and as an unheated–unmixed anaerobic digester. Septic tanks provide long settling times (6–8 hours or more) but do not separate decomposing solids from the wastewater flow. When the tank becomes full, solids will be discharged with the flow. The process is suitable for small facilities (i.e., schools, motels, homes, etc.) but, due to the long detention times and lack of control, it is not suitable for larger applications.

Two-Story (Imhoff) Tank

The *two-story or Imhoff tank*, named for German engineer Karl Imhoff (1876–1965), is similar to a septic tank in the removal of settleable solids and the anaerobic digestion of solids. The difference is that the two-story tank consists of a settling compartment where sedimentation is accomplished, a lower compartment where settled solids and digestion takes place, and gas vents. Solids removed from the wastewater by settling pass from the settling compartment into the digestion compartment through a slot in the bottom of the settling compartment. The design of the slot prevents solids from returning to the settling compartment. Solids decompose anaerobically in the digestion section. Gases produced as a result of the solids decomposition are released through the gas vents running along each side of the settling compartment.

Plain Settling Tanks (Clarifiers)

The plain settling tank or clarifier optimizes the settling process. Sludge is removed from the tank for processing in other downstream treatment units. Flow enters the tank, is slowed and distributed evenly across the width and depth of the unit, passes through the unit, and leaves over the effluent weir. Detention time within the primary settling tank is 1–3 hours (2 hours on average). Sludge removal is accomplished frequently on either continuous or intermittent basis. Continuous removal requires additional sludge treatment processes to remove the excess water resulting from removal of sludge which contains less than 2–3 percent solids. Intermittent sludge removal requires the sludge be pumped from the tank on a schedule frequent enough to prevent large clumps of solids rising to the surface but infrequent enough to obtain 4–8 percent solids in the sludge withdrawn.

Scum must be removed from the surface of the settling tank frequently. This is normally a mechanical process but may require manual start-up. The system should be operated frequently enough to prevent excessive buildup and scum carryover but not so frequent as to cause hydraulic overloading of the scum removal system. Settling tanks require housekeeping and maintenance. Baffles (prevent floatable solids, scum, from leaving the tank), scum troughs, scum collectors, effluent troughs, and effluent weirs require frequent cleaning to prevent heavy biological growths and solids accumulations. Mechanical equipment must be lubricated and maintained as specified in the manufacturer's recommendations or in accordance with procedures listed in the plant O & M Manual.

Process control sampling and testing is used to evaluate the performance of the settling process. Settleable solids, dissolved oxygen, pH, temperature, total suspended solids, and BOD_5, as well as sludge solids and volatile matter, testing are routinely carried out.

Secondary Treatment

The main purpose of *secondary treatment* (sometimes referred to as biological treatment) is to provide biochemical oxygen demand (BOD) removal beyond what is achievable by primary treatment. There are three commonly used approaches, all of which take advantage of the ability of microorganisms to convert organic wastes (via biological treatment), into stabilized, low-energy compounds. Two of these approaches, the *trickling filter* (and/or its variation, the *rotating biological contactor* (RBC)) and the *activated sludge* process, sequentially follow normal primary treatment. The third, *ponds* (oxidation ponds or lagoons), however, can provide equivalent results without preliminary treatment. In this section, we present a brief overview of the secondary treatment process followed by a detailed discussion of wastewater treatment ponds (used primarily in smaller treatment plants), trickling filters, and RBCs. We then shift focus to the activated sludge process—the secondary treatment process, which is used primarily in large installations and is the main focus of the handbook.

Secondary treatment refers to those treatment processes which use biological processes to convert dissolved, suspended, and colloidal organic wastes to more stable solids which can either be removed by settling or discharged to the environment without causing harm. Exactly what is secondary treatment? As defined by the Clean Water Act (CWA), secondary treatment produces an effluent with no more than 30 mg/L BOD_5 and 30 mg/L total suspended solids.

Note: The CWA also states that ponds and trickling filters will be included in the definition of secondary treatment even if they do not meet the effluent quality requirements continuously.

Most secondary treatment processes decompose solids aerobically producing carbon dioxide, stable solids, and more organisms. Because solids are produced, all of the biological processes must include some form of solids removal (settling tank, filter, etc.). Secondary treatment processes can be separated into two large categories: fixed film systems and suspended growth systems.

Fixed film systems are processes that use a biological growth (biomass or slime) which is attached to some form of media. Wastewater passes over or around the media and the slime. When the wastewater and slime are in contact, the organisms remove and oxidize the organic solids. The media may be stone, redwood, synthetic materials or any other substance that is durable (capable of withstanding weather conditions for many years), provides a large area for slime growth while providing open space for ventilation, and is not toxic to the organisms in the biomass. Fixed film devices include trickling filters and rotating biological contactors. *Suspended growth systems* are processes that use a biological growth, which is mixed with the wastewater. Typical suspended growth systems consist of various modifications of the activated sludge process.

Treatment Ponds

Wastewater treatment can be accomplished using *ponds* (aka, *lagoons*). Ponds are relatively easy to build and manage, they accommodate large fluctuations in flow, and they can also provide treatment that approaches conventional systems (producing a highly purified effluent) at much lower cost. It is the cost (the economics) that drives many managers to decide on the pond option. The actual degree of treatment provided depends on the type and number of ponds used. Ponds can be used as the sole type of treatment or they can be used in conjunction with other forms of wastewater treatment—that is, other treatment processes followed by a pond or a pond followed by other treatment processes.

Stabilization ponds (aka, treatment ponds) have been used for treatment of wastewater for over 3,000 years. The first recorded construction of a pond system in the United States was at San Antonio, Texas, in 1901. Today, over 8,000 wastewater treatment ponds are in place, involving more than 50 percent of the wastewater treatment facilities in the United States (CWNS, 2000). Facultative ponds account for 62 percent, aerated ponds 25 percent, anaerobic 0.04 percent, and total containment 12 percent of the pond treatment systems. They treat a variety of wastewaters from domestic wastewater to complex industrial wastes, and they function under a wide range of weather conditions, from tropical to arctic. Ponds can be used alone or in combination with other wastewater treatment processes. As our understanding of pond operating mechanisms has increased, different types of ponds have been developed for application in specific types of wastewater under local environmental conditions. This handbook focuses on municipal wastewater treatment pond systems.

While the tendency in the United States has been for smaller communities to build ponds, in other parts of the world, including Australia, New Zealand, Mexico and Latin America, Asia, and Africa, treatment ponds have been built for large cities. As a result, our understanding of the biological, biochemical, physical, and climatic factors that interact to transform the organic compounds, nutrients, and pathogenic organisms found in sewage into less harmful chemicals and unviable organisms (i.e., dead or sterile) has grown since 1983. A wealth of experience has been built up as civil, sanitary, or environmental engineers, operators, public works managers, and public health and environmental agencies have gained more experience with these systems. While some of this information makes its way into technical journals and text books, there is a need for a less formal presentation of the subject for those working in the field every day (USEPA, 2011).

Ponds are designed to enhance the growth of natural ecosystems that are either anaerobic (providing conditions for bacteria that grow in the absence of oxygen (O_2) environments), aerobic (promoting the growth of O_2 producing and/or requiring organs, such as algae and bacteria), or facultative, which is a combination of the two. Ponds are managed to reduce concentrations of biochemical oxygen demand, TSS, and coliform numbers (fecal or total) to meet water-quality requirements.

Types of Ponds

Ponds can be classified based on their location in the system, by the type of wastes they receive, and by the main biological process occurring in the pond. First we look at the types of ponds according to their location and the type of wastes they receive: raw sewage stabilization ponds, oxidation ponds, and polishing ponds. Then, in the following section, we look at ponds classified by the type of processes occurring within the pond: *aerobic ponds*, *anaerobic ponds*, *facultative ponds*, and *aerated ponds*.

Ponds Based on Location and Types of Wastes They Receive

Raw Sewage Stabilization Pond

The raw sewage stabilization pond is the most common type of pond. With the exception of screening and shredding, this type of pond receives no prior treatment. Generally, raw sewage stabilization ponds are designed to provide a minimum of 45 days detention time and to receive no more than 30 pounds of BOD_5 per day per acre. The quality of the discharge is dependent on the time of the year. Summer months produce high BOD_5 removal but excellent suspended-solids removals. The pond consists of an influent structure, pond berm or walls, and an effluent structure designed to permit selection of the best quality effluent. Normal operating depth of the pond is 3–5 ft. The process occurring in the pond involves bacteria decomposing the organics in the wastewater (aerobically and anaerobically) and algae using the products of the bacterial action to produce oxygen (photosynthesis). Because this type of pond is the most commonly used in wastewater treatment, the process that occurs within the pond is described in greater detail in the following text.

When wastewater enters the stabilization pond several processes begin to occur. These include settling, aerobic decomposition, anaerobic decomposition, and photosynthesis. Solids in the wastewater will settle to the bottom of the pond. In addition to the solids in the wastewater entering the pond, solids, which are produced by the biological activity, will also settle to the bottom. Eventually, this will reduce the detention time and the performance of the pond. When this occurs (20–30 years normal) the pond will have to be replaced or cleaned.

Bacteria and other microorganisms use the organic matter as a food source. They use oxygen (aerobic decomposition), organic matter and nutrients to produce carbon dioxide, water, and stable solids, which may settle out, and more organisms. Carbon dioxide is an essential component of the photosynthesis process occurring near the surface of the pond. Organisms also use the solids that settled out as food material; however, the oxygen levels at the bottom of the pond are extremely low so the process used is anaerobic decomposition. The organisms use the organic matter to produce gases (hydrogen sulfide, methane, etc.) which are dissolved in the water, stable solids, and more organisms. Near the surface of the pond a population of green algae will develop which can use the carbon dioxide produced by the bacterial population, nutrients, and sunlight to produce more algae and oxygen which is dissolved into the water. The dissolved oxygen is then used by organisms in the aerobic decomposition process.

When compared with other wastewater treatment systems involving biological treatment, a stabilization pond treatment system is the simplest to operate and maintain. Operation and maintenance activities include collecting and testing samples for dissolved oxygen (D.O), and pH, removing weeds and other debris (scum) from the pond, mowing the berms, repairing erosion, and removing burrowing animals.

Note: Dissolved oxygen and pH levels in the pond will vary throughout the day. Normal operation will result in very high D.O. and pH levels due to the natural processes occurring.

Note: When operating properly the stabilization pond will exhibit a wide variation in both dissolved oxygen and pH. This is due to the photosynthesis occurring in the system.

Oxidation Pond

An oxidation pond, which is normally designed using the same criteria as the stabilization pond, receives flows that have passed through a stabilization pond or primary settling tank. This type of pond provides biological treatment, additional settling and some reduction in the number of fecal coliform present.

Polishing Pond

A polishing pond, which uses the same equipment as a stabilization pond, receives flow from an oxidation pond or from other secondary treatment systems. Polishing ponds remove additional BOD_5, solids and fecal coliform, and some nutrients. They are designed to provide 1–3 days detention time and normally operate at a depth of 5–10 ft. Excessive detention time or too shallow a depth will result in alga growth, which increases influent, suspended-solids concentrations.

Biochemistry in a Pond

Photosynthesis

Photosynthesis is the process whereby organisms use solar energy to fix CO_2 and obtain the reducing power to convert it into organic compounds. In wastewater ponds, the dominant photosynthetic organisms include algae, cyanobacteria, and purple sulfur bacteria (Pipes, 1961; Pearson, 2005).

Photosynthesis may be classified as oxygenic or anoxygenic, depending on the source of reducing power used by a particular organism. In oxygenic photosynthesis, water serves as the source of reducing power, with O_2 as a by-product. The equation representing oxygenic photosynthesis is:

$$H_2O + \text{sunlight} \rightarrow 1/2O_2 + 2H^+ + 2e^-$$

Oxygenic photosynthetic algae and cyanobacteria convert CO_2 to organic compounds, which serve as the major source of chemical energy for other aerobic organisms. Aerobic bacteria need the O_2 produced to function in their role as primary consumers in degrading complex organic waste material.

Anoxygenic photosynthesis does not produce O_2 and, in fact, occurs in the complete absence of O_2. The bacteria involved in anoxygenic photosynthesis are largely strict anaerobes, unable to function in the presence of O_2. They obtain energy by reducing inorganic compounds. Many photosynthetic bacteria utilize reduced *S* compounds or element *S* in anoxygenic photosynthesis according to the following equation:

$$H_2S \rightarrow S^o + 2H^+ + 2e^-$$

Respiration

Respiration is a physiological process by which organic compounds are oxidized into CO_2 and water. Respiration is also an indicator of cell material synthesis. It is a complex process that consists of many interrelated biochemical reactions (Pearson, 2005). Aerobic respiration, common to species of bacteria, algae, protozoa, invertebrates, and higher plants and animals, may be represented by the following equation:

$$C_2H_{12}O_6 + 6O_2 + \text{enzymes} \rightarrow 6CO_2 + 6H_2O + \text{new cells}$$

The bacteria involved in aerobic respiration are primarily responsible for degradation of waste products.

In the presence of light, respiration and photosynthesis can occur simultaneously in algae. However, the respiration rate is low compared to the photosynthesis rate, which results in a net consumption of CO_2 and production of O_2. In the absence of light, on the other hand, algal respiration continues while photosynthesis stops, resulting in a net consumption of O_2 and production of CO_2 (USEPA, 2011).

Trickling Filters

Trickling filters have been used to treat wastewater since the 1890s. It was found that if settled wastewater was passed over rock surfaces, slime grew on the rocks and the water became cleaner. Today we still use this principle but, in many installations, instead of rocks we use plastic media. In most wastewater treatment systems, the *trickling filter* follows primary treatment and includes a secondary settling tank or clarifier. Trickling filters are widely used for the treatment of domestic and industrial wastes. The process is a fixed film biological treatment method designed to remove BOD_5 and suspended solids.

A trickling filter consists of a rotating distribution arm that sprays and evenly distributes liquid wastewater over a circular bed of fist-sized rocks, other coarse, rough materials, or synthetic media. The spaces between the media allow air to circulate easily so that aerobic conditions can be maintained. The spaces also allow wastewater to trickle down through, around and over the media. A layer of biological slime that absorbs and consumes the wastes trickling through the bed covers the media material. The organisms aerobically decompose the solids producing more organisms and stable wastes, which either become part of the slime or are discharged back into the wastewater flowing over the media. This slime consists mainly of bacteria, but it may also include algae, protozoa, worms, snails, fungi, and insect larvae. The accumulating slime occasionally sloughs off (*sloughings*) individual media materials and is collected at the bottom of the filter, along with the treated wastewater, and passed on to the secondary settling tank where it is removed. The overall performance of the trickling filter is dependent on hydraulic and organic loading, temperature, and recirculation.

Rotating Biological Contactors

The *rotating biological contactor* is a biological treatment system (see Figure 10.9) and is a variation of the attached growth idea provided by the trickling filter. Still relying on microorganisms that grow on the surface of a medium, the RBC is instead a *fixed film* biological treatment device—the basic biological process, however, is similar to that occurring in the trickling filter. An RBC consists of a series of closely spaced (mounted side by side), circular, plastic (synthetic) disks, that are typically about 3.5 m in diameter and attached to a rotating horizontal shaft. Approximately 40 percent of each disk is submersed in a tank containing the wastewater to be treated. As the RBC rotates, the attached biomass film (zoogleal slime) that grows on the surface of the disk move into and out of the wastewater. While submerged in the wastewater, the microorganisms absorb organics; while they are rotated out of the wastewater, they are supplied with needed oxygen for aerobic decomposition. As the zoogleal slime reenters the wastewater, excess solids and waste products are stripped off the media as sloughings. These sloughings are transported with the wastewater flow to a settling tank for removal.

Modular RBC units are placed in series. Simply because a single contactor is not sufficient to achieve the desired level of treatment, the resulting treatment achieved exceeds conventional secondary treatment. Each individual contactor is called a stage and the group is known as a train. Most RBC systems consist of two or more trains with three or more stages in each. The key advantage in using RBCs instead of trickling filters is that RBCs are easier to operate under varying load conditions, since it is easier to keep the solid medium wet at all times. Moreover, the level of nitrification, which can be achieved by an RBC system, is significant—especially when multiple stages are employed.

RBC Equipment

The equipment that makes up an RBC includes the rotating biological contactor (the media: either standard or high density), a center shaft, drive system, tank, baffles, housing or cover, and a settling tank. The *rotating biological contactor* consists of circular sheets of synthetic material (usually plastic) which are mounted side by side on a shaft. The sheets (media) contain large amounts of surface area for growth of the biomass. The *center shaft* provides the support for the disks of media and must be strong enough to support the weight of the media and the biomass. Experience has indicated a major problem has been the collapse of the support shaft. The *drive system* provides the motive force to rotate the disks and shaft. The drive system may be mechanical or air driven or a combination of each. When the drive system does not provide uniform movement of the RBC, major operational problems can arise.

The *tank* holds the wastewater that the RBC rotates in. It should be large enough to permit variation of the liquid depth and detention time. *Baffles* are required to permit proper adjustment of the loading applied to each stage of the RBC process. Adjustment can be made to increase or decrease the submergence of the RBC. RBC stages are normally enclosed in some type of protective structure (*cover*) to prevent loss of biomass due to severe weather changes (snow, rain, temperature, wind, sunlight, etc.). In many instances this housing greatly restricts access to the RBC. The *settling tank* is provided to remove the sloughing material created by the biological activity and is similar in design to the primary settling tank. The settling tank provides 2–4 hours detention times to permit settling of lighter biological solids.

RBC Operation

During normal operation, operator vigilance is required to observe the RBC movement, slime color, and appearance. However, if the unit is covered, observations may be limited to that portion of the media, which can be viewed through the access door. Slime color and appearance can indicate process condition, for example:

- Gray, shaggy slime growth indicates normal operation
- Reddish brown, golden shaggy growth indicates nitrification
- White chalky appearance indicates high sulfur concentrations
- No slime indicates severe temperature or pH changes

Sampling and testing should be conducted daily for dissolved oxygen content and pH. BOD_5 and suspended-solids testing should also be accomplished to aid in assessing performance.

RBC Expected Performance

The RBC normally produces a high-quality effluent with BOD_5 at 85–95 percent and suspended-solids removal at 85–95 percent. The RBC treatment process may also significantly reduce (if designed for this purpose) the levels of organic nitrogen and ammonia nitrogen.

Activated Sludge

The biological treatment systems discussed to this point (ponds, trickling filters, and rotating biological contactors) have been around for years. The trickling filter, for example, has been around and successfully used since the late 1800s. The problem with ponds, trickling filters and RBCs is that they are temperature sensitive, remove less BOD and, trickling filters, for example, cost more to build than the activated sludge systems that were later developed.

Note: Although trickling filters and other systems cost more to build than activated sludge systems, it is important to point out that activated sludge systems cost more to operate because of the need for energy to run pumps and blowers.

The activated sludge process follows primary settling. The basic components of an activated sludge sewage treatment system include an aeration tank and a secondary basin, settling basin, or clarifier. Primary effluent is mixed with settled solids recycled from the secondary clarifier and is then introduced into the aeration tank. Compressed air is injected continuously into the mixture through porous diffusers located at the bottom of the tank, usually along one side.

Wastewater is fed continuously into an aerated tank, where the microorganisms metabolize and biologically flocculate the organics. Microorganisms (activated sludge) are settled from the aerated mixed liquor under quiescent conditions in the final clarifier and are returned to the aeration tank. Left uncontrolled, the number of organisms would eventually become too great; therefore, some must periodically be removed (wasted). A portion of the concentrated solids from the bottom of the settling tank must be removed from the process (waste activated sludge or WAS). Clear supernatant from the final settling tank is the plant effluent.

Activated Sludge Terminology

To better understand the discussion of the activated sludge process presented in the following sections, you must understand the terms associated with the process. Some of these terms have been used and defined earlier in the text, but we list them here again to refresh your memory. Review these terms and remember them. They are used throughout the discussion.

- *Adsorption*—taking in or reception of one substance into the body of another by molecular or chemical actions and distribution throughout the absorber.
- *Activated*—to speed up reaction. When applied to sludge, it means that many aerobic bacteria and other microorganisms are in the sludge particles.
- *Activated sludge*—a floc or solid formed by the microorganisms. It includes organisms, accumulated food materials, and waste products from the aerobic decomposition process.
- *Activated sludge process*—a biological wastewater treatment process in which a mixture or influent and activated sludge is agitated and aerated. The activated sludge is subsequently separated from the treated mixed liquor by sedimentation and is returned to the process as needed. The treated wastewater overflows the weir of the settling tank in which separation from the sludge takes place.
- *Adsorption*—the adherence of dissolved, colloidal, or finely divided solids to the surface of solid bodies when they are brought into contact.

- *Aeration*—mixing air and a liquid by one of the following methods: spraying the liquid in the air; diffusing air into the liquid; or agitating the liquid to promote surface adsorption of air.
- *Aerobic*—a condition in which "free" or dissolved oxygen is present in the aquatic environment. Aerobic organisms must be in the presence of dissolved oxygen to be active.
- *Bacteria*—single-cell plants that play a vital role in stabilization of organic waste.
- *Biochemical oxygen demand*—a measure of the amount of food available to the microorganisms in a particular waste. It is measured by the amount of dissolved oxygen used up during a specific time period (usually five days, expressed as BOD_5).
- *Biodegradable*—from "degrade" (to wear away or break down chemically) and "bio" (by living organisms). Put it all together, and you have a "substance, usually organic, which can be decomposed by biological action."
- *Bulking*—a problem in activated sludge plants that results in poor settleability of sludge particles.
- *Coning*—a condition that may be established in a sludge hopper during sludge withdrawal, when part of the sludge moves toward the outlet while the remainder tends to stay in place. Development of a cone or channel of moving liquids surrounded by relatively stationary sludge.
- *Decomposition*—generally, in waste treatment, decomposition refers to the changing of waste matter into simpler, more stable forms that will not harm the receiving stream.
- *Diffuser*—a porous plate or tube through which air is forced and divided into tiny bubbles for distribution in liquids. Commonly made of carborundum, aluminum, or silica sand.
- *Diffused air aeration*—a diffused air activated sludge plant takes air, compresses it, then discharges the air below the water surface to the aerator through some type of air diffusion device.
- *Dissolved oxygen*—atmospheric oxygen dissolved in water or wastewater, usually abbreviated as DO.

Note: The typical required DO for a well-operated activated sludge plant is between 2.0 mg/L and 2.5 mg/L.

- *Facultative*—facultative bacteria can use either molecular (dissolved) oxygen or oxygen obtained from food materials. In other words, facultative bacteria can live under aerobic or anaerobic conditions.
- *Filamentous bacteria*—organisms that grow in thread or filamentous form.
- *Food-to-microorganisms ratio*—a process control calculation used to evaluate the amount of food (BOD or COD) available per pound of mixed liquor volatile suspended solids. This may be written as F/M ratio.
- *Fungi*—multicellular aerobic organisms.

- *Gould sludge age*—a process control calculation used to evaluate the amount of influent suspended solids available per pound of mixed liquor suspended solids.
- *Mean cell residence time* (MCRT)—the average length of time mixed liquor suspended-solids particle remains in the activated sludge process. This is usually written as MCRT and may also be referred to as *sludge retention rate* (STR).
- *Mixed liquor suspended solids* (MLSS)—the suspended-solids concentration of the mixed liquor. Many references use this concentration to represent the amount of organisms in the liquor. Many references use this concentration to represent the amount of organisms in the activated sludge process. This is usually written MLSS.
- *Mixed liquor volatile suspended solids* (MLVSS)—the organic matter in the mixed liquor suspended solids. This can also be used to represent the amount of organisms in the process. This is normally written as MLVSS.
- *Nematodes*—microscopic worms that may appear in biological waste treatment systems.
- *Nutrients*—substances required to support plant organisms. Major nutrients are carbon, hydrogen, oxygen, sulfur, nitrogen, and phosphorus.
- *Protozoa*—single-cell animals that are easily observed under the microscope at a magnification of 100x. Bacteria and algae are prime sources of food for advanced forms of protozoa.
- *Return activated sludge*—the solids returned from the settling tank to the head of the aeration tank. This is normally written as RAS.
- *Rising sludge*—rising sludge occurs in the secondary clarifiers or activated sludge plant when the sludge settles to the bottom of the clarifier, is compacted, and then rises to the surface in relatively short time.
- *Rotifiers*—multicellular animals with flexible bodies and cilia near their mouths used to attract food. Bacteria and algae are their major source of food.
- *Secondary treatment*—a wastewater treatment process used to convert dissolved or suspended materials into a form that can be removed.
- *Settleability*—a process control test used to evaluate the settling characteristics of the activated sludge. Readings taken at 30–60 minutes are used to calculate the settled sludge volume (SSV) and the sludge volume index (SVI).
- *Settled sludge volume*—the volume of mL/L (or percent) occupied by an activated sludge sample after 30–60 minutes of settling. Normally written as SSV with a subscript to indicate the time of the reading used for calculation (SSV_{30} or SSV_{60}).
- *Shock load*—the arrival at a plant of a waste toxic to organisms, in sufficient quantity or strength to cause operating problems, such as odor or sloughing off of the growth of slime on the trickling filter media. Organic overloads also can cause a shock load.
- *Sludge volume index*—a process control calculation used to evaluate the settling quality of the activated sludge.

- *Solids*—material in the solid state

  Dissolved—solids present in solution. Solids that will pass through a glass fiber filter.

  Fixed—also known as the inorganic solids. The solids that are left after a sample is ignited at 550°C (centigrade) for 15 minutes.

  Floatable solids—solids that will float to the surface of still water, sewage, or other liquid. Usually composed of grease particles, oils, light plastic material, etc. Also called *scum*.

  Non-settleable—finely divided suspended solids that will not sink to the bottom in still water, sewage, or other liquid in a reasonable period, usually two hours. Non-settleable solids are also known as colloidal solids.

  Suspended—the solids that will not pass through a glass fiber filter.

  Total—the solids in water, sewage, or other liquids; it includes the suspended solids and dissolved solids.

  Volatile—the organic solids. Measured as the solids that are lost on ignition of the dry solids at 550°C.
- *Waste activated sludge*—the solids being removed from the activated sludge process. This is normally written as WAS.

Activated Sludge Process: Equipment

The equipment requirements for the activated sludge process are more complex than other processes discussed. Equipment includes an *aeration tank*, *aeration*, *system-settling tank*, *return sludge*, and *waste sludge system*. These are discussed in the following.

Aeration Tank

The *aeration tank* is designed to provide the required detention time (depends on the specific modification) and ensure that the activated sludge and the influent wastewater are thoroughly mixed. Tank design normally attempts to ensure no dead spots are created.

Aeration

Aeration can be mechanical or diffused. Mechanical aeration systems use agitators or mixers to mix air and mixed liquor. Some systems use sparge ring to release air directly into the mixer. Diffused aeration systems use pressurized air released through diffusers near the bottom of the tank. Efficiency is directly related to the size of the air bubbles produced. Fine bubble systems have a higher efficiency. The diffused air system has a blower to produce large volumes of low-pressure air (5–10 psi), air lines to carry the air to the aeration tank, headers to distribute the air to the diffusers which release the air into the wastewater.

Settling Tank

Activated sludge systems are equipped with plain *settling tanks* designed to provide 2–4 hours hydraulic detention time.

Return Sludge

The return sludge system includes pumps, a timer or variable speed drive to regulate pump delivery, and a flow measurement device to determine actual flow rates.

Waste Sludge

In some cases the *waste activated sludge* withdrawal is accomplished by adjusting valves on the return system. When a separate system is used it includes pump(s), timer or variable speed drive and a flow measurement device.

Oxidation Ditches

An oxidation ditch is modified extended aeration activated sludge biological treatment process that utilizes longs solids retention times (SRTs) to remove biodegradable organics. Oxidation ditches are typically complete mix systems, but they can be modified to approach plug flow conditions. (***Note***: as conditions approach plug flow, diffused air must be used to provide enough mixing. The system will also no longer operate as an oxidation ditch). Typical oxidation ditch treatment systems consist of a single or multi-channel configuration within a ring, oval, or horseshoe-shaped basin. As a result, oxidation ditches are called "racetrack type" reactors. Horizontally or vertically mounted aerators provide circulation, oxygen transfer, and aeration in the ditch.

Preliminary treatment, such as bar screens and grit removal, normally precedes the oxidation ditch. Primary settling prior to an oxidation ditch is sometimes practiced but is not typical in this design. Tertiary filters may be required after clarification, depending on the effluent requirements. Disinfection is required and reaeration may be necessary prior to final discharge. Flow to the oxidation ditch is aerated and mixed with return sludge from a secondary clarifier.

Surface aerators, such as brush rotors, disc aerators, draft tube aerators, or fine bubble diffusers, are used to circulate the mixed liquor. The mixing process entrains oxygen into the mixed liquor to foster microbial growth and the motive velocity ensures contact of microorganisms with the incoming wastewater. The aeration sharply increases the dissolved oxygen (DO) concentration but decreases as biomass uptake oxygen as the mixed liquor travels through the ditch. Solids are maintained in suspension as the mixed liquor travels through the ditch. Solids are maintained in suspension as the mixed liquor circulated around the ditch. If design SRT's are selected for nitrification, a high degree of nitrification will occur. Oxidation ditch effluent is usually settled in a separate secondary clarifier. An anaerobic tank may be added prior to the ditch to enhance biological phosphorus removal.

An oxidation ditch may also be operated to achieve partial denitrification. One of the most common design modifications for enhanced nitrogen removal is known as the Modified Ludzack-Ettinger (MLE) process. In this process, an anoxic tank is added upstream of the ditch along with mixed liquor recirculation for the aerobic zone to the tank to achieve higher levels of denitrification. In the aerobic basin, autotrophic bacteria (nitrifiers) convert ammonia nitrogen to nitrite-nitrogen and then to nitrate nitrogen. In the anoxic zone, heterotrophic bacteria convert nitrate nitrogen to nitrogen gas which is released into the atmosphere. Some mixed liquor from the aerobic basin is recirculated to the anoxic zone to provide mixed liquor with a high concentration of nitrate nitrogen to the anoxic zone.

Several manufacturers have developed modifications to the oxidation ditch design to remove nutrient sin conditions cycled or phased between the anoxic and aerobic states. While the mechanics of operation differ by manufacturer, in general, the process consists of two separate aeration basins, the first anoxic and the second aerobic. Wastewater and return activated sludge (RAS) are introduced into the first reactor which operates under anoxic conditions. Mixed liquor then flows into the second reactor operating under aerobic conditions. The process is then reversed and the second reactor begins to operate under anoxic conditions.

With regard to applicability, the oxidation ditch process is a fully demonstrated secondary wastewater treatment technology, applicable in any situation where activated sludge treatment (conventional or extend aeration) is appropriate. Oxidation ditches are applicable in plants that require nitrification because the basins can be sized using an appropriate SRT to achieve nitrification at the mixed liquor minimum temperature. This technology is very effective in small installations, small communities, and isolated institutions, because it requires more land than conventional treatment plants (USEPA 2000a).

There are currently more than 9,000 municipal oxidation ditch installations in the United States (Spellman, 2007). Nitrification to less than 1 mg/L ammonia nitrogen consistently occurs when ditches are designed and operated for nitrogen removal.

Disinfection of Wastewater

Like drinking water, liquid wastewater effluent is disinfected. Unlike drinking water, wastewater effluent is disinfected not to directly (direct end-of-pipe connection) protects a drinking water supply but instead is treated to protect public health in general. This is particularly important when the secondary effluent is discharged into a body of water used for swimming or water supply for a downstream water supply. In the treatment of water for human consumption, treated water is typically chlorinated (although ozonation is also currently being applied in many cases). Chlorination is the preferred disinfection in potable water supplies because of chlorine's unique ability to provide a residual. This chlorine residual is important because when treated water leaves the waterworks facility and enters the distribution system, the possibility of contamination is increased. The residual works to continuously disinfect water right up to the consumer's tap.

In this section, we discuss basic chlorination and dechlorination. In addition, we describe ultraviolet (UV) irradiation, ozonation, bromine chlorine, and no disinfection.

Chlorine Disinfection

Chlorination for disinfection, as shown in Figure 10.8, follows all other steps in conventional wastewater treatment. The purpose of chlorination is to reduce the population of organisms in the wastewater to levels low enough to ensure that pathogenic organisms will not be present in sufficient quantities to cause disease when discharged.

Note: Chlorine gas is heavier than (vapor density of 2.5). Therefore, exhaust from a chlorinator room should be taken from floor level.

Note: The safest action to take in the event of a major chlorine container leak is to call the fire department.

Note: You might wonder why it is that chlorination of critical waters such as natural trout streams is not normal practice. This practice is strictly prohibited because chlorine and its by-products (i.e., chloramines) are extremely toxic to aquatic organisms.

Chlorination Terminology

Remember that there are several terms used in discussion of disinfection by chlorination. Because it is important for the operator to be familiar with these terms, we repeat key terms again.

- *Chlorine*—a strong oxidizing agent which has strong disinfecting capability. A yellow-green gas which is extremely corrosive, and is toxic to humans in extremely low concentrations in air.
- *Contact Time*—the length of time the disinfecting agent and the wastewater remain in contact.
- *Demand*—the chemical reactions, which must be satisfied before a residual or excess chemical will appear.
- *Disinfection*—refers to the selective destruction of disease-causing organisms. All the organisms are not destroyed during the process. This differentiates disinfection from sterilization, which is the destruction of all organisms.
- *Dose*—the amount of chemical being added in milligrams/liter.
- *Feed Rate*—the amount of chemical being added in pounds per day.
- *Residual*—the amount of disinfecting chemical remaining after the demand has been satisfied.
- *Sterilization*—the removal of all living organisms.

Chlorine Facts

- Elemental chlorine (Cl_2—gaseous) is a yellow-green gas, 2.5 times heavier than air.
- The most common use of chlorine in wastewater treatment is for disinfection. Other uses include odor control and activated sludge bulking control. Chlorination takes place prior to the discharge of the final effluent to the receiving waters (see Figure 10.8).
- Chlorine may also be used for nitrogen removal, through a process called *breakpoint chlorination*. For nitrogen removal, enough chlorine is added to the wastewater to convert all the ammonium nitrogen gas. To do this, approximately 10 mg/L of chlorine must be added for every 1 mg/L of ammonium nitrogen in the wastewater.
- For disinfection, chlorine is fed manually or automatically into a chlorine contact tank or basin, where it contacts flowing wastewater for at least 30

minutes to destroy disease-causing microorganisms (pathogens) found in treated wastewater.

- Chorine may be applied as a gas, a solid, or in liquid hypochlorite form.
- Chorine is a very reactive substance. It has the potential to react with many different chemicals (including ammonia), as well as with organic matter. When chlorine is added to wastewater, several reactions occur:
    (1) Chlorine will react with any reducing agent (i.e., sulfide, nitrite, iron, and thiosulfate) present in wastewater. These reactions are known as *chlorine demand*. The chlorine used for these reactions is not available for disinfection.
    (2) Chlorine also reacts with organic compounds and ammonia compounds to form chlor-organics and chloramines. Chloramines are part of the group of chlorine compounds that have disinfecting properties and show up as part of the chlorine residual test.
    (3) After all of the chlorine demands are met, addition of more chlorine will produce free residual chlorine. Producing free residual chlorine in wastewater requires very large additions of chlorine.

Hypochlorite Facts

Hypochlorite, though there are some minor hazards associated with its use (skin irritation, nose irritation, and burning eyes), is relatively safe to work with. It is normally available in dry form as a white powder, pellet, or tablet or in liquid form. It can be added directly using a dry chemical feeder or dissolved and fed as a solution.

Note: In most wastewater treatment systems, disinfection is accomplished by means of combined residual.

Ultraviolet Irradiation

Although ultraviolet (UV) disinfection was recognized as a method for achieving disinfection in the late nineteenth century, its application virtually disappeared with the evolution of chlorination technologies. However, in recent years, there has been resurgence in its use in the wastewater field, largely as a consequence of concern for discharge of toxic chlorine residual. Even more recently, UV has gained more attention because of the tough new regulations on chlorine use imposed by both OSHA and the USEPA. Because of this relatively recent increased regulatory pressure, many facilities are actively engaged in substituting chlorine for other disinfection alternatives. Moreover, UV technology itself has made many improvements, which now makes UV attractive as a disinfection alternative. Ultraviolet light has very good germicidal qualities and is very effective in destroying microorganisms. It is used in hospitals, biological testing facilities, and many other similar locations. In wastewater treatment, the plant effluent is exposed to ultraviolet light of a specified wavelength and intensity for a specified contact period. The effectiveness of the process is dependent upon:

- UV light intensity
- Contact time
- Wastewater quality (turbidity)

For any one treatment plant, disinfection success is directly related to the concentration of colloidal and particulate constituents in the wastewater.

The Achilles' heel of UV for disinfecting wastewater is turbidity. If the wastewater quality is poor, the ultraviolet light will be unable to penetrate the solids and the effectiveness of the process decreases dramatically. For this reason, many states limit the use of UV disinfection to facilities that can reasonably be expected to produce an effluent containing ≤ 30 mg/L or less of BOD_5 and total suspended solids.

The main components of a UV disinfection system are mercury arc lamps, a reactor, and ballasts. The source of UV radiation is either the low-pressure or medium-pressure mercury arc lamp with low or high intensities. Note that in the operation of UV systems, UV lamps must be readily available when replacements are required. The best lamps are those with a stated operating life of at least 7,500 hours and those that do not produce significant amounts of ozone or hydrogen peroxide. The lamps must also meet technical specifications for intensity, output, and arc length. If the UV light tubes are submerged in the wastestream, they must be protected inside quartz tubes, which not only protect the lights but also make cleaning and replacement easier.

Contact tanks must be used with UV disinfection. They must be designed with the banks of UV lights in a horizontal position, either parallel or perpendicular to the flow or with banks of lights placed in a vertical position perpendicular to the flow.

Note: The contact tank must provide, at a minimum, 10-second exposure time.

We stated earlier that turbidity problems are the result of using UV in wastewater treatment—and this is the case. However, if turbidity is its Achilles' heel, then the need for increased maintenance (as compared to other disinfection alternatives) is the toe of the same foot. UV maintenance requires that the tubes be cleaned on a regular basis or as needed. In addition, periodic acid washing is also required to remove chemical buildup.

Routine monitoring is required. Monitoring to check on bulb burnout, buildup of solids on quartz tubes, and UV light intensity is necessary.

Note: UV light is extremely hazardous to the eyes. Never enter an area where UV lights are in operation without proper eye protection. Never look directly into the ultraviolet light.

Ozonation

Ozone is a strong oxidizing gas that reacts with most organic and many inorganic molecules. It is produced when oxygen molecules separate, collide with other oxygen atoms, and form a molecule consisting of three oxygen atoms. For high-quality effluents, ozone is a very effective disinfectant. Current regulations for domestic treatment systems limit use of ozonation to filtered effluents unless the system's effectiveness can be demonstrated prior to installation.

Note: Effluent quality is the key performance factor for ozonation.

For ozonation of wastewater, the facility must have the capability to generate pure oxygen along with an ozone generator. A contact tank with ≥ 10-minute contact time at design average daily flow is required. Off-gas monitoring for process control is also required. In addition, safety equipment capable of monitoring ozone in the atmosphere and a ventilation system to prevent ozone levels exceeding 0.1 ppm is required.

The actual operation of the ozonation process consists of monitoring and adjusting the ozone generator and monitoring the control system to maintain the required ozone concentration in the off-gas. The process must also be evaluated periodically using biological testing to assess its effectiveness.

Note: Ozone is an extremely toxic substance. Concentrations in air should not exceed 0.1 ppm. It also has the potential to create an explosive atmosphere. Sufficient ventilation and purging capabilities should be provided.

Note: Ozone has certain advantages over chlorine for disinfection of wastewater: (1) ozone increases DO in the effluent; (2) ozone has a briefer contact time; (3) ozone has no undesirable effects on marine organisms; and (4) ozone decreases turbidity and odor.

Bromine Chloride

Bromine chloride is a mixture of bromine and chlorine. It forms hydrocarbons and hydrochloric acid when mixed with water. Bromine chloride is an excellent disinfectant that reacts quickly and normally does not produce any long-term residuals.

Note: Bromine chloride is an extremely corrosive compound in the presence of low concentrations of moisture.

The reactions occurring when bromine chloride is added to the wastewater are similar to those occurring when chlorine is added. The major difference is the production of bromamine compounds rather than chloramines. The bromamine compounds are excellent disinfectants but are less stable and dissipate quickly. In most cases, the bromamines decay into other, less toxic compounds rapidly and are undetectable in the plant effluent. The factors that affect performance are similar to those affecting the performance of the chlorine disinfection process. Effluent quality, contact time, etc., have a direct impact on the performance of the process.

No Disinfection

In a very limited number of cases, treated wastewater discharges without disinfection are permitted. These are approved on a case-by-case basis. Each request must be evaluated based upon the point of discharge, the quality of the discharge, the potential for human contact, and many other factors.

Advanced Wastewater Treatment

Advanced wastewater treatment is defined as the method(s) and/or process(es) that remove more contaminants (suspended and dissolved substances) from wastewater than are taken out by conventional biological treatment. Put another way, advanced wastewater treatment is the application of a process or system that follows secondary treatment or that includes phosphorus removal or nitrification in conventional secondary treatment.

Advanced wastewater treatment is used to augment conventional secondary treatment because secondary treatment typically removes only 85–95 percent of the biochemical oxygen demand and total suspended solids in raw sanitary sewage. Generally, this leaves 30 mg/L or less of BOD and TSS in the secondary effluent. To meet stringent water-quality standards, this level of BOD and TSS in secondary effluent may not prevent violation of water-quality standards—the plant may not make permit. Thus, advanced wastewater treatment is often used to remove additional pollutants from treated wastewater.

In addition to meeting or exceeding the requirements of water-quality standards, treatment facilities use advanced wastewater treatment for other reasons as well. For example, sometimes, conventional secondary wastewater treatment is not sufficient to protect the aquatic environment. In a stream, for example, when periodic flow events occur, the stream may not provide the amount of dilution of effluent needed to maintain the necessary dissolved oxygen (DO) levels for aquatic organism survival.

Secondary treatment has other limitations. It does not significantly reduce the effluent concentration of nitrogen and phosphorus (important plant nutrients) in sewage. An over-abundance of these nutrients can over-stimulate plant and algae growth such that they create water-quality problems. For example, if discharged into lakes, these nutrients contribute to algal blooms and accelerated eutrophication (lake aging). Also, the nitrogen in the sewage effluent may be present mostly in the form of ammonia compounds. If in high enough concentration, ammonia compounds are toxic to aquatic organisms. Yet another problem with these compounds is that they exert a *nitrogenous* oxygen demand in the receiving water, as they convert to nitrates. This process is called nitrification.

Note: The term *tertiary treatment* is commonly used as a synonym for advanced wastewater treatment. However, these two terms do not have precisely the same meaning. Tertiary suggests a third step that is applied after primary and secondary treatment.

Advanced wastewater treatment can remove more than 99 percent of the pollutants from raw sewage and can produce an effluent of almost potable (drinking) water quality. However, obviously, advanced treatment is not cost-free. The cost of advanced treatment, for operation and maintenance as well as for retrofit of present conventional processes, is very high (sometimes doubling the cost of secondary treatment). Therefore, a plan to install advanced treatment technology calls for careful study—the benefit-to-cost ratio is not always big enough to justify the additional expense.

Even considering the expense, application of some form of advanced treatment is not uncommon. These treatment processes can be physical, chemical, or biological. The specific process used is based upon the purpose of the treatment and the quality of the effluent desired.

Chemical Treatment

The purpose of chemical treatment is to remove:

- Biochemical oxygen demand
- Total suspended solids
- Phosphorus

- Heavy metals
- Other substances that can be chemically converted to a settleable solid

Chemical treatment is often accomplished as an "add-on" to existing treatment systems or by means of separate facilities specifically designed for chemical addition. In each case, the basic process necessary to achieve the desired results remains the same:

- Chemicals are thoroughly mixed with the wastewater.
- The chemical reactions that occur form solids (coagulation).
- The solids are mixed to increase particle size (flocculation).
- Settling and/or filtration (separation) then remove the solids.

The specific chemical used depends on the pollutant to be removed and the characteristics of the wastewater. Chemicals may include the following:

- Lime
- Alum (aluminum sulfate)
- Aluminum salts
- Ferric or ferrous salts
- Polymers
- Bioadditives

Microscreening

Microscreening (also called *microstraining*) is an advanced treatment process used to reduce suspended solids. The microscreens are composed of specially woven steel wire fabric mounted around the perimeter of a large revolving drum. The steel wire cloth acts as a fine screen, with openings as small as 20 μm (or millionths of a meter)—small enough to remove microscopic organisms and debris. The rotating drum is partially submerged in the secondary effluent, which must flow into the drum then outward through the microscreen. As the drum rotates, captured solids are carried to the top where a high-velocity water spray flushes them into a hopper or backwash tray mounted on the hollow axle of the drum. Backwash solids are recycled to plant influent for treatment. These units have found greatest application in treatment of industrial waters and final polishing filtration of wastewater effluents. The expected performance for suspended-solids removal is 95–99 percent, but the typical suspended-solids removal achieved with these units is about 55 percent. The normal range is 10–80 percent.

According to Metcalf and Eddy (2003), the functional design of the microscreen unit involves the following considerations: (1) the characterization of the suspended solids with respect to the concentration and degree of flocculation; (2) the selection of unit design parameter values that will not only ensure capacity to meet maximum hydraulic loadings with critical solids characteristics but also provide desired design performance over the expected range of hydraulic and solids loadings; and (3) the provision of backwash and cleaning facilities to maintain the capacity of the screen.

Filtration

The purpose of *filtration* processes used in advanced treatment is to remove suspended solids. The specific operations associated with a filtration system are dependent on the equipment used. A general description of the process follows.

Wastewater flows to a filter (gravity or pressurized). The filter contains single, dual, or multimedia. Wastewater flows through the media, which removes solids. The solids remain in the filter. Backwashing the filter as needed removes trapped solids. Backwash solids are returned to the plant for treatment. Processes typically remove 95–99 percent of the suspended matter.

Membrane Bioreactors

In this section we present a more in depth discussion of membrane bioreactors because the technologies most commonly used for performing secondary treatment of municipal wastewater rely on microorganisms suspended in the wastewater to treat it. Although these technologies work well in many situations, they have several drawbacks, including the difficulty of growing the right types of microorganisms and the physical requirement of a large site. The use of microfiltration membrane bioreactors (MBRs), a technology that has become increasingly used in the past ten years, overcomes many of the limitations of conventional systems. These systems have the advantage of combining a suspended growth biological reactor with solids removal via filtration. The membranes can be designed for and operated in small spaces and with high removal efficiency of contaminants such as nitrogen, phosphorus, bacteria, biochemical oxygen demand, and total suspended solids. The membrane filtration system in effect can replace the secondary clarifier and sand filters in a typical activated sludge treatment system. Membrane filtration allows a higher biomass concentration to be maintained, thereby allowing smaller bioreactors to be used (USEPA, 2007a).

Membrane Filtration

Membrane filtration involves the flow of water-containing pollutants across a membrane. Water permeates through the membrane into a separate channel for recovery. Because of the cross-flow movement of water and the waste constituents, materials left behind do not accumulate at the membrane surface but are carried out of the system for later recovery or disposal. The water passing through the membrane is called the *permeate*, while the water with the more concentrated materials is called the *concentrate* or *retentate*.

Membranes are constructed of cellulose or other polymer material, with a maximum pore size set during the manufacturing process. The requirement is that the membranes prevent passage of particles the size of microorganisms, or about 1 micron (0.001 mm), so that they remain in the system. This means that MBR systems are good for removing solid material, but the removal of dissolved wastewater components must be facilitated by using additional treatment steps.

Membranes can be configured in a number of ways. For MBR applications, the two configurations most often used are hollow fibers grouped in bundles or as flat plates. The hollow fiber bundles are connected by manifolds in units that are designed for easy changing and servicing.

Biological Nitrification

Biological *nitrification* is the first basic step of *biological nitrification-denitrification.*

In nitrification, the secondary effluent is introduced into another aeration tank, trickling filter, or biodisc. Because most of the carbonaceous BOD has already been removed, the microorganisms that drive in this advanced step are the nitrifying bacteria *Nitrosomonas* and *Nitrobacter.* In nitrification, the ammonia nitrogen is converted to nitrate nitrogen, producing a *nitrified effluent.* At this point, the nitrogen has not actually been removed, only converted to a form that is not toxic to aquatic life and that does not cause an additional oxygen demand. The nitrification process can be limited (performance affected) by alkalinity (requires 7.3 parts alkalinity to 1.0 part ammonia nitrogen); pH; dissolved oxygen availability; toxicity (ammonia or other toxic materials); and process mean cell residence time (sludge retention time). As a general rule, biological nitrification is more effective and achieves higher levels of removal during the warmer times of the year.

Biological Denitrification

Biological denitrification removes nitrogen from the wastewater. When bacteria come in contact with a nitrified element in the absence of oxygen, they reduce the nitrates to nitrogen gas, which escapes the wastewater. The denitrification process can be done in either an anoxic activated sludge system (suspended growth) or a column system (fixed growth). The denitrification process can remove up to 85 percent or more of nitrogen. After effective biological treatment, little oxygen demanding material is left in the wastewater when it reaches the denitrification process. The denitrification reaction will only occur if an oxygen demand source exists when no dissolved oxygen is present in the wastewater. An oxygen demand source is usually added to reduce the nitrates quickly. The most common demand source added is soluble BOD or methanol. Approximately 3 mg/L of methanol is added for every 1 mg/L of nitrate nitrogen. Suspended growth denitrification reactors are mixed mechanically, but only enough to keep the biomass from settling without adding unwanted oxygen. The submerged filters of different types of media may also be used to provide denitrification. A fine media downflow filter is sometimes used to provide both denitrification and effluent filtration. A fluidized sand bed where wastewater flows upward through a media of sand or activated carbon at a rate to fluidize the bed may also be used. Denitrification bacteria grow on the media.

Carbon Adsorption

The main purpose of *carbon adsorption* used in advanced treatment processes is the removal of refractory organic compounds (non-BOD_5) and soluble organic materials that are difficult to eliminate by biological or physical/chemical treatment. In the carbon adsorption process, wastewater passes through a container filled either with carbon powder or carbon slurry. Organics adsorb onto the carbon (i.e., organic molecules are attracted to the activated carbon surface and are held there) with sufficient contact time. A carbon system usually has several columns or basins used as contactors. Most contact chambers are either open concrete gravity-type systems or steel pressure containers applicable to either upflow or downflow operation. With use, carbon loses its adsorptive capacity. The carbon must then be regenerated or replaced with fresh carbon. As head loss develops in carbon contactors, they are backwashed with clean effluent in much the same way the effluent filters are backwashed. Carbon used for adsorption may be in a granular form or in a powdered form.

Note: Powdered carbon is too fine for use in columns and is usually added to the wastewater, then later removed by coagulation and settling.

Land Application

The application of secondary effluent onto a land surface can provide an effective alternative to the expensive and complicated advanced treatment methods discussed previously and the biological nutrient removal (BNR) system discussed later. A high-quality polished effluent (i.e., effluent with high levels of TSS, BOD, phosphorus, and nitrogen compounds as well as refractory organics are reduced) can be obtained by the natural processes that occur as the effluent flows over the vegetated ground surface and percolates through the soil. Limitations are involved with land application of wastewater effluent. For example, the process needs large land areas. Soil type and climate are also critical factors in controlling the design and feasibility of a land treatment process.

Biological Nutrient Removal

Nitrogen and phosphorus are the primary causes of cultural eutrophication (i.e., nutrient enrichment due to human activities) in surface waters. The most recognizable manifestations of this eutrophication are algal blooms that occur during the summer. Chronic symptoms of over-enrichment include low dissolved oxygen, fish kills, murky water, and depletion of desirable flora and fauna. In addition, the increase in algae and turbidity increases the need to chlorinate drinking water, which, in turn, leads to higher levels of disinfection by-products that have been shown to increase the risk of cancer (USEPA, 2007c). Excessive amounts of nutrients can also stimulate the activity of microbes, such as *Pfisteria*, which may be harmful to human health (USEPA, 2001d).

Approximately 25 percent of all water body impairments are due to nutrient-related causes (e.g., nutrients, oxygen depletion, algal growth, ammonia, harmful algal blooms, biological integrity, and turbidity) (USEPA, 2007d). In efforts to reduce the number of nutrient impairments, many point source discharges have received more stringent effluent limits for nitrogen and phosphorus. To achieve these new, lower effluent limits, facilities have begun to look beyond traditional treatment technologies.

Recent experience has reinforced the concept that biological nutrient removal systems are reliable and effective in removing nitrogen and phosphorus. The process is based upon the principle that, under specific conditions, microorganisms will remove more phosphorus and nitrogen than is required for biological activity; thus, treatment can be accomplished without the use of chemicals. Not having to use and therefore having to purchase chemicals to remove nitrogen and phosphorus potentially has numerous cost-benefit implications. In addition, because chemicals are not required to be used, chemical waste products are not produced, reducing the need to handle and dispose of waste. Several patented processes are available for this purpose. Performance depends on the biological activity and the process employed.

Enhanced Biological Nutrient Removal (EBNR)

Removing phosphorus from wastewater in secondary treatment processes has evolved into innovative *enhanced biological nutrient removal* technologies. An ENBR treatment process promotes the production of phosphorus accumulating organisms which utilize more phosphorus in their metabolic processes than a conventional secondary biological treatment process (USEPA, 2007b). The average total phosphorus concentration in raw domestic wastewater is usually 6–8 mg/L and the total phosphorus concentration in municipal wastewater after conventional secondary treatment is routinely reduced to 3–4 mg/L. Whereas, EBNR incorporated into the secondary treatment system can often reduce total phosphorus concentrations to 0.3 mg/L and less. Facilities using EBNR significantly reduced the amount of phosphorus to be removed through the subsequent chemical addition and tertiary filtration process. This improved the efficiency of the tertiary process and significantly reduced the costs of chemicals used to remove phosphorus. Facilities using EBNR reported that their chemical dosing was cut in half after EBNR was installed to remove phosphorus (USEPA, 2007b).

Treatment provided by these WWTPs also removes other pollutants which commonly affect water quality to very low levels (USEPA, 2007b). Biochemical oxygen demand and total suspended solids are routinely less than 2 mg/L and fecal coliform bacteria less than 10 fcu/100 mL. Turbidity of the final effluent is very low which allows for effective disinfection using ultraviolet light, rather than chlorination. Recent studies report that wastewater treatment plants using EBNR also significantly reduced the amount of pharmaceuticals and personal health care products from municipal wastewater, as compared to the removal accomplished by conventional secondary treatment.

Epstein & Alpert (1984) make the point that if the biosolids cake is higher in solids content, it reduces the need for space, fuel, labor, equipment, and size of the receiving facility, e.g., a composting facility.

Probably one of the best summarizations of the various reasons why it is important to dewater biosolids is given by Metcalf & Eddy (1991) in the following: (1) the costs of transporting biosolids to the ultimate disposal site is greatly reduced when biosolids volume is reduced; (2) dewatered biosolids allow for easier handling; (3) dewatering biosolids (reduction in moisture content) allows for more efficient incineration; (4) if composting is the beneficial reuse choice, dewatered biosolids decrease the amount and therefore the cost of bulking agents; (5) with the USEPA's new 503 rule, dewatering biosolids may be required to render the biosolids less offensive; and (6) when landfilling is the ultimate disposal option, dewatering biosolids is required to reduce leachate production.

Again, the point being made here is that the importance of adequately dewatering biosolids for proper disposal/reuse can't be overstated.

The unit processes that are most often used for dewatering biosolids are (1) vacuum filtration, (2) pressure filtration, (3) centrifugation, and (4) drying beds. The biosolids cake produced by common dewatering processes has a consistency similar to dry, crumbly, bread pudding (Spellman, 1996). This non-fluid dewatered dry, crumbly cake product is easily handled, non-offensive, and can be land applied manually and by conventional agricultural spreaders (Outwater, 1994).

Rotary Vacuum Filtration

Rotary vacuum filters have also been used for many years to dewater sludge. The vacuum filter includes filter media (belt, cloth, or metal coils), media support (drum), vacuum system, chemical feed equipment, and conveyor belt(s) to transport the dewatered solids. In operation, chemically treated solids are pumped to a vat or tank in which a rotating drum is submerged. As the drum rotates, a vacuum is applied to the drum. Solids collect on the media and are held there by the vacuum as the drum rotates out of the tank. The vacuum removes additional water from the captured solids. When solids reach the discharge zone, the vacuum is released and the dewatered solids are discharged onto a conveyor belt for disposal. The media is then washed prior to returning to the start of the cycle.

The three principal types of rotary vacuum filters are rotary drum, coil, and belt. The *rotary drum* filter consists of a cylindrical drum rotating partially submerged in a vat or pan of conditioned sludge. The drum is divided length-wise into a number of sections that are connected through internal piping to ports in the valve body (plant) at the hub. This plate rotates in contact with a fixed valve plate with similar parts, which are connected to a vacuum supply, a compressed air supply, and an atmosphere vent. As the drum rotates, each section is thus connected to the appropriate service.

The *coil type* vacuum filter uses two layers of stainless-steel coils arranged in corduroy fashion around the drum. After a de-watering cycle, the two layers of springs leave the drum bed and are separated from each other so that the cake is lifted off the lower layer and is discharged from the upper layer. The coils are then washed and reapplied to the drum. The coil filter is used successfully for all types of sludges; however, sludges with extremely fine particles or ones that are resistant to flocculation de-water poorly with this system.

The media on a *belt filter* leaves the drum surface at the end of the drying zone and passes over a small-diameter discharge roll to aid cake discharge. Washing of the media occurs next. Then the media are returned to the drum and to the vat for another cycle. This type of filter normally has a small-diameter curved bar between the point where the belt leaves the drum and the discharge roll. This bar primarily aids in maintaining belt dimensional stability.

Pressure Filtration

Pressure filtration differs from vacuum filtration in that the liquid is forced through the filter media by a positive pressure instead of a vacuum. Several types of presses are available, but the most commonly used types are plate-and-frame presses and belt presses. *Filter presses* include the belt or plate-and-frame types. The belt filter includes two or more porous belts, rollers, and related handling systems for chemical makeup and feed, and supernatant and solids collection and transport.

The plate-and-frame filter consists of a support frame, filter plates covered with porous material, hydraulic or mechanical mechanism for pressing plates together, and related handling systems for chemical makeup and feed, and supernatant and solids collection and transport. In the plate-and-frame filter, solids are pumped (sandwiched) between plates. Pressure (200–250 psi) is applied to the plates and water is "squeezed" from the solids. At the end of the cycle the pressure is released and as the plates separate the solids drop out onto a conveyor belt for transport to storage or disposal. Performance factors for plate-and-frame presses include feed sludge characteristics, type and amount of chemical conditioning, operating pressures, and the type and amount of precoat.

The belt filter uses a coagulant (polymer) mixed with the influent solids. The chemically treated solids are discharged between two moving belts. First water drains from the solids by gravity. Then, as the two belts move between a series of rollers, pressure "squeezes" additional water out of the solids. The solids are then discharged onto a conveyor belt for transport to storage/disposal. Performance factors for the belt press include sludge feed rate, belt speed, belt tension, belt permeability, chemical dosage, and chemical selection.

Filter presses have lower operation and maintenance costs than vacuum filters or centrifuges. They typically produce a good quality cake and can be batch operated. However, construction and installation costs are high. Moreover, chemical addition is required and the presses must be operated by skilled personnel.

Centrifugation

Centrifuges of various types have been used in dewatering operations for at least 30 years and appear to be gaining in popularity. Depending on the type of centrifuge used, in addition to centrifuge pumping equipment for solids feed and centrate removal, chemical makeup and feed equipment and support systems for removal of dewatered solids are required.

Sludge Incineration

Not surprisingly, incinerators produce the maximum solids and moisture reductions. The equipment required depend on whether the unit is a multiple hearth or fluid-bed incinerator. Generally, the system will require a source of heat to reach ignition temperature, solids feed system and ash handling equipment. It is important to note that the system must also include all required equipment (e.g., scrubbers) to achieve compliance with air pollution control requirements. Solids are pumped to the incinerator. The solids are dried then ignited (burned). As they burn the organic matter is converted to carbon dioxide and water vapor and the inorganic matter is left behind as ash or "fixed" solids. The ash is then collected for reuse of disposal.

Land Application of Biosolids

The purpose of land application of biosolids is to dispose of the treated biosolids in an environmentally sound manner by recycling nutrients and soil conditioners. In order to be land applied, wastewater biosolids must comply with state and federal biosolids management/disposal regulations. Biosolids must not contain materials that are dangerous to human health (i.e., toxicity, pathogenic organisms, etc.) or dangerous to the environment (i.e., toxicity, pesticides, heavy metals, etc.). Treated biosolids are land applied by either direct injection or application and plowing in (incorporation).

NPDES Permits

In the United States, all treatment facilities which discharge to State waters must have a discharge permit issued by the State Water Control Board or other appropriate State agency. This permit is known on the national level as the National Pollutant Discharge Elimination System (NPDES) permit and on the state level as the (State) Pollutant Discharge Elimination System (state-PDES) permit. The permit states the specific conditions which must be met to legally discharge treated wastewater to State waters. The permit contains general requirements (applying to every discharger) and specific requirements (applying only to the point source specified in the permit). A general permit is a discharge permit, which covers a specified class of dischargers. It is developed to allow dischargers with the specified category to discharge under specified conditions. All discharge permits contain general conditions. These conditions are standard for all dischargers and cover a broad series of requirements. Read the general conditions of the treatment facility's permit carefully. Permit tees must retain certain records.

BIBLIOGRAPHY

Albrecht, R., 1987. How to succeed in compost marketing. *BioCycle*, 28(9): 26–27.

Alexander, R., 1991. Sludge compost use on athletic fields. *BioCycle*, 32(7): 69–71.

American Public Health Association (APHA), 1992. *Standard Methods for the Examination of Water and Wastewater*, 18th ed. Washington, DC.

Anderson, J.B., & Zwieg, H.P., 1962. Biology of waste stabilization ponds. *Southwest Water Works Journal*, 44(2): 15–18.

Assenzo, J.R., & Reid, G.W., 1966. Removing nitrogen and phosphorus by bio-oxidation ponds in central Oklahoma. *Water and Sewage Works*, 13(8): 294–299.

Benedict, A.H., Epstein, E., & English, J.N., 1986. Municipal sludge composting technology evaluation. *Journal WPCF*, 58(4): 279–289.

Brockett, O.D., 1976. Microbial reactions in facultative ponds-1. The anaerobic nature of oxidation pond sediments. *Water Research*, 10(1): 45–49.

Bryan, F.L., 1974. *Diseases Transmitted in Food Contaminated with Wastewater*. EPA 660/2-74-041, June 1974. Washington, DC: U.S. Environmental Protection Agency.

Burnett, C.H., 1992. Small cities + warm climates = windrow composting. *Presented at the Water Environment Federation 65th Annual Conference & Exposition*, New Orleans, LA. September, 20–24.

Burnett, G.W., & Schuster, G.S., 1973. *Pathogenic Microbiology*. St Louis, MO: C.V. Mosby Company.

CDC, 2013. *Parasites—Taeniasis*. Atlanta, GA: Centers for Disease Control and Prevention.

CDC, 2020. *Parasites—Ascariasis*. Atlanta, GA: Centers for Disease Control and Prevention.

Cheremisinoff, P.N., 1995. Gravity separation for efficient solids removal. *The National Environmental Journal*, 5(6): 29–32.

Cheremisinoff, P.N., & Young, R.A., 1981. *Pollution Engineering Practice Handbook*. Ann Arbor, MI: Ann Arbor Science Publishers, Inc.

Corbitt, R.A., 1990. *Standard Handbook of Environmental Engineering*. New York: McGraw-Hill, Inc.
Craggs, R., 2005. Nutrients. In *Pond Treatment Technology*. A. Hilton (ed.). London, UK: IWA Publishing.
Crawford, G., Daigger, G., Fisher, J., Blair, S., & Lewis, R., 2005. Parallel operation of large membrane bioreactors at Traverse City. In *Proceedings of the Water Environment Federation 78th Annual Conference & Exposition*, Washington, DC, CD-ROM. October 29–November 2, 2005.
Crawford, G., Fernandez, A., Shawwa, A., & Daigger, G., 2002. Competitive bidding and evaluation of membrane bioreactor equipment—Three large plant case studies. In *Proceedings of the Water Environment Federation 75th Annual Conference & Exposition*, Chicago, IL, CD-ROM. September 28–October 2, 2002.
Crawford, G., Thompson, D., Lozier, J., Daigger, G., & Fleischer, E., 2000. Membrane bioreactors—A designer's perspective. In *Proceedings of the Water Environment Federation 73rd Annual Conference & Exposition on Water Quality and Wastewater Treatment*, Anaheim, CA, CD-ROM. October 14–18, 2000.
Crites, R., & Tchobanoglous, G., 1998. *Small and Decentralized Wastewater Management Systems*. Boston, MA: WCB McGraw-Hill, Inc.
Crites, R.W., Middlebrooks, E.J., & Reed, S.C., 2006. *Natural Wastewater Treatment Systems*. Boca Raton, FL: CRC, Taylor and Francis Group.
CWHS (Clean Watersheds Needs Survey), 2000. *Report to Congress*. EPA-832-R-10-002. Washington, DC.
Dean, R.S., & Smith, J.E., 1973. The properties of sludges. In *Proceedings of the Joint Conference on Recycling Municipal Sludges and effluents on Land*, pp. 39–47, Champaign, IL. Washington, DC: National Association of State Land-Grant Colleges. July 1973.
Emrick, J., & Abraham, K., 2002. Long-term BNR operations—Cold in Montana! In *Proceedings of the Water Environmental Federation 75th Annual Technical Exhibition & Conference*, Chicago, IL. September 28–October 2, 2002.
Engineering Science, Inc., April 1987. *Monterey Wastewater Reclamation Study for Agriculture—Final Report*. Berkeley, CA: Engineering Science, Inc.
EPA, 1991. *Preliminary Risk Assessment for Parasites in Municipal Sewage Sludge Applied to Land*. EPA 600/6-91/001. March 1991. Washington, DC: U.S. Environmental Protection Agency.
EPA, 1992. *Technical Support Document for Reduction of Pathogens and Vector Attraction in Sewage Sludge*. EPA R-93-004. Washington, DC: U.S. Environmental Protection Agency.
EPA, 1992a. *Manual Guidelines for Water Reuse*. EPA 625/R-92/004. Washington, DC: U.S. Environmental Protection Agency.
Epstein, E., 1994. Composting and bioaerosols. *BioCycle*, 35(1): 51–58.
Epstein, E., 1998. Design and operations of composting facilities: Public health aspect. Accessed 11/26/12 @ http://www.http://redptech.com/tch15.htm.
Epstein, E., & Alpert, J.E., 1984. Sludge dewatering and compost economics. *BioCycle*, 25(10): 31–34.
Epstein, E., & Epstein, J., 1989. Public health issues and composting. *BioCycle*, 30(8): 50–53.
Feachem, R.G., Bradley, D.L., Garelick, H., & Mara, D.D., 1980. *Technology for Water Supply and Sanitation—Health Aspects of Excreta and Sullage Management: A State-of-the-Art Review*. Washington, DC: World Bank.
Finstein, M.S., Miller, F.C., Hogan, J.A., & Strom, P.F., 1987. Analysis of EPA guidance on composting sludge. *BioCycle*, 28(1): 20–26.

Finstein, M.S., Miller, F.C., & Strom, P.F., 1986. Monitoring and valuating composting process performance. *Journal of WPCF*, 58: 272–278.

Fleischer, E.J., Broderick, T.A., Daigger, G.T., Fonseca, A.D., Holbrook, R.D., & Murthy, S.N., 2005. Evaluation of membrane bioreactor process capabilities to meet stringent effluent nutrient discharge requirements. *Water Environment Research*, 77: 162–178.

Fleischer, E.J., Broderick, T.A., Daigger, G.T., Lozier, J.C., Wollmann, A.M., & Fonseca, A.D., 2001. Evaluating the next generation of water reclamation processes. In *Proceedings of the Water Environment Federation 74th Annual Conference & Exposition*, Atlanta, GA, CD-ROM. October 13–17, 2001.

Gallert, C., & Winter, J., 2005. *Bacterial Metabolism in Wastewater Treatment Systems. Environmental Biotechnology*. H.J. Jordening & J. Winter (eds.). New York: Wiley VCH.

Gannett, F., 2012. Refinement of nitrogen removal from municipal wastewater treatment plants. Prepared for the Maryland Department of the Environment. Accessed @ http://www.mde.state.md.us/assets/emovals/BRE%20Gannet%20Flemin-GMB%200 presentation.pdf.

Gaudy, A.F., Jr., & Gaudy, E.T., 1980. *Microbiology for Environmental Scientists and Engineers*. New York: McGraw Hill.

Gerba, C.P., 1983. Using Sludge in Land Applications. In *Utilization of Municipal Wastewater and Sludge on Land*, pp. 147–195. A.L. Page, T. L. Gleason, J. E. Smith, I.K. Iskander, and L.E. Sommers (eds.). Riverside, CA: University of California.

Gloyna, E.F., 1976. Facultative waste stabilization pond design. In *Ponds as a Waste Treatment Alternative*. Gloyna, E.F., Malina, J.F., Jr., & Davis, E.M. (eds.). Water Resources Symposium No. 9, University of Texas Press, Austin, TX.

Grady, C.P.L., Jr., Daigger, G.T., Lover, N.G., & Filipe, C.D.M., 2011. *Biological Wastewater Treatment*, 3rd ed. Boca Raton, FL: CRC Press.

Grolund, E., 2002. *Microalgae at Wastewater Treatment in Cold Climates*. Department of Environmental Engineering. SE 971 87 LULEA Sweden, Lic Thesis 2002:35.

Haug, R.T., 1980. *Compost Engineering: Principles and Practices*. Lancaster, PA: Technomic Publishing, Inc.

Haug, R.T., 1986. Composting process design criteria: Part III. *BioCycle*, 27(10): 53–57.

Haug, R.T., & Davis, B., 1981. Composting results in Los Angeles. *BioCycle*, 22(6): 19–24.

Hay, J.C., 1996. Pathogen destruction and biosolids composting. *BioCycle, Journal of Waste Recycling*, 37(6): 67–72.

Hermanowicz, S.W., Jenkins, D., Merlo, R.P., & Trussell, R.S., 2006. *Effects of Biomass Properties on Submerged Membrane Bioreactor (SMBR) Performance and Solids Processing*. Document No. 01-CTS-19UR. Arlington, VA: Water Environment Federation.

Jagger, J., 1967. *Introduction to Research in Ultraviolet Photobiology*. Englewood Cliffs, NJ: Prentice-Hall, Inc.

Knudson, G.B., 1985. Photoreactivation of UV-irradiated *Legionella pneumoplila* and other *Legionella* species. *Applied and Environmental Microbiology*, 49: 975–980.

Kordon, C, 1992. *The Language of the Cell*. New York: McGraw-Hill.

Linden, K.G., Shin, G.A., Faubert, G., Cairns, W., & Sobsey, M.D., 2002. UV disinfection of *Giardia Lamblia* cysts in water. *Environmental Science and Technology*, 36: 2519–2522.

Logsdon, G.S., Thurman, V.C., Frindt, E.S., & Stoeker, J.G., 1985. Evaluating sedimentation and various filter media for the removal of Giardia cysts. *Journal ASSA*, 77(2): 61.

Lue-Hing, C., Zenz, D.R., & Kuchenrither, R., 1992. *Municipal Sewage Sludge Management: Processing, Utilization, and Disposal*. Lancaster, PA: Technomic Publishing, Inc.

Lynch, J.M., & Poole, N.J., 1979. *Microbial Ecology, A Conceptual Approach*. New York: John Wiley & Sons.
McGhee, T.J., 1991. *Water Supply and Sewerage*. New York: McGraw-Hill, Inc.
Metcalf & Eddy, Inc., 1991. *Wastewater Engineering: Treatment, Disposal, Reuse*, 3rd ed. New York: McGraw-Hill.
Metcalf & Eddy, Inc., 2003. *Wastewater Engineering: Treatment, Disposal, Reuse*, 4th ed. New York: McGraw-Hill.
Metcalf & Eddy, 2003. *Wastewater Engineering: Treatment, Disposal and reuse*. New York: McGraw-Hill.
Middlebrooks, E.J., Middlebrooks, C.H., Reynolds, J.H., Watters, G.Z., Reed, S.C., & George, D.B., 1982. *Wastewater Stabilization Lagoon Design, Performance and Upgrading*. New York: Macmillan Publishing, Co, Inc.
Middlebrooks, E.J., & Pano, A., 1983. Nitrogen removal in aerated lagoons. *Water Research*, 17(10): 1369–1378.
Millner, P. (ed.), 1995. Bioaerosols and composting. *BioCycle*, 36(1): 48–54.
Natvik, O., Dawson, B., Emrick, J., and Murphy, S., 2003. BNR "Then" and "Now"—A case study—Kalispell advanced wastewater treatment plant. In *Proceedings of the Water Environment Federation 76th Annual Technical Exhibition & Conference*, Los Angeles, CA. October 11–15, 2003.
NEIWPCC, 1988. *Guides for the Design of Wastewater Treatment Works TR-16*. Wilmington, MA: New England Interstate Water Pollution Control Commission.
Oguma, K., Katayama, H., Mitani, H., Morita, S., Hirata, T., & Ohgaki, S., 2001. Determination of pyrimidine dimmers in *Escherichia coli* and *Cryptosporidium parvum* during UV light inactivation, photoreactivation, and dark repair. *Applied and Environmental Microbiology*, 67: 4630–4637.
Oswald, W.J., 1990a. Advanced integrated wastewater pond systems: Supplying water and saving the environment for six billion people. In *Proceedings of the ASCE Convention, Environmental Engineering Division*, San Francisco, CA. November 5–8.
Oswald, W.J., 1990b. Sistemas Avanzados De Lagunas Integradas Para Tratamiento De Aguas Servidas (SALI). In *Proceedings of the ASCE Convention, Environmental Engineering Division*, San Francisco, CA. November 5–8.
Oswald, W.J., 1996. *A Syllabus on Advanced Integrated Pond Systems®*. Berkeley, CA: University of California.
Outwater, A.B., 1994. *Reuse of Sludge and Minor Wastewater Residuals*. Boca Raton, FL: Lewis Publishers.
Pano, A., & Middlebrooks, E.J., 1982. Ammonia nitrogen removal in facultative waste water stabilization ponds. *Journal (Water Pollution Control Federation)*, 54(4): 2148.
Park, J., 2012. Biological nutrient removal theories and design. Available @ http://www.dnr.state.wi,us/org/water/wm/ww/biophos/bnr_remvoal.htm.
Paterson, C., & Curtis, T., 2005. Physical and chemical environments. In *Pond Treatment Technology*. A. Shilton (ed.). London, UK: IWA Publishing.
Pearson, H., 2005. Microbiology of waste stabilisation ponds. In *Pond Treatment Technology*. A. Shilton (ed.). London, UK: IWA Publishing.
Peot, C., 1998. Compost use in wetland restoration. Design for success. In *Proceedings s of the 12th Annual Residual and Biosolids Management Conference*. Water Environment Federation, Alexandra, VA. October 22, 1998.
Pipes, W.O., Jr., 1961. Basic biology of stabilization ponds. *Water and Sewage Works*, 108(4): 131–136.
Rauth, A.M., 1965. The physical state of viral nucleic acid and the sensitivity of viruses to ultraviolet light. *Biophysical Journal*, 5: 257–273.

Richard, M., & Bowman, D., 1991. Troubleshooting the aerated and facultative waste treatment lagoon. *Presented at the USEPA's Natural/Constructed Wetlands Treatment System Workshop*, Denver, CO.
Sawyer, C.N., McCarty, P.L., & Parkin, G.F., 1994. *Chemistry for Environmental Engineering.* New York: McGraw Hill.
Shilton, A., (ed.), 2005. *Pond Treatment Technology.* London: IWA Publishing.
Shin, G.A., Linden, K.G., Arrowood, M.J., Faubert, G., & Sosbey, M.D., 2001. DNA repair of UV-irradiated *Cryptosporidium parvum* oocysts and *Giardia lamblia* cysts. In *Proceedings of the First International Ultraviolet Association Congress*, Washington, DC. June 14–16.
Singleton, P., & Sainsbury, D., 1994. *Dictionary of Microbiology & Molecular Biology*, 2nd ed. New York: John Wiley & Sons.
Sloan Equipment, 1999. *Aeration Products.* Owings Mills, MD.
Sopper, W.E., 1993. *Municipal Sludge Use in Land Reclamation.* Boca Raton, FL: Lewis Publishers.
Spellman, F.R., 1996. *Stream Ecology and Self-Purification.* Boca Raton, FL: CRC Press.
Spellman, F.R., 2000. *Microbiology for Water and Wastewater Operators.* Boca Raton, FL: CRC Press.
Spellman, F.R., 2007. *The Science of Water*, 2nd ed. Boca Raton, FL: CRC Press.
Tchobanoglous, G., Theisen, H., & Vigil, S.A., 1993. *Integrated Solid Waste Management.* New York: McGraw-Hill, Inc.
Toomey, W.A., 1994. Meeting the challenge of yard trimmings diversion. *BioCycle*, 35(5): 55–58.
Ullrich, A.H., 1967. Use of wastewater stabilization ponds in two different systems. *JWPCF*, 39(6): 965–977.
USEPA, 1975. *Process Design Manual for Nitrogen Control.* EPA-625/1-75-007. Cincinnati, OH: Center for Environmental Research Information.
USEPA, 1977a. *Operations Manual for Stabilization Ponds.* EPA-430/9-77-012, NTIS No. PB-279443. Washington, DC: Office of Water Program Operations.
USEPA, 1977b. *Upgrading Lagoons.* EPA-625/4-73-001, NTIS No. PB 259974. Cincinnati, OH: Center for Environmental Research Information.
USEPA, 1989. *Technical Support Document for Pathogen Reducing in Sewage Sludge.* NTIS No. PB89-136618. Springfield, VA: National Technical Information Service.
USEPA, 1993. *Manual: Nitrogen Control.* EPA-625/R-93/010, Cincinnati, OH.
USEPA, 1997. *Innovative Uses of Compost: Disease Control for Plants and Animals.* Washington, DC: United States Environmental Protection Agency.
USEPA, 1999a. *Wastewater Technology Fact Sheet: Ultraviolet Disinfection.* Washington, DC: United States Environmental Protection Agency.
USEPA, 1999b. *Wastewater Technology Fact Sheet: Ozone Disinfection.* Washington, DC: United States Environmental Protection Agency.
USEPA, 2000a. *Clean Watersheds Needs Survey 2000, Report to Congress*, EPA-832-R-10-002. Washington, DC: U.S. Environmental Protection Agency.
USEPA, 2000b. *Wastewater Technology Fact Sheet Package Plants.* Washington, DC: United States Environmental Protection Agency.
USEPA, 2001. *Memorandum: Development and Adoption of Nutrient Criteria into Water Quality Standards.* Available @ http://oaspub.epa.gov/waters/national_rept.control #TOP_IMP.
USEPA, 2006. *UV Disinfection Guidance Manual.* Washington, DC: United States Environmental Protection Agency.
USEPA, 2007a. *Wastewater Management Fact Sheet: Membrane Bioreactors.* Washington, DC: United States Environmental Protection Agency.

USEPA, 2007b. *Advanced Wastewater Treatment to Achieve Low Concentration of Phosphorus.* Washington, DC: Environmental Protection Agency.
USEPA, 2007c. *Biological Nutrient Removal Processes and Costs.* Washington, DC: United States Environmental Protection Agency.
USEPA, 2007d. *National Section 303(d) List Fact Sheet.* Available @ http://iaspub.epa.gov/waters/national_rept.control.
USEPA, 2007e. *Innovative Uses of Compost: Disease Control for Plants and Animals.* EPA/530-F-97-044. Washington, DC: Office of Solid Waste and Emergency Response, U.S. EPA.
USEPA, 2008. *Municipal Nutrient Removal Technologies Reference Document Volume 2—Appendices.* Washington, DC: United States Environmental Protection Agency.
USEPA, 2011. *Principles of Design and Operations of Wastewater treatment Pond Systems for Plant operators, Engineers, and Managers.* Washington, DC: U.S. Environmental Protection Agency.
Vasconcelos, V.M., & Pereira, E., 2001. Cyanobacteria diversity and toxicity in a wastewater treatment plant (Portugal). *Water Research*, 35(5): 1354–1357.
Vesilind, P.A., 1980. *Treatment and Disposal of Wastewater Sludges*, 2nd ed. Ann Arbor, MI: Ann Arbor Science Publishers, Inc.
Wallis-Lage, C., Hemken, B., et al., 2006. MBR plants: Larger and more complicated. *Presented at the Water Reuse Association's 21st Annual Water Reuse Symposium*, Hollywood, CA. September 2006.
WEF (Water Environment Federation), 1985. *Operation of Extended Aeration Package Plants. Manual of Practice NO. OM-7.* Alexandria, VA: WEF.
WEF (Water Environment Federation), 1995. *Wastewater Residuals Stabilization. Manual of Practice FD-9.* Alexandria, VA: Water Environment Federation.
WEF (Water Environment Federation), 1998. *Design of Municipal Wastewater Treatment Plants, Manual of Practice No. 8*, 4th ed. Vol. 2. Alexandria, VA: WEF.
Wilbur, C., & Murray, C., 1990. Odor source evaluation. *BioCycle*, 31(3): 68–72.
Yeager, J.G., & Ward, R.I., 1981. Effects of moisture content on long-term survival and regrowth of bacteria in wastewater sludge. *Applied and Environmental Microbiology*, 41(5): 1117–1122.

Part 4

Wastewater-Based Epidemiology

11 Fundamentals of Biomonitoring, Monitoring, Sampling, and Testing*

In January, we take our nets to a no-name stream in the foothills of the Blue Ridge Mountains of Virginia to do a special kind of macroinvertebrate monitoring—looking for "winter stoneflies." Winter stoneflies have an unusual life cycle. Soon after hatching in early spring, the larvae bury themselves in the streambed. They spend the summer lying dormant in the mud, thereby avoiding problems like overheated streams, low oxygen concentrations, fluctuating flows, and heavy predation. In later November, they emerge, grow quickly for a couple of months, and then lay their eggs in January.

January monitoring of winter stoneflies helps in interpreting the results of spring and fall macroinvertebrate surveys. In spring and fall, a thorough benthic survey is conducted, based on *Protocol II* of the USEPA's "Rapid Bioassessment Protocols for Use in Streams and Rivers." Some sites on various rural streams have poor diversity and sensitive families. Is the lack of macroinvertebrate diversity because of specific warm-weather conditions, high water temperature, low oxygen, or fluctuating flows, or is some toxic contamination present? In the January screening, if winter stoneflies are plentiful, seasonal conditions were probably to blame for the earlier results; if winter stoneflies are absent, the site probably suffers from toxic contamination (based on our rural location, probably emanating from non-point sources) that is present year-round. Though different genera of winter stoneflies are found in our region (southwestern Virginia), Allocapnia is sought because it is present even in the smallest streams.

—F.R. Spellman

***Note*:** Although this book's focus is on using wastewater-based epidemiology and data obtained via measurements made from findings, none of this data is available without sampling, testing, and analyzing the sample taken. Thus, in this chapter, basic sampling, testing, and analysis of water/wastewater samples is presented first, primarily to introduce the reader to some of the different methods employed to test

* Based on F.R. Spellman (2007). *The Science of Water*, 2nd ed. Boca Raton, FL: CRC Press.

water and wastewater to ensure water quality. A discussion of wastewater-based epidemiology used to measure and locate possible outbreaks of bacterial infections, protozoan contamination, and viral episodes, including COVID-19 follows.

WHAT IS BIOMONITORING AND BIOSURVEYING?

The life in, and physical characteristics of, a stream ecosystem provide insight into the historical and current status of its quality. The assessment of a water body ecosystem based on organisms (including pathogenic organisms) living in it is called *biomonitoring*. The assessment of the system based on its physical characteristics is called a habitat assessment. Biomonitoring and habitat assessments are two tools that stream ecologists use to assess the water quality of a stream.

Biosurveying is used to assess ecological resources, such as rivers, streams, lakes, and wetlands. Biosurveying involves collection and analysis of animal and/or plant samples which serve as bioindicators. Studies are conducted according to published procedures to ensure consistency in data collection and analysis, and to compare findings to established metrics. One advantage of biosurveys is that when properly conducted, they can identify pollution problems that are expensive and difficult to detect using other methods.

In general, biological monitoring the use of organisms (called assemblages)—periphytons, fish, and macroinvertebrates—uses the assemblages to assess environmental condition. Biological observation is more representative as it reveals cumulative effects as opposed to chemical observation, which is representative only at the actual time of sampling.

Again, the presence of different assemblages of organisms is used in conducting biological assessments and/or biosurveys. In selecting the appropriate assemblage {not sure what goes here} for a particular biomonitoring situation, the advantages of using each assemblage must be considered along with the objectives of the program. Some of the advantages of using periphytons (algae), benthic macroinvertebrates, and fish in a biomonitoring program are presented in this section.

Important Note: Periphytons are a complex matrix of benthic attached algae, cyanobacteria, heterotrophic microbes, and detritus that are attached to submerged surfaces in most aquatic ecosystems.

ADVANTAGES OF USING PERIPHYTON

1. Algae generally have rapid reproduction rates and very short life cycles, making them valuable indicators of short-term impacts.
2. As primary producers, algae are most directly affected by physical and chemical factors.
3. Sampling is easy, is inexpensive, requires few people, and creates minimal impact on resident biota.
4. Relatively standard methods exist for revaluation of functional and non-taxonomic structural (biomass, chlorophyll measurements) characteristics of algal communities.

5. Algal assemblages are sensitive to some pollutants which may not visibly affect other aquatic assemblages or may only affect other organisms at higher concentrations (i.e., herbicides) (Carins and Dickson 1971; American Public Health Association et al. 1971; Patrick 1973; Rodgers et al. 1979; Weitzel 1979; Karr 1981; USEPA 1983).

Advantages of Using Fish

1. Fish are good indicators of long-term (several years) effects and broad habitat conditions because they are relatively long-lived and mobile (Karr et al. 1986)
2. Fish assemblages include a range of species that represent a variety of trophic levels (omnivores, herbivores, insectivores, planktivores, piscivores, etc.). They tend to integrate effects of lower trophic levels; thus, fish assemblage structure is reflective of integrated environmental health.
3. Fish are at the top of the aquatic food web and are consumed by humans, making them important for assessing contamination.
4. Fish are relatively easy to collect and identify to the species level. Most specimens can be sorted and identified in the field by experienced fisheries professionals, and subsequently released unharmed.
5. Environmental requirements of most fish are comparatively well known. Life history information is extensive for many species, and information on fish distributions is commonly available.
6. Aquatic life uses (water quality standards) are typically characterized in terms of fisheries (cold water, cool water, warm water, sport, forage). Monitoring fish provides a direct evaluation of "fishability" and fish "propagation," which emphasizes the importance of fish to anglers and commercial fishermen.
7. Fish account for nearly half of the endangered vertebrate species and subspecies in the United States (Warren and Burr, 1994).

Advantages of Using Macroinvertebrates

Benthic macroinvertebrates are the larger organisms such as aquatic insects, insect larvae, and crustaceans that live in the bottom portions of a waterway for part of their life cycle. They are ideal for use in biomonitoring, as they are ubiquitous, relatively sedentary, and long-lived. They provide a cross-section of the situation, as some species are extremely sensitive to pollution, while others are more tolerant. However, like toxicity testing, biomonitoring does not tell you why animals are present or absent. As mentioned, benthic macroinvertebrates are excellent indicators for several reasons:

1. Biological communities reflect overall ecological integrity (i.e., chemical, physical, and biological integrity). Therefore, biosurvey results directly assess the status of a waterbody relative to the primary goal of the Clean Water Act (CWA).

2. Biological communities integrate the effects of different stressors and thus provide a broad measure of their aggregate impact.
3. Because they are ubiquitous, communities integrate the stressors over time and provide an ecological measure of fluctuating environmental conditions.
4. Routine monitoring of biological communities can be relatively inexpensive because they are easy to collect and identify.
5. The status of biological communities is of direct interest to the public as a measure of a particular environment.
6. Where criteria for specific ambient impacts do not exist (e.g., nonpoint sources that degrade habitats), biological communities may be the only practical means of evaluation.

Benthic macroinvertebrates act as continuous monitors of the water they live in. Unlike chemical monitoring, which provides information about water quality at the time of measurement (a snapshot), biological monitoring can provide information about past and/or episodic pollution (a videotape). This concept is analogous to miners who took canaries into deep mines with them to test for air quality. If the canary died, the miners knew the air was bad and they had to leave the mine. Biomonitoring a water body ecosystem uses the same theoretical approach. Aquatic macroinvertebrates are subject to pollutants in the water body. Consequently, the health of the organisms reflects the quality of the water they live in. If the pollution levels reach a critical concentration, certain organisms will migrate away, fail to reproduce, or die, eventually leading to the disappearance of those species at the polluted site. Normally, these organisms will return if conditions improve in the system (Bly & Smith, 1994).

Biomonitoring (and the related term, Bioassessment) surveys are conducted before and after an anticipated impact to determine the effect of the activity on the water body habitat. Moreover, surveys are performed periodically to monitor water body habitats and watch for unanticipated impacts. Finally, biomonitoring surveys are designed to reference conditions or to set biocriteria (serve as monitoring thresholds to signal future impacts, regulatory actions, etc.) for determining that an impact has occurred (Camann, 1996).

***Note*:** The primary justification for bioassessment and monitoring is that degradation of water body habitats affects the biota using those habitats. Therefore, the living organisms themselves provide the most direct means of assessing real environmental impacts. Although the focus of this text is on macroinvertebrate protocols, the periphyton and fish protocols are discussed briefly in the following before an in-depth discussion of the macroinvertebrate protocols.

PERIPHYTON PROTOCOLS

Benthic algae (periphyton or phytobenthos) are primary producers, and an important foundation of many stream food webs (Bahls, 1993; Stevenson 1996). These organisms also stabilize substrata and serve as habitat for many other organisms. Because benthic algal assemblages are attached to substrate, their characteristics are affected by physical, chemical, and biological disturbances that occur in the stream reach during the time in which the assemblage developed.

Diatoms in particular are useful ecological indicators because they are found in abundance in most lotic ecosystems. Diatoms and many other algae can be identified to species by experienced algologists. The great numbers of species provide multiple, sensitive indicators of environmental change and the specific conditions of their habitat. Diatom species are differentially adapted to a wide range of ecological conditions.

Periphyton indices of biotic integrity have been developed and tested in several regions (Kentucky Department of Environmental Protection 1993; Hill 1997). Since the ecological tolerances for many species are known, changes in community composition can be used to diagnose the environmental stressors affecting ecological health as well as to assess biotic integrity (Stevenson, 1998; Stevenson and Pan, 1999). Periphyton protocols may be used by themselves, but they are most effective when used with one or more of the other assemblages and protocols. They should be used with habitat and benthic macroinvertebrate assessments particularly, because of the close relation between periphyton and these elements of stream ecosystems.

Presently, few states have developed protocols for periphyton assessment. Montana, Kentucky, and Oklahoma have developed periphyton bioassessment programs. Other states are exploring the possibility of developing periphyton programs. Algae have been widely used to monitor water quality in rivers in Europe, where many different approaches have been used for sampling and data analysis.

FISH PROTOCOLS

Monitoring of the fish assemblage is an integral component of many water quality management programs, and its importance is reflected in the aquatic life use-support designations of many states. Assessments of the fish assemblage must measure the overall structure and function of the ichthyofaunal community to adequately evaluate biological integrity and protect surface water resource quality. Fish bioassessment data quality and comparability are assured through the utilization of qualified fisheries professionals and consistent methods.

In the fish protocol, the principal evaluation mechanism utilizes the technical framework of the Index of Biotic Integrity (IBI)—a fish assemblage assessment approach developed by Karr (1981). The IBI incorporates the zoogeographic, ecosystem, community, and population aspects of the fish assemblage into a single ecologically based index. Calculation and interpretation of the IBI involves a sequence of activities including fish sample collection; data tabulation; and regional modification and calibration of metrics and expectation values.

The fish protocol involves careful, standardized field collection, species identification and enumeration, and analyses using aggregated biological attributes or quantification of numbers of key species. The role of experienced fisheries scientists in the adaptation and application of the bioassessment protocol and the taxonomic identification of fishes cannot be overemphasized. The fish bioassessment protocols survey yields an objective discrete measure of the condition of the fish assemblage.

Although the fish survey can usually be completed in the field by qualified fish biologists, difficult species identifications will require laboratory confirmation. Data provided by the fish bioassessment protocols can serve to assess use attainment, develop biological criteria, prioritize sites for further evaluation, provide a reproducible impact assessment, and evaluate status and trends of the fish assemblage.

Fish collection procedures must focus on a multi-habitat approach—sampling habitats in relative proportion to their local representation (as determined during site reconnaissance). Each sample reach should contain riffle, run, and pool habitat, when available. Whenever possible, the reach should be sampled sufficiently upstream of any bridge or road crossing to minimize the hydrological effects on overall habitat quality. Wadeability and accessibility may ultimately govern the exact placement of the sample reach. A habitat assessment is performed, and physical/chemical parameters measured concurrently with fish sampling to document and characterize available habitat specifics within the sample reach.

MACROINVERTEBRATE PROTOCOLS

Benthic macroinvertebrates, by indicating the extent of oxygenation of a stream, may be regarded as indicators of the intensity of pollution from organic waste. The responses of aquatic organisms in water bodies to large quantities of organic wastes are well documented. They occur in a predictable cyclical manner. For example, upstream from the discharge point, a stream can support a wide variety of algae, fish, and other organisms, but in the section of the water body where oxygen levels are low (below 5 ppm), only a few types of worms survive. As stream flow courses downstream, oxygen levels recover, and those species that can tolerate low rates of oxygen (such as gar, catfish, and carp) begin to appear. In a stream, eventually, at some further point downstream, a clean-water zone reestablishes itself, and a more diverse and desirable community of organisms returns. Due to this characteristic pattern of alternating levels of dissolved oxygen (in response to the dumping of large amounts of biodegradable organic material), a stream, as stated above, goes through a cycle called an oxygen sag curve. Its state can be determined using the biotic index as an indicator of oxygen content.

The Biotic Index

The biotic index is a systematic survey of macroinvertebrates organisms. Because the diversity of species in a stream is often a good indicator of the presence of pollution, the biotic index can be used to correlate with stream quality. Observation of types of species present or missing is used as an indicator of stream pollution. The biotic index, used in the determination of the types, species, and numbers of biological organisms present in a stream, is commonly used as an auxiliary to BOD determination in determining stream pollution. The biotic index is based on two principles:

1. A large dumping of organic waste into a stream tends to restrict the variety of organisms at a certain point in the stream.

2. As the degree of pollution in a stream increases, key organisms tend to disappear in a predictable order. The disappearance of particular organisms tends to indicate the water quality of the stream.

There are several different forms of the biotic index. In Great Britain, for example, the Trent Biotic Index (TBI), the Chandler score, the Biological Monitoring Working Party (BMWP) score, and the Lincoln Quality Index (LQI) are widely used. Most of the forms use a biotic index that ranges from 0 to 10. The most polluted stream, which therefore contains the smallest variety of organisms, is at the lowest end of the scale (0); the clean streams are at the highest end (10). A stream with a biotic index of greater than 5 will support game fish; on the other hand, a stream with a biotic index of less than 4 will not support game fish.

As mentioned, because they are easy to sample, macroinvertebrates have predominated in biological monitoring. Macroinvertebrates are a diverse group. They demonstrate tolerances that vary between species. Thus, discrete differences tend to show up, containing both tolerant and sensitive indicators. Macroinvertebrates can be easily identified using identification keys that are portable and easily used in field settings. Present knowledge of macroinvertebrate tolerances and response to stream pollution is well documented. In the United States, for example, the Environmental Protection Agency (EPA) has required states to incorporate narrative biological criteria into its water quality standards since 1993. The National Park Service (NPS) has collected macroinvertebrate samples from American streams since 1984. Through their sampling effort, the NPS has been able to derive quantitative biological standards (Huff, 1993).

The biotic index provides a valuable measure of pollution. This is especially the case for species that are very sensitive to lack of oxygen. An example of an organism that is commonly used in biological monitoring is the stonefly. Stonefly larvae live underwater and survive best in well-aerated, unpolluted waters with clean gravel bottoms. When stream water quality deteriorates due to organic pollution, stonefly larvae cannot survive. The degradation of stonefly larvae has an exponential effect upon other insects and fish that feed off the larvae; when the stonefly larvae disappear, so in turn do many insects and fish (O'Toole 1986).

Table 11.1 shows a modified version of the BMWP biotic index. Considering that the BMWP biotic index indicates ideal stream conditions, it considers the sensitivities of different macroinvertebrate species which are represented by diverse populations and are excellent indicators of pollution. These aquatic macroinvertebrates are organisms that are large enough to be seen by the unaided eye. Moreover, most aquatic macroinvertebrates species live for at least a year; and they are sensitive to stream water quality both on a short-term basis and over the long term. For example, mayflies, stoneflies, and caddisflies are aquatic macroinvertebrates that are considered clean-water organisms; they are generally the first to disappear from a stream if water quality declines and are, therefore, given a high score. On the other hand, tubificid worms (which are tolerant to pollution) are given a low score.

In Table 11.1, a score from 1 to 10 is given for each family present. A site score is calculated by adding the individual family scores. The site score or total score is then

TABLE 11.1
BMWP Score System (Modified for Illustrative Purposes)

Families	Common-name examples	Score
Heptageniidae	Mayflies	10
Leuctridae	Stoneflies	9–10
Aeshnidae	Dragonflies	8
Polycentropidae	Caddisflies	7
Hydrometridae	Water Strider	6–7
Gyrinidae	Whirligig beetle	5
Chironomidae	Mosquitoes	2
Oligochaeta	Worms	1

divided by the number of families recorded to derive the Average Score Per Taxon (ASPT). High ASPT scores result due to such taxa as stoneflies, mayflies, and caddisflies being present in the stream. A low ASPT score is obtained from streams that are heavily polluted and dominated by tubificid worms and other pollution-tolerant organism. From Table 11.1, it can be seen that those organisms having high scores, especially mayflies and stoneflies, are the most sensitive, and others, such as dragonflies and caddisflies, are very sensitive to any pollution (deoxygenation) of their aquatic environment.

As noted earlier, the benthic macroinvertebrate biotic index employs the use of certain benthic macroinvertebrates to determine (to gauge) the water quality (relative health) of a water body (stream or river). Benthic macroinvertebrates are classified into three groups based on their sensitivity to pollution. The number of taxa in each of these groups are tallied and assigned a score. The scores are then summed to yield a score that can be used as an estimate of the quality of the water body life.

Metrics within the Benthic Macroinvertebrates

Table 11.2 provides a sample index of macroinvertebrates and their sensitivity to pollution. The three groups based on their sensitivity to pollution are:

Group One—Indicators of Poor Water Quality
Group Two—Indicators of Moderate Water Quality
Group Three—Indicators of Good Water Quality

In summary, it can be said that unpolluted streams normally support a wide variety of macroinvertebrates and other aquatic organisms with relatively few of any one kind. Any significant change in the normal population usually indicates pollution.

BIOLOGICAL SAMPLING IN STREAMS

A few years ago, we were preparing to perform benthic macroinvertebrate sampling protocols in a wadeable section in one of the countless reaches of the Yellowstone

TABLE 11.2
Sample Index of Macroinvertebrates

Group one (Sensitive)	Group two (Somewhat Sensitive)	Group three (Tolerant)
Stonefly larva	Alderfly larva	Aquatic worm
Caddisfly larva	Damselfly larva	Midgefly larva
Water penny larva	Cranefly larva	Blackfly larva
Riffle beetle adult	Beetle adult	Leech
Mayfly larva	Dragonfly larva	Snails
Gilled snail	Sowbugs	

River, WY. It was autumn; windy, and cold. Before we stepped into the slow-moving frigid waters, we stood for a moment at the bank and took in the surroundings. The pallet of autumn is austere in Yellowstone. The coniferous forests east of the Mississippi lack the bronzes, the coppers, the peach-tinted yellows, and the livid scarlets that set the mixed stands of the East aflame. All we could see in that line was the quaking aspen and its gold. This autumnal gold, which provides the closest thing to eastern autumn in the West, is mined from the narrow, rounded crowns of Populus tremuloides. The aspen trunks stand stark white and antithetical against the darkness of the firs and pines, the shiny pale gold leaves sensitive to the slightest rumor of wind. Agitated by the slightest hint of breeze, the gleaming upper surfaces bounced the sun into our eyes. Each tree scintillated, like a show of gold coins in freefall. The aspens' bright, metallic flash seemed, in all their glittering motion, to make a valiant dying attempt to fill the spectrum of fall.

Because they were bright and glorious, we did not care that they could not approach the colors of an eastern autumn. While nothing is comparable to experiencing leaf-fall in autumn along the Appalachian Trail, the fact that this autumn was not the same simply did not matter. This spirited display of gold against dark green lightened our hearts and eased the task that was before us, warming the thought of the bone-chilling water and all. With the aspens gleaming gold against the pines and firs, it simply did not seem to matter. Notwithstanding the glories of nature alluded to above, one should not be deceived. Conducting biological sampling in a water body is not only the nuts and bolts of biological sampling, but it is also very hard and important work.

Biological Sampling: Planning

When planning a biological sampling outing, it is important to determine the precise objectives. One important consideration is to determine whether sampling will be accomplished at a single point or at isolated points. Additionally, the frequency of sampling must be determined. That is, will sampling be accomplished at hourly, daily, weekly, monthly, or even longer intervals? Whatever frequency of sampling is chosen, the entire process will probably continue over a protracted period (i.e., preparing for biological sampling in the field might take several months from the initial planning stages to the time when actual sampling occurs). An experienced freshwater ecologist should be centrally involved in all aspects of planning.

In *Monitoring Water Quality: Intensive Stream Bioassay*, 08/18/2000, the USEPA recommends that the following issues should be considered in planning the sampling program:

Availability of reference conditions for the chosen area
Appropriate dates to sample in each season
Appropriate sampling gear
Availability of laboratory facilities
Sample storage
Data management
Appropriate taxonomic keys, metrics, or measurement for macroinvertebrate analysis
Habitat assessment consistency
A USGS topographical map
Familiarity with safety procedures

When the initial objectives (issues) have been determined and the plan devised, then the sampler can move to other important aspects of the sampling procedure. Along with the items just mentioned, it is imperative that the sampler understands what biological sampling is all about.

Sampling is one of the most basic and important aspects of water quality management (Tchobanoglous & Schroeder, 1985). Biological sampling allows for rapid and general water quality classification. Rapid classification is possible because quick and easy crosschecking between stream biota and a standard stream biotic index is possible. Biological sampling is typically used for general water quality classification in the field because sophisticated laboratory apparatus is usually not available. Additionally, stream communities often show a great deal of variation in basic water quality parameters such as DO BOD, suspended solids, and coliform bacteria. This occurrence can be observed in eutrophic lakes that may vary from oxygen saturation to less than 0.5 mg/L in a single day, and the concentration of suspended solids may double immediately after a heavy rain. Moreover, the sampling method chosen must consider the differences in the habits and habitats of the aquatic organisms.

The first step toward accurate measurement of a stream's water quality is to make sure that the sampling targets those organisms (i.e., macroinvertebrates) that are most likely to provide the information that is being sought. Second, it is essential that representative samples be collected. Laboratory analysis is meaningless if the sample collected was not representative of the aquatic environment being analyzed. As a rule, samples should be taken at many locations, as often as possible. If, for example, you are studying the effects of sewage discharge into a stream, you should first take at least six samples upstream of the discharge, six samples at the discharge, and at least six samples at several points below the discharge for two to three days (the six-six sampling rule). If these samples show wide variability, then the number of samples should be increased. On the other hand, if the initial samples exhibit little variation, then a reduction in the number of samples may be appropriate (Kittrell, 1969).

When planning the biological sampling protocol (using biotic indices as the standards) remember that when the sampling is to be conducted in a stream, findings are based on the presence or absence of certain organisms. Thus, the absence of these organisms must be a function of pollution and not of some other ecological problem. The preferred (favored in this text) aquatic group for biological monitoring in stream is the macroinvertebrates, which are usually retained by 30 mesh sieves (pond nets).

Sampling Stations

After determining the number of samples to be taken, sampling stations (locations) must be determined. Several factors determine where the sampling stations should be set up. These factors include stream habitat types, the position of the wastewater effluent outfalls, the stream characteristics, stream developments (dams, bridges, navigation locks, and other man-made structures), the self-purification characteristics of the stream, and the nature of the objectives of the study (Velz, 1970). The stream habitat types used in this discussion are those that are macroinvertebrate assemblage in stream ecosystems. Some combination of these habitats would be sampled in a multi-habitat approach to benthic sampling (Barbour et al. 1997):

1. Cobble (hard substrate)—Cobble is prevalent in the riffles (and runs), which are a common feature throughout most mountain and piedmont streams. In many high-gradient streams, this habitat type will be dominant. However, riffles are not a common feature of most coastal or other low-gradient streams. Sample shallow areas with coarse substrates (mixed gravel, cobble or larger) by holding the bottom of the dip net against the substrate and dislodging organisms by kicking (this is where the "designated kicker," your sampling partner, comes into play) the substrate for 0.5 m upstream of the net.
2. Snags—Snags and other woody debris that have been submerged for a relatively long period (not recent deadfall) provides excellent colonization habitat. Sample submerged woody debris by jabbing in medium-sized snag material (sticks and branches). The snag habitat may be kicked first to help to dislodge organisms, but only after placing the net downstream of the snag. Accumulated woody material in pool areas is considered snag habitat. Large logs should be avoided because they are generally difficult to sample adequately.
3. Vegetated banks—When lower banks are submerged and have roots and emergent plants associated with them, they are sampled in a fashion similar to snags. Submerged areas of undercut banks are good habitats to sample. Sample banks with protruding roots and plants by jabbing into the habitat. Bank habitat can be kicked first to help dislodge organisms, but only after placing the net downstream.

4. Submerged macrophytes—Submerged macrophytes are seasonal in their occurrence and may not be a common feature of many streams, particularly those that are high gradient. Sample aquatic plants that are rooted on the bottom of the stream in deep water by drawing the net through the vegetation from the bottom to the surface of the water (maximum of 0.5 m each jab). In shallow water, sample by bumping or jabbing the net along the bottom in the rooted area, avoiding sediments where possible.
5. Sand (and other fine sediment)—Usually the least productive macroinvertebrate habitat in streams, this habitat may be the most prevalent in some streams. Sample banks of unvegetated or soft soil by bumping the net along the surface of the substrate rather than dragging the net through soft substrate; this reduces the amount of debris in the sample.

It is usually impossible to go out and count each and every macroinvertebrate present in a waterway. This would be comparable to counting different sized grains of sand on the beach. Thus, in a biological sampling program (i.e., based on our experience), the most common sampling methods are the transect and the grid. Transect sampling involves taking samples along a straight line either at uniform or at random intervals (see Figure 11.1). The transect involves the cross-section of a lake or stream or the longitudinal section of a river or stream. The transect sampling method allows for a more complete analysis by including variations in habitat.

In grid sampling, an imaginary grid system is placed over the study area. The grids may be numbered, and random numbers are generated to determine which grids should be sampled (see Figure 11.2). This type of sampling method allows for quantitative analysis because the grids are all of a certain size. For example, to sample a stream for benthic macroinvertebrates, grids that are 0.25 m^2 may be used. Then, the weight or number of benthic macroinvertebrates per square meter can be determined.

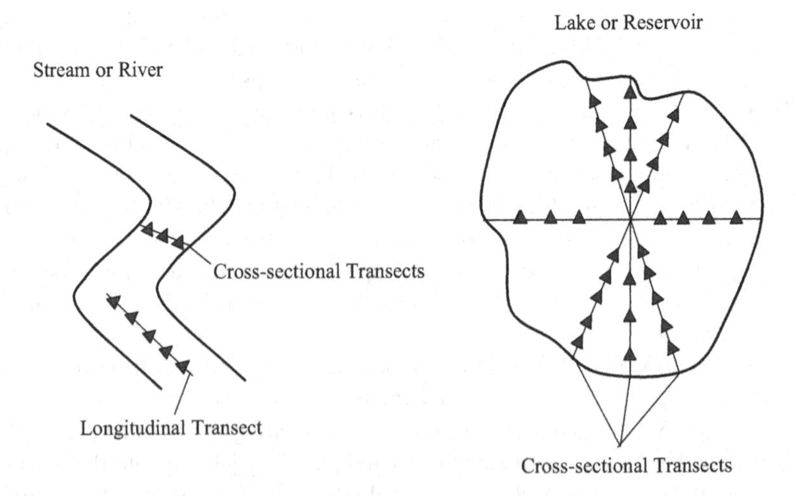

FIGURE 11.1 Transect sampling.

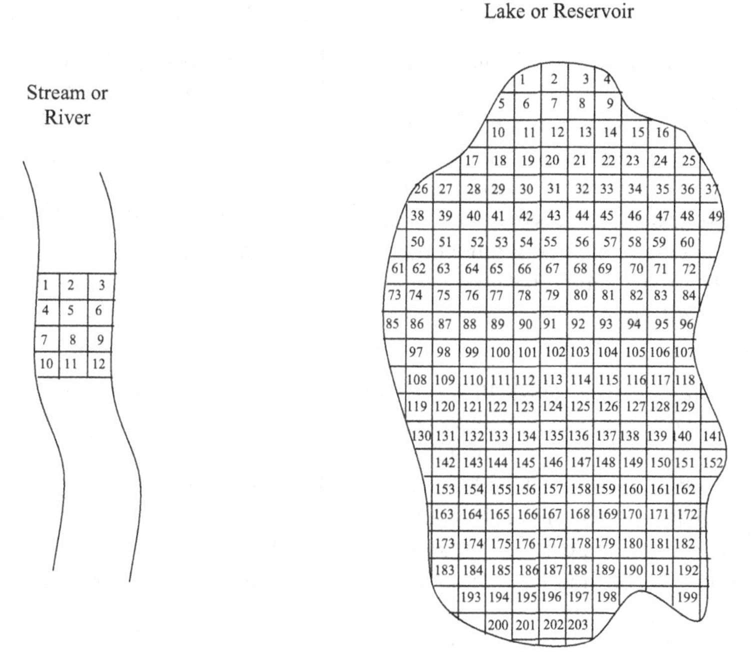

FIGURE 11.2 Grid sampling.

Random sampling requires that each possible sampling location have an equal chance of being selected. Numbering all sampling locations, and then using a computer, calculator, or a random numbers table to collect a series of random numbers can do this. An illustration of how to put the random numbers to work is provided in the following example. Given a pond that has 300 grid units, find 8 random sampling locations using the following sequence of random numbers taken from a standard random numbers table: 101, 209, 007, 018, 099, 100, 017, 069, 096, 033, 041, 011. The first eight numbers of the sequence could be selected and only grids would be sampled to obtain a random sample. In grid sampling to collect samples a sieve bucket (see Figure 11.3) is often used.

FIGURE 11.3 Sieve bucket.

Macroinvertebrate Sampling in Rocky-Bottom Streams

Rocky-bottom streams are defined as those with bottoms made up of gravel, cobbles, and boulders in any combination. They usually have definite riffle areas. As mentioned, riffle areas are fairly well oxygenated and, therefore, are prime habitats for benthic macroinvertebrates. In these streams, we use the rocky-bottom sampling method. This method of macroinvertebrate sampling is used in streams that have riffles and gravel/cobble substrates. Three samples are to be collected at each site, and a composite sample is obtained (i.e., one large total sample).

Step 1—A site should have already been located on a map, with its latitude and longitude indicated.

A. Samples will be taken in three different spots within a 100-yard stream site. These spots may be three separate riffles; one large riffle with different current velocities; or, if no riffles are present, three run areas with gravel or cobble substrate. Combinations are also possible (if, for example, your site has only one small riffle and several run areas). Mark off the 100-yard stream site. If possible, it should begin at least 50 yards upstream of any human-made modification of the channel, such as a bridge, dam, or pipeline crossing. Avoid walking in the stream because this might dislodge macroinvertebrates and alter sampling results.
B. Sketch the 100-yard sampling area. Indicate the location of the three sampling spots on the sketch. Mark the most downstream site as Site 1, the middle site as Site 2, and the upstream site as Site 3.

Step 2—Get into place.

A. Always approach sampling locations from the downstream end and sample the site farthest downstream first (Site 1). This prevents biasing of the second and third collections with dislodged sediment of macroinvertebrates. Always use a clean kick-seine, relatively free of mud and debris from previous uses. Fill a bucket about one third full of stream water and fill your spray bottle.
B. Select a 3-foot-by-3-foot riffle area for sampling at Site 1. One member of the team, the net holder, should position the net at the downstream end of this sampling area. Hold the net handles at a 45-degree angle to the water's surface. Be sure that the bottom of the net fits tightly against the streambed so that no macroinvertebrates escape under the net. You may use rocks from the sampling area to anchor the net against the stream bottom. Do not allow any water to flow over the net.

Step 3—Dislodge the macroinvertebrates.

A. Pick up any large rocks in the 3-foot by 3-foot sampling area and rub them thoroughly over the partially filled bucket so that any macroinvertebrates clinging to the rocks will be dislodged into the bucket. Then place each cleaned rock outside of the sampling area. After sampling is completed, rocks can be returned to the stretch of stream they came from.

B. The member of the team designated as the "kicker" should thoroughly stir up the sampling areas with their feet, starting at the upstream edge of the 3-foot by 3-foot sampling area and working downstream, moving toward the net. All dislodged organisms will be carried by the stream flow into the net. Be sure to disturb the first few inches of stream sediment to dislodge burrowing organisms. As a guide, disturb the sampling area for about three minutes, or until the area is thoroughly worked over.

C. Any large rocks used to anchor the net should be thoroughly rubbed into the bucket as above.

Step 4—Remove the net.

A. Next, remove the net without allowing any of the organisms it contains to wash away. While the net holder grabs the top of the net handles, the kicker grabs the bottom of the net handles and the net's bottom edge. Remove the net from the stream with a forward scooping motion.

B. Roll the kick net into a cylinder shape and place it vertically in the partially filled bucket. Pour or spray water down the net to flush its contents into the bucket. If necessary, pick debris and organisms from the net by hand. Release back into the stream any fish, amphibians, or reptiles caught in the net.

Step 5—Collect the second and third samples.

A. Once all of the organisms have been removed from the net, repeat the steps above at Sites 2 and 3. Put the samples from all three sites into the same bucket. Combining the debris and organisms from all three sites into the same bucket is called compositing.

Note: If your bucket is nearly full of water after you have washed the net clean, let the debris and organisms settle to the bottom. Then, cup the net over the bucket and pour the water through the net into a second bucket. Inspect the water in the second bucket to be sure no organisms came through.

Step 6—Preserve the sample.

A. After collecting and compositing all three samples, it is time to preserve the sample. All team members should leave the stream and return to a relatively flat section of the stream bank with their equipment. The next step will be to remove large pieces of debris (leaves, twigs, and rocks) from the sample. Carefully remove the debris one piece at a time. While holding the material over the bucket, use the forceps, spray bottle, and your hands to pick, rub,

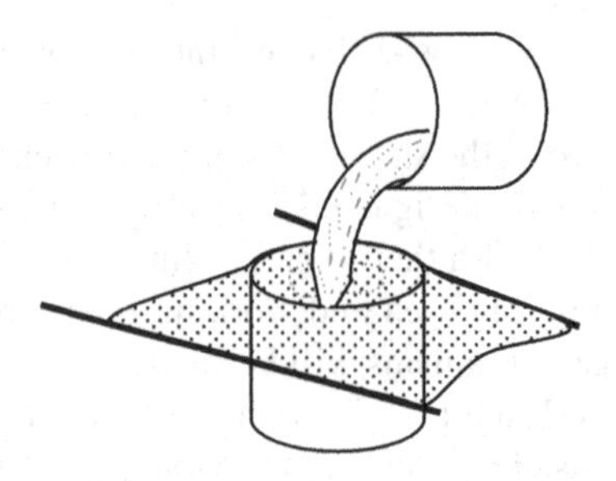

FIGURE 11.4 Pouring sample water through the net.

and rinse the leaves, twigs, and rocks to remove any attached organisms. Use a magnifying lens and forceps to find and remove small organisms clinging to the debris. When satisfied that the material is clean, discard it back into the stream.

B. The water will have to be drained before transferring the material to the jar. This process will require two team members. Place the kick net over the second bucket, which has not yet been used and should be completely empty. One team member should push the center of the net into bucket #2, creating a small indentation or depression. Then, hold the sides of the net closely over the mouth of the bucket. The second person can now carefully pour the remaining contents of bucket #1 onto a small area of the net to drain the water and concentrate the organisms. Use care when pouring so that organisms are not lost over the side of the net (see Figure 11.4). Use the spray bottle, forceps, sugar scoop, and gloved hands to remove all material from bucket #1 onto the net. When you are satisfied that bucket #1 is empty, use your hands and the sugar scoop to transfer the material from the net into the empty jar. Bucket #2 captured the water and any organisms that might have fallen through the netting during pouring. As a final check, repeat the process above, but this time, pour bucket #2 over the net, into bucket #1. Transfer any organisms on the net into the jar.
C. Now, fill the jar (so that all material is submerged) with the alcohol from the second jar. Put the lid tightly back onto the jar and gently turn the jar upside down two or three times to distribute the alcohol and remove air bubbles.
D. Complete the sampling station ID tag. Be sure to use a pencil, not a pen, because the ink will run in the alcohol! The tag includes your station number, the stream, and location (e.g., upstream from a road crossing), date, time, and the names of the members of the collecting team. Place the ID tag into the sample container, writing side facing out, so that identification can be seen clearly.

Rocky-Bottom Habitat Assessment

The habitat assessment (including measuring general characteristics and local land use) for a rocky-bottom stream is conducted in a 100-yard section of stream that includes the riffles from which organisms were collected.

Step 1—Delineate the habitat assessment boundaries.

A. Begin by identifying the most downstream riffle that was sampled for macroinvertebrates. Using tape measure or twine, mark off a 100-yard section extending 25 yards below the downstream riffle and about 75 yards upstream.
B. Complete the identifying information of the field data sheet for the habitat assessment site. On the stream sketch, be as detailed as possible, and be sure to note which riffles were sampled.

Step 2—Describe the general characteristics and local land use on the field sheet.

1. For safety reasons as well as to protect the stream habitat, it is best to estimate the following characteristics rather than actually wading into the stream to measure them.

A. Water appearance can be a physical indicator of water pollution.
   1. Clear—colorless, transparent
   2. Milky—cloudy-white or gray, not transparent; might be natural or due to pollution
   3. Foamy—might be natural or due to pollution, generally detergents or nutrients (foam that is several inches high and does not brush apart easily is generally due to pollution)
   4. Turbid—cloudy brown due to suspended silt or organic material
   5. Dark brown—might indicate that acids are being released into the stream due to decaying plants
   6. Oily sheen—multicolored reflection might indicate oil floating in the stream, although some sheens are natural
   7. Orange—might indicate acid drainage
   8. Green—might indicate that excess nutrients are being released into the stream
B. Water odor can be a physical indicator of water pollution
   1. None or natural smell
   2. Sewage—might indicate the release of human waste material
   3. Chlorine—might indicate that a sewage treatment plant is over-chlorinating its effluent
   4. Fishy—might indicate the presence of excessive algal growth or dead fish
   5. Rotten eggs—might indicate sewage pollution (the presence of a natural gas)
C. Water temperature can be particularly important for determining whether the stream is suitable as habitat for some species of fish and macroinvertebrates that have distinct temperature requirements. Temperature also has a direct effect on the amount of dissolved oxygen available to aquatic organisms. Measure temperature by submerging a thermometer

for at least two minutes in a typical stream run. Repeat once and average the results.

D. The width of the stream channel can be determined by estimating the width of the streambed that is covered by water from bank to bank. If it varies widely along the stream, estimate an average width.
E. Local land use refers to the part of the watershed within one-quarter mile upstream of and adjacent to the site. Note which land uses are present, as well as which ones seem to be having a negative impact on the stream. Base observations on what can be seen, what was passed on the way to the stream, and, if possible, what is noticed when leaving the stream.

Step 3—Conduct the habitat assessment. The following information describes the parameters that will be evaluated for rocky-bottom habitats. Use these definitions when completing the habitat assessment field data sheet. The first two parameters should be assessed directly at the riffle(s) or run(s) that were used for the macroinvertebrate sampling. The last eight parameters should be assessed in the entire 100-yard section of the stream.

A. Attachment sites for macroinvertebrates are essentially the amount of living space or hard substrates (rocks, snags) available for aquatic insects and snails. Many insects begin their life underwater in streams and need to attach themselves to rocks, logs, branches, or other submerged substrates. The greater the variety and number of available living spaces or attachment sites, the greater the variety of insects in the stream. Optimally, cobble should predominate, and boulders and gravel should be common. The availability of suitable living spaces for macroinvertebrates decreases as cobble becomes less abundant and boulders, gravel, or bedrock become more prevalent.
B. Embeddedness refers to the extent to which rocks (gravel, cobble, and boulders) are surrounded by, covered, or sunken into the silt, sand, or mud of the stream bottom. Generally, as rocks become embedded, fewer living spaces are available to macroinvertebrates and fish for shelter, spawning, and egg incubation.

   Note: To estimate the percent of embeddedness, observe the amount of silt or finer sediments overlaying and surrounding the rocks. If kicking does not dislodge the rocks or cobbles, they might be greatly embedded.
C. Shelter for fish includes the relative quantity and variety of natural structures in a stream, such as fallen trees, logs, and branches; cobble and large rock; and undercut banks that are available to fish for hiding, sleeping, or feeding. A wide variety of submerged structures in the stream provide fish with many living spaces; the more living spaces in a stream, the more types of fish the stream can support.
D. Channel alteration is a measure of large-scale changes in the shape of the stream channel. Many streams in urban and agricultural areas have been straightened, deepened (e.g., dredged), or diverted into concrete channels,

often for flood control purposes. Such streams have far fewer natural habitats for fish, macroinvertebrates, and plants than do naturally meandering streams. Channel alteration is present when the stream runs through a concrete channel, when artificial embankments, riprap, and other forms of artificial bank stabilization or structures are present; when the stream is very straight for significant distances; when dams, bridges, and flow-altering structures such as combined sewer overflow (CSO) are present; when the stream is of uniform depth due to dredging; and when other such changes have occurred. Signs that indicate the occurrence of dredging include straightened, deepened, and otherwise uniform stream channels, as well as the removal of streamside vegetation to provide dredging equipment access to the stream.

E. Sediment deposition is a measure of the amount of sediment that has been deposited in the stream channel and the changes to the stream bottom that have occurred as a result of the deposition. High levels of sediment deposition create an unstable and continually changing environment that is unsuitable for many aquatic organisms. Sediments are naturally deposited in areas where the stream flow is reduced, such as in pools and bends, or where flow is obstructed. These deposits can lead to the formation of islands, shoals, or point bars (sediments that build up in the stream, usually at the beginning of a meander) or can result in the complete filling of pools. To determine whether these sediment deposits are new, look for vegetation growing on them: new sediments will not yet have been colonized by vegetation.

F. Stream velocity and depth combinations are important for the maintenance of healthy aquatic communities. Fast water increases the amount of dissolved oxygen in the water, keeps pools from being filled with sediment, and helps food items like leaves, twigs, and algae move more quickly through the aquatic system. Slow water provides spawning areas for fish and shelters macroinvertebrates that might be washed downstream in higher stream velocities. Similarly, shallow water tends to be more easily aerated (i.e., it holds more oxygen), but deeper water stays cooler longer. Thus, the best stream habitat includes all of the following velocity/depth combinations and can maintain a wide variety of organisms.

slow (<1 ft/sec), shallow (<1.5 ft)
slow, deep
fast, deep
fast, shallow

Measure stream velocity by marking off a 10 ft section of stream run and measuring the time it takes an orange, stick, or other floating biodegradable object to float the 10 ft. Repeat five times, in the same 10 ft section, and determine the average time. Divide the distance (10 ft) by the average time (seconds) to determine the velocity in feet per second.

Measure the stream depth by using a stick of known length and taking readings at various points within your stream site, including riffles, runs, and pools. Compare velocity and depth at various points within the 100-yard site to see how many of the combinations are present.

G. Channel flow status is the percent of the existing channel that is filled with water. The flow status changes as the channel enlarges or as flow decreases because of dams and other obstructions, diversions for irrigation, or drought. When water does not cover much of the streambed, the living area for aquatic organisms is limited.

   Note: For the following parameters, evaluate the conditions of the left and right stream banks separately. Define the "left" and "right" banks by standing at the downstream end of the study stretch and look upstream. Each bank is evaluated on a scale of 0–10.

H. Bank vegetation protection measures the amount of the stream bank that is covered by natural (i.e., growing wild and not obviously planted) vegetation. The root system of plants growing on stream banks helps hold soil in place, reducing erosion. Vegetation on banks provides shade for fish and macroinvertebrates and serves as a food source by dropping leaves and other organic matter into the stream. Ideally, a variety of vegetation should be present, including trees, shrubs, and grasses. Vegetation disruption can occur when the grasses and plants on the stream banks are mowed or grazed, or when the trees and shrubs are cut back or cleared.

I. Condition of banks measures erosion potential and whether the stream banks are eroded. Steep banks are more likely to collapse and suffer from erosion than are gently sloping banks and are, therefore, considered to have erosion potential. Signs of erosion include crumbling, unvegetated banks, exposed tree roots, and exposed soil.

J. The riparian vegetative zone is defined as the width of natural vegetation from the edge of the stream bank. The riparian vegetative zone is a buffer zone to pollutants entering a stream from runoff. It also controls erosion and provides stream habitat and nutrient input into the stream.

   Note: A wide, relatively undisturbed riparian vegetative zone reflects a healthy stream system; narrow, far less useful riparian zones occur when roads, parking lots, fields, lawns, and other artificially cultivated areas, bare soil, rock, or buildings are near the stream bank. The presence of "old fields" (i.e., previously developed agricultural fields allowed to revert to natural conditions) should rate higher than fields in continuous or periodic use. In arid areas, the riparian vegetative zone can be measured by observing the width of the area dominated by riparian or water-loving plants, such as willows, marsh grasses, and cottonwood trees.

Macroinvertebrate Sampling in Muddy-Bottom Streams

In muddy-bottom streams, as in rocky-bottom streams, the goal is to sample the most productive habitat available and look for the widest variety of organisms. The most productive habitat is the one that harbors a diverse population of pollution-sensitive macroinvertebrates. Samples should be obtained using a D-frame net (see Figure 11.5)

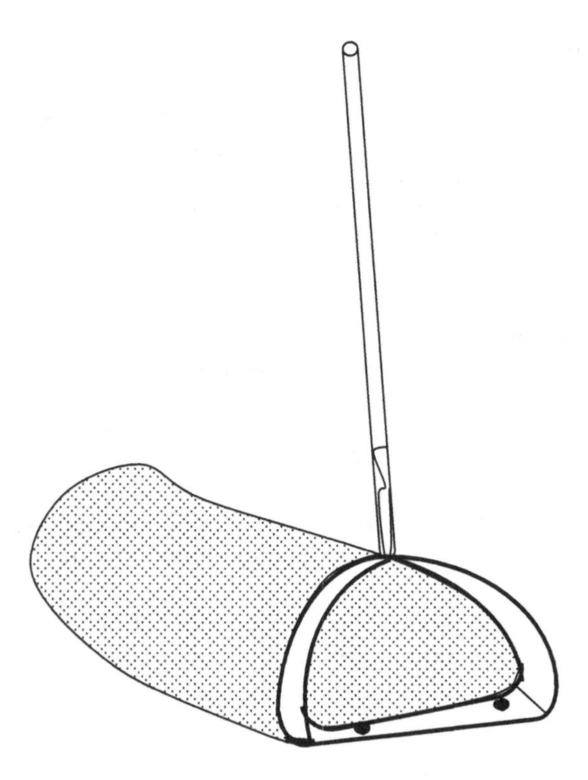

FIGURE 11.5 D-frame aquatic net.

to jab at the habitat and scoop up the organisms that are dislodged. The idea is to collect a total sample that consists of 20 jabs taken from a variety of habitats. Use the following method of macroinvertebrate sampling in streams that have muddy-bottom substrates.

Step 1—Determine which habitats are present.

1. Muddy-bottom streams usually have four habitats: vegetated banks margins, snags and logs, aquatic vegetation beds and decaying organic matter, and silt/sand/gravel substrate. It is generally best to concentrate sampling efforts on the most productive habitat available, yet to sample other principal habitats if they are present. This ensures that you will secure as wide a variety of organisms as possible. Not all habitats are present in all streams or are present in significant amounts. If the sampling areas have not been preselected, determine which of the following habitats are present.

Note: Avoid standing in the stream while making habitat determinations.

A. Vegetated bank margins consist of overhanging bank vegetation and submerged root mats attached to banks. The bank margins may also contain submerged, decomposing leaf packs trapped in root wads or lining the streambanks. This is generally a highly productive habitat in a muddy stream, and it is often the most abundant type of habitat.

B. Snags and logs consist of submerged wood, primarily dead trees, logs, branches, roots, cypress knees, and leaf packs lodged between rocks or logs. This is also a very productive muddy-bottom stream habitat.

C. Aquatic vegetation beds and decaying organic matter consist of beds of submerged, green/leafy plants that are attached to the stream bottom. This habitat can be as productive as vegetated bank margins and snags and logs.

D. Silt/sand/gravel substrate includes sandy, silty, or muddy stream bottoms; rocks along the stream bottom; and/or wetted gravel bars. This habitat may also contain algae-covered rocks (Aufwuchs). This is the least productive of the four muddy-bottom stream habitats, and it is always present in one form or another (e.g., silt, sand, mud, or gravel might predominate).

Step 2—Determine how many times to jab in each habitat type. The sampler's goal is to jab 20 times. The D-frame net (see Figure 11.5) is 1 ft wide, and a jab should be approximately 1 ft in length. Thus, 20 jabs equal 20 ft^2 of combined habitat.

A. If all 4 habitats are present in plentiful amounts, jab the vegetated banks 10 times and divide the remaining 10 jabs by the amount of the remaining 3 habitats.

B. If 3 habitats are present in plentiful amounts, and 1 is absent, jab the silt/sand/gravel substrate, the least productive habitat, 5 times and divide the remaining 15 jabs between the other 2 more productive habitats.

C. If only 2 habitats are preset in plentiful amounts, the silt/sand/gravel substrate will most likely be one of those habitats. Jab the silt/sand/gravel substrate 5 times and the more productive habitat 15 times.

D. If some habitats are plentiful and others are sparse, sample the sparse habitats to the extent possible, even if you can take only one or two jabs. Take the remaining jabs from the plentiful habitat(s). This rule also applies if you cannot reach a habitat because of unsafe stream conditions. Jab 20 times.

Note: Because the sampler might need to make an educated guess to decide how many jabs to take in each habitat type, it is critical that the sampler note on the field data sheet, how many jabs were taken in each habitat. This information can be used to help characterize the findings.

Step 3—Get into place.

1. Outside and downstream of the first sampling location (first habitat), rinse the dip net and check to make sure it does not contain any macroinvertebrates or debris from the last time it was used. Fill a bucket approximately one third with clean stream water. Also, fill the spray bottle with clean stream water. This bottle will be used to wash the net between jabs and after sampling is completed.

Note: This method of sampling requires only one person to disturb the stream habitats. While one person is sampling, a second person should stand outside the sampling area, holding the bucket and spray bottle. After every few jabs, the sampler should hand the net to the second person, who then can rinse the contents of the net into the bucket.

Step 4—Dislodge the macroinvertebrates.

1. Approach the first sample site from downstream, and sample while walking upstream. Sample in the four habitat types as follows:
   - A. Sample vegetated bank margins by jabbing vigorously, with an upward motion, brushing the net against vegetation and roots along the bank. The entire jab motion should occur underwater.
   - B. To sample snags and logs, hold the net with one hand under the section of submerged wood being sampled. With the other hand (which should be gloved), rub about 1 ft^2 of area on the snag or log. Scoop organisms, bark, twigs, or other organic matter dislodged into the net. Each combination of log rubbing, and net scooping is one jab.
   - C. To sample aquatic vegetation beds, jab vigorously, with an upward motion, against or through the plant bed. The entire jab motion should occur underwater.
   - D. To sample a silt/sand/gravel substrate, place the net with one edge against the stream bottom and push it forward about a foot (in an upstream direction) to dislodge the first few inches of silt, sand, gravel, or rocks. To avoid gathering a net full of mud, periodically sweep the mesh bottom of the net back and forth in the water, making sure that water does not run over the top of the net. This will allow fine silt to rinse out of the net. When 20 jabs have been completed, rinse the net thoroughly in the bucket. If necessary, pick any clinging organisms from the net by hand, and put them in the bucket.

Step 5—Preserve the sample.

1. Look through the material in the bucket, and immediately return any fish, amphibians, or reptiles to the stream. Carefully remove large pieces of debris (leaves, twigs, and rocks) from the sample. While holding the material over the bucket, use the forceps, spray bottle, and your hands to pick, rub, and rinse the leaves, twigs, and rocks to remove any attached organisms. Use the magnifying lens and forceps to find and remove small organisms clinging to the debris. When satisfied that the material is clean, discard it back into the stream.
2. Drain the water before transferring material to the jar. This process will require two people. One person should place the net into the second bucket, like a sieve (this bucket, which has not yet been used, should be completely empty) and hold it securely. The second person can now carefully pour the remaining contents of bucket #1 onto the center of the net to drain the water

Station ID Tag

Station # ________________________________

Stream ________________________________

Location ________________________________

Date/Time ________________________________

Team Members: ________________________________

____

FIGURE 11.6 Station ID Tag.

and concentrate the organisms. Use care when pouring so that organisms are not lost over the side of the net. Use the spray bottle, forceps, sugar scoop, and gloved hands to remove all the material from bucket #1 onto the net. When satisfied that bucket #1 is empty, use your hands and the sugar scoop to transfer all the material from the net into the empty jar. The contents of the net can also be emptied directly into the jar by turning the net inside out into the jar. Bucket #2 captured the water and any organisms that might have fallen through the netting. As a final check, repeat the process above, but this time, pour bucket #2 over the net, into bucket #1. Transfer any organisms on the net into the jar.

3. Fill the jar (so that all material is submerged) with alcohol. Put the lid tightly back onto the jar and gently turn the jar upside down two or three times to distribute the alcohol and remove air bubbles.
4. Complete the sampling station ID tag (see Figure 11.6). Be sure to use a pencil, not a pen, because the ink will run in the alcohol. The tag should include your station number, the stream, and location (e.g., upstream form a road crossing), date, time, and the names of the members of the collecting crew. Place the ID tag into the sample container, writing side facing out, so that identification can be seen clearly.

   Note: To prevent samples from being mixed up, samplers should place the ID tag inside the sample jar.

Muddy-Bottom Assessment

1. The following information describes the parameters to be evaluated for muddy-bottom habitats. Use these definitions when completing the habitat assessment field data sheet.

A. Shelter for fish and attachment sites for macroinvertebrates are essentially the amount of living space and shelter (rocks, snags, and undercut banks) available for fish, insects, and snails. Many insects attach themselves to rocks, logs, branches, or other submerged substrates. Fish can hide or feed in these areas. The greater the variety and number of available shelter sites or attachment sites, the greater the variety of fish and insects in the stream.

Note: Many of the attachment sites result from debris falling into the stream from the surrounding vegetation. When debris first falls into the water, it is termed new fall, and it has not yet been "broken down" by microbes (conditioned) for macroinvertebrate colonization. Leaf material or debris that is conditioned is called old fall. Leaves that have been in the stream for some time lose their color, turn brown or dull yellow, become soft and supple with age, and might be slimy to the touch. Woody debris becomes blackened or dark in color; smooth bark becomes coarse and partially disintegrated, creating holes and crevices. It might also be slimy to the touch.

B. Pool substrate characterization evaluates the type and condition of bottom substrates found in pools. Pools with firmer sediment types (e.g., gravel and sand) and rooted aquatic plants support a wider variety of organisms than do pools with substrates dominated by mud or bedrock and no plants. In addition, a pool with one uniform substrate type will support far fewer types of organisms than will a pool with a wide variety of substrate types.
C. Pool variability rates the overall mixture of pool types found in the stream according to size and depth. The four basic types of pools are large-shallow, large-deep, small-shallow, and small-deep. A stream with many pool types will support a wide variety of aquatic species. Rivers with low sinuosity (few bends) and monotonous pool characteristics do not have sufficient quantities and types of habitats to support a diverse aquatic community.
D. Channel alteration
E. Sediment deposition
F. Channel sinuosity evaluates the sinuosity or meandering of the stream. Streams that meander provide a variety of habitats (such as pools and runs) and stream velocities and reduce the energy from current surges during storm events. Straight stream segments are characterized by even stream depth and unvarying velocity, and they are prone to flooding. To evaluate this parameter, imagine how much longer the stream would be if it were straightened out.
G. Channel flow status
H. Bank vegetative protection
I. Condition of banks
J. Riparian vegetative zone width

Note: Whenever stream sampling is to be conducted, it is a good idea to have a reference collection on hand. A reference collection is a sample of locally found macroinvertebrates that have been identified, labeled, and preserved in alcohol. The program advisor, along with a professional biologist/entomologist, should assemble the reference collection, properly identify all samples, preserve them in vials, and label them. This collection may then be used as a training tool and, in the field, as an aid in macroinvertebrate identification.

Post-Sampling Routine

After completing the stream characterization and habitat assessment, make sure that all of the field data sheets have been completed properly and that the information is legible. Be sure to include the site's identifying name and the sampling date on each sheet. This information will function as a quality control element. Before leaving the stream location, make sure that all sampling equipment/devices have been collected and rinsed properly. Double check to see that sample jars are tightly closed and properly identified. All samples, field sheets, and equipment should be returned to the team leader at this point. Keep a copy of the field data sheet(s) for comparison with future monitoring trips and for personal records. The next step is to prepare for macroinvertebrate laboratory work. This step includes all the work needed to set up a laboratory for processing samples into subsamples and identifying macroinvertebrates to the family level. A professional biologist/entomologist/freshwater ecologist or the professional advisor should supervise the identification procedure. (The actual laboratory procedures after the sampling and collecting phase are beyond the scope of this text.)

Sampling Devices

In addition to the sampling equipment mentioned previously, it may be desirable to employ, depending on stream conditions, other sampling devices. Additional sampling devices commonly used, and discussed in the following sections, include dissolved oxygen and temperature monitors, sampling nets (including the D-frame aquatic net), sediment samplers (dredges), plankton samplers, and Secchi disks. The methods described below are approved by the USEPA. Coverage that is more detailed is available in APHA (1998).

Dissolved Oxygen and Temperature Monitor

As mentioned, the dissolved oxygen (DO) content of a stream sample can provide the investigator with vital information, as DO content reflects the stream's ability to maintain aquatic life.

The Winkler DO with Azide Modification Method

The Winkler DO with azide modification method is commonly used to measure DO content. The Winkler method is best suited for clean waters. The Winkler method can be used in the field but is better suited for laboratory work where better accuracy may be achieved. The Winkler method adds a divalent manganese solution followed by a strong alkali to a 300 mL BOD bottle of stream water sample. Any DO rapidly oxidizes an equivalent amount of divalent manganese to basic hydroxides of higher balance states. When the solution is acidified in the presence of iodide, oxidized manganese again reverts to the divalent state, and iodine, equivalent to the original

DO content of the sample, is liberated. The amount of iodine is then determined by titration with a standard, usually thiosulfate, solution.

Fortunately, for the field biologist, this is the age of miniaturized electronic circuit components and devices; thus, it is not too difficult to obtain portable electronic measuring devices for DO and temperature that are of quality construction and have better than moderate accuracy. These modern electronic devices are usually suitable for laboratory and field use. The device may be subjected to severe abuse in the field; therefore, the instrument must be durable, accurate, and easy to use. Several quality DO monitors are available commercially.

When using a DO monitor, it is important to calibrate (standardize) the meter prior to use. Calibration procedures can be found in Standard Methods (latest edition) or in manufacturer's instructions for the meter to be used. Determining the air temperature, the DO at saturation for that temperature, and then adjusting the meter so that it reads the saturation value usually accomplishes meter calibration. After calibration, the monitor is ready for use. As mentioned, all recorded measurements, including water temperatures and DO readings, should be entered in a field notebook.

Sampling Nets

A variety of sampling nets are available for use in the field. The two-person seine net shown in Figure 11.7 is 20 ft long and 4 ft deep with a 1/8-inch mesh and is utilized to collect a variety of organisms. Two people, each holding one end, walk upstream, and small organisms are gathered in the net. Dip nets are used to collect organisms in shallow streams. The Surber sampler (collects macroinvertebrates stirred up from the bottom; see Figure 11.8) can be used to obtain a quantitative sample (number of organisms/square feet). It is designed for sampling riffle areas in streams and rivers up to a depth of about 450 mm (18 inches). It consists of two folding stainless-steel frames set at right angles to each other. The frame is placed on the bottom, with the net extending downstream. Using your hand or a rake, all sediment enclosed by the frame is dislodged. All organisms are caught in the net and transferred to another vessel for counting.

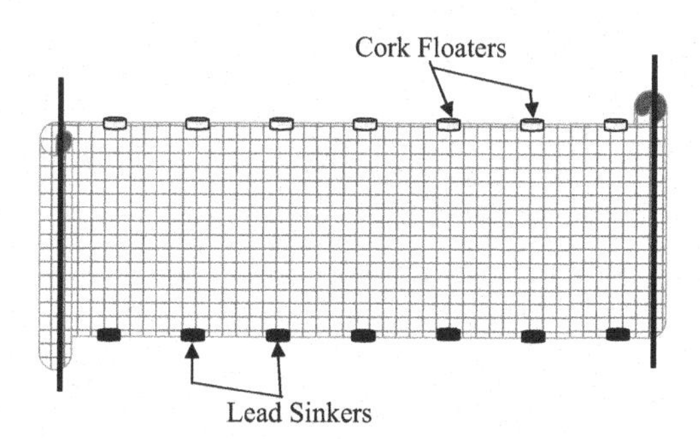

FIGURE 11.7 Two-person seine net.

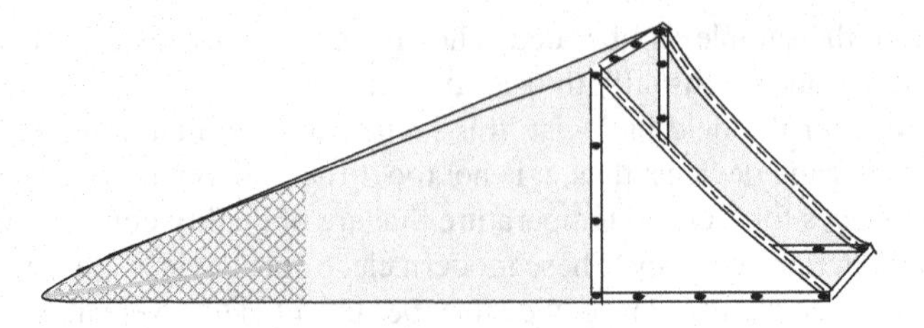

FIGURE 11.8 Surber sampler.

The D-frame aquatic dip net (see Figure 11.5) is ideal for sweeping over vegetation or for use in shallow streams.

Sediment Samplers (Dredges)

A sediment sampler or dredge is designed to obtain a sample of the bottom material in a slow-moving stream and the organisms in it. The simple homemade dredge shown in Figure 11.9 works well in water too deep to sample effectively with handheld tools. The homemade dredge is fashioned from a #3 coffee can and a smaller can with a tight-fitting plastic lid (peanut cans work well). To use the homemade dredge, first invert it under water so the can fills with water and no air is trapped. Then, lower the dredge as quickly as possible with the "down" line. The idea is to bury the open end of the coffee can in the bottom. Then, quickly pull the "up" line to bring the can to the surface with a minimum loss of material. Dump the contents into a sieve or observation pan to sort. It works best in bottoms composed of sediment, mud, sand, and small gravel. The bottom sampling dredge can be used for a number of different analyses. Because the bottom sediments represent a good area in which to find macroinvertebrates and benthic algae, the communities of organisms living

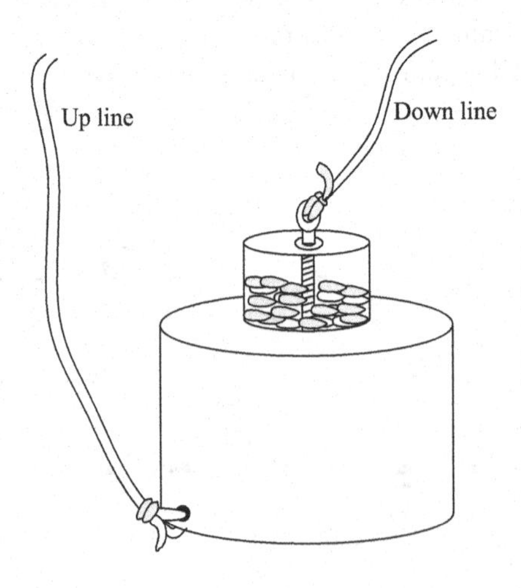

FIGURE 11.9 Homemade dredge.

on or in the bottom can be easily studied quantitatively and qualitatively. A chemical analysis of the bottom sediment can be conducted to determine what chemicals are available to organisms living in the bottom habitat.

Plankton Sampler

(More detailed information on plankton sampling can be found in AWRI, 2000.) Plankton (meaning to drift) are distributed through the stream and, in particular, in pool areas. They are found at all depths and are comprised of plant (phytoplankton) and animal (zooplankton) forms. Plankton show a distribution pattern that can be associated with the time of day and seasons. The three fundamental sizes of plankton are nanoplankton, microplankton, and macroplankton. The smallest are nanoplankton that ranges in size from 5–60 microns (millionth of a meter). Because of their small size, most nanoplankton will pass through the pores of a standard sampling net. Special fine mesh nets can be used to capture the larger nanoplankton. Most planktonic organisms fall into the microplankton or net plankton category. The sizes range from the largest nanoplankton to about 2 mm (thousandths of a meter). Nets of various sizes and shapes are used to collect microplankton. The nets collect the organism by filtering water through fine meshed cloth. The plankton nets on the vessels are used to collect microplankton. The third group of plankton, as associated with size, is called macroplankton. They are visible to the naked eye. The largest can be several meters long.

The plankton net or sampler (see Figure 11.10) is a device that makes it possible to collect phytoplankton and zooplankton samples. For quantitative comparisons of different samples, some nets have a flow meter used to determine the amount of water passing through the collecting net.

The plankton net or sampler provides a means of obtaining samples of plankton from various depths so that distribution patterns can be studied. Considering the depth of the water column that is sampled allows you to make quantitative determinations. The net can be towed to sample plankton at a single depth (horizontal tow) or lowered into the water to sample the water column (vertical tow). Another possibility is oblique tows where the net is lowered to a predetermined depth and raised at a constant rate as the vessel moves forward.

The plankton net or sampler provides a means of obtaining samples of plankton from various depths so that distribution patterns can be studied. Considering the depth of the water column that is sampled allows you to make quantitative determinations. The net can be towed to sample plankton at a single depth (horizontal tow) or lowered into the water to sample the water column (vertical tow). Another possibility is oblique tows where the net is lowered to a predetermined depth and raised at a constant rate as the vessel moves forward.

After towing and removal from the stream, the sides of the net are rinsed to dislodge the collected plankton. If a quantitative sample is desired, a certain quantity of water is collected. If the plankton density is low, then the sample may be concentrated using a low-speed centrifuge or some other filtering device. A definite volume of the sample is studied under the compound microscope for counting and identification of plankton.

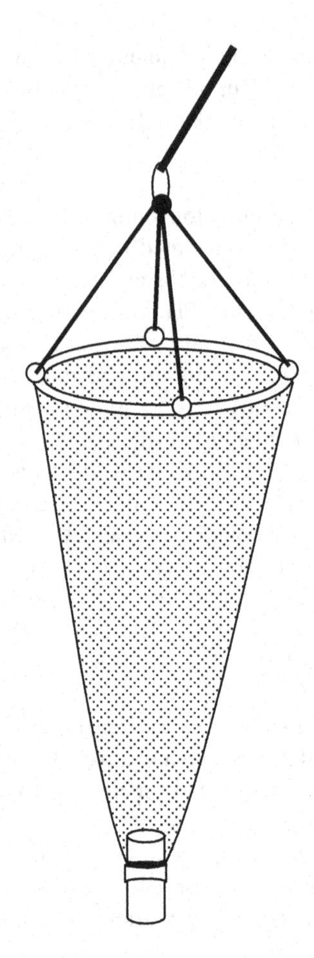

FIGURE 11.10 Plankton net.

Secchi Disk

For determining water turbidity or degree of visibility in a stream, a Secchi disk is often used (Figure 11.11). The Secchi disk originated with Father Pietro Secchi, an astrophysicist and scientific advisor to the Pope, who was requested to measure transparency in the Mediterranean Sea by the head of the Papal Navy. Secchi used some white disks to measure the clarity of water in the Mediterranean in April of 1865. Various sizes of disks have been used since that time, but the most frequently used disk is an 8-inch-diameter metal disk painted in alternate black-and-white quadrants. The disk shown in Figure 11.11 is 20 cm in diameter; it is lowered into the stream using the calibrated line. To use the Secchi disk properly, it should be lowered into the stream water until no longer visible. At the point where it is no longer visible, a measurement of the depth is taken. This depth is called the Secchi disk transparency

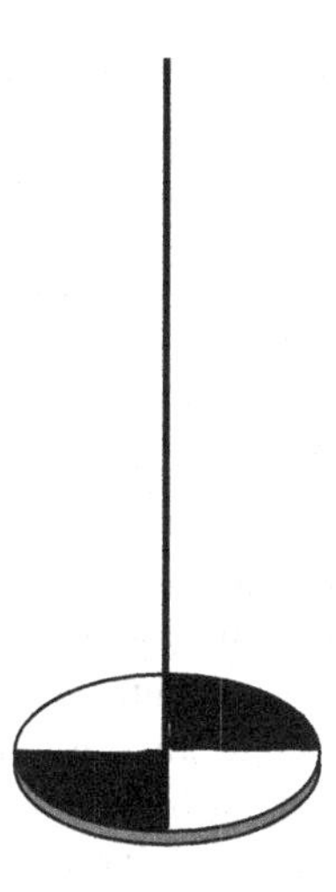

FIGURE 11.11 Secchi disk.

light extinction coefficient. The best results are usually obtained after early morning and before late afternoon.

Miscellaneous Sampling Equipment

Several other sampling tools/devices are available for use in sampling a stream. For example, consider the standard sand-mud sieve. Generally made of heavy-duty galvanized 1/8-inch mesh screen supported by a water-sealed 24×15×3-inch wood frame, this device is useful for collecting burrowing organisms found in soft bottom sediments. Moreover, no stream sampling kit would be complete without a collecting tray, collecting jars of assorted sizes, heavy-duty plastic bags, small pipets, large two-ounce pipets, fine mesh straining net, and black china marking pencil. In addition, depending upon the quantity of material to be sampled, it is prudent to include several 3- and 5-gallon collection buckets in the stream sampling field kit.

THE BOTTOMLINE ON BIOLOGICAL SAMPLING

This discussion has stressed the practice of biological monitoring, employing the use of biotic indices as key measuring tools. We emphasized biotic indices not only for their simplicity of use but also for the relative accuracy they provide, although their development and use can sometimes be derailed. The failure of a monitoring protocol to assess environmental conditions accurately, or to protect running waters, usually stems from conceptual, sampling, or analytical pitfalls. Biotic indices can be combined with other tools for measuring the condition of ecological systems in ways that enhance or hinder their effectiveness. The point is, like any other tool, they can be misused. However, that biotic indices can be, and are, misused does not mean that the indices' approach itself is useless. Thus, to ensure that the biotic indices approach is not useless, it is important for the practicing freshwater ecologist and water sampler to remember a few key guidelines:

1. Sampling everything is not the goal. As Botkin (1990) note, biological systems are complex and unstable in space and time, and samplers often feel compelled to study all components of this variation. Complex sampling programs proliferate. However, every study need not explore everything. Freshwater samplers and monitors should avoid the temptation to sample all the unique habitats and phenomena that make freshwater monitoring so interesting. Concentration should be placed on the central components of a clearly defined research agenda (a sampling/monitoring protocol)—detecting and measuring the influences of human activities on the water body's ecological system.
2. With regard to the influence of human activities on the water body's ecological system, we must see protecting biological condition as a central responsibility of water resource management. One thing is certain, until biological monitoring is seen as essential to track attainment of that goal and biological criteria as enforceable standards mandated by the Clean Water Act, life in the nation's freshwater systems will continue to decline.

Biomonitoring is only one of several tools available to the water practitioner. No matter the tool employed, all results depend upon proper biomonitoring techniques. Biological monitoring must be designed to obtain accurate results—present approaches need to be strengthened. In addition, "the way it's always been done" must be reexamined and efforts must be undertaken to do what works to keep freshwater systems alive. We can afford nothing less.

PREPARATION AND SAMPLING CONSIDERATIONS

The sections that follow detail specific equipment considerations and analytical procedures for each of the most common water quality parameters. Sampling devices should be corrosion resistant, easily cleaned, and capable of collecting desired samples safely and in accordance with test requirements. Whenever possible, assign a sampling device to each sampling point. Sampling equipment must be cleaned on a regular schedule to avoid contamination.

Note: Some tests require special equipment to ensure the sample is representative. For example, fecal bacteria sampling requires special equipment and/or procedures to prevent collection of nonrepresentative samples.

Cleaning Procedures

Reused sample containers and glassware must be cleaned and rinsed before the first sampling run, and after each run by following Method A or Method B described below. The most suitable method depends on the parameter being measured.

Method A: General Preparation of Sampling Containers

Use the following method when preparing all sample containers and glassware for monitoring conductivity, total solids, turbidity, pH, and total alkalinity. Wearing latex gloves:

1. Wash each sample bottle or piece of glassware with a brush and phosphate-free detergent.
2. Rinse three times with cold tap water.
3. Rinse three times with distilled or deionized water.

Method B: Acid Wash Procedures

Use this method when preparing all sample containers and glassware for monitoring nitrates and phosphorus. Wearing latex gloves:

1. Wash each sample bottle or piece of glassware with a brush and phosphate-free detergent.
2. Rinse three times with cold tap water.
3. Rinse with 10 percent hydrochloric acid.
4. Rinse three times with deionized water.

Sample Types

Two types of samples are commonly used for water quality monitoring: grab samples and composite samples. The type of sample used depends on the specific test, the reason the sample is being collected, and the applicable regulatory requirements.

Grab samples are taken all at once, at a specific time and place. They are representative only of the conditions at the time of collection. Grab samples must be used to determine pH, total residual chlorine (TRC), dissolved oxygen (DO), and fecal coliform concentrations. Grab samples may also be used for any test which does not specifically prohibit their use.

Note: Before collecting samples for any test procedure, it is best to review the sampling requirements of the test.

Composite samples consist of a series of individual grab samples collected over a specified period in proportion to flow. The individual grab samples are mixed together in proportion to the flow rate at the time the sample was collected to form the composite sample. This type of sample is taken to determine average conditions in a large volume of water whose properties vary significantly over the course of a day.

Collecting Samples from a Stream

In general, sample away from the stream bank in the main current. Never sample stagnant water. The outside curve of the stream is often a good place to sample, because the main current tends to hug this bank. In shallow stretches, carefully wade into the center current to collect the sample. A boat is required for deep sites. Try to maneuver the boat into the center of the main current to collect the water sample. When collecting a water sample for analysis in the field or at the lab, follow the steps below.

Whirl-pak® Bags

To collect water samples using Whirl-pak bags, use the following procedures:

1. Label the bag with the site number, date, and time.
2. Tear off the top of the bag along the perforation above the wire tab just before sampling. Avoid touching the inside of the bag. If you accidentally touch the inside of the bag, use another one.
3. Wading—Try to disturb as little bottom sediment as possible. In any case, be careful not to collect water that contains bottom sediment. Stand facing upstream. Collect the water samples in front of you.

   Boat—Carefully reach over the side and collect the water sample on the upstream side of the boat.
4. Hold the two white pull-tabs in each hand and lower the bag into the water on your upstream side with the opening facing upstream. Open the bag midway, between the surface and the bottom by pulling the white pull-tabs. The bag should begin to fill with water. You may need to "scoop" water into the bag by drawing it through the water upstream and away from you. Fill the bag no more than three fourths full!
5. Lift the bag out of the water. Pour out excess water. Pull on the wire tabs to close the bag. Continue holding the wire tabs and flip the bag over at least four to five times quickly to seal the bag. Do not try to squeeze the air out of the top of the bag. Fold the ends of the bag, being careful not to puncture the bag. Twist them together, forming a loop.
6. Fill in the bag number and/or site number on the appropriate field data sheet. This is important. It is the only way the lab specialist will know which bag goes with which site.
7. If samples are to be analyzed in a lab, place the sample in the cooler with ice or cold packs. Take all samples to the lab.

Screw-Cap Bottles

To collect water samples using screw-cap bottles, use the following procedures (see Figure 11.12).

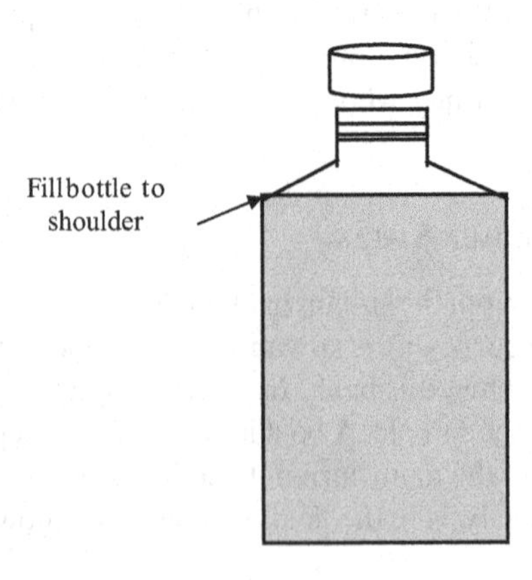

FIGURE 11.12 Sampling bottle. Filling to shoulder assures collecting enough sample. Do not overfill.

1. Label the bottle with the site number, date, and time.
2. Remove the cap from the bottle just before sampling. Avoid touching the inside of the bottle or the cap. If you accidentally touch the inside of the bottle, use another one.
3. Wading—Try to disturb as little bottom sediment as possible. In any case, be careful not to collect water that has sediment from bottom disturbance. Stand facing upstream. Collect the water sample on your upstream side, in front of you. You may also tape your bottle to an extension pole to sample from deeper water.

   Boat—Carefully reach over the side and collect the water sample on the upstream side of the boat.
4. Hold the bottle near its base and plunge it (opening downward) below the water surface. If you are using an extension pole, remove the cap, turn the bottle upside down, and plunge it into the water, facing upstream. Collect a water sample 8–12 inches beneath the surface, or midway between the surface and the bottom if the stream reach is shallow.
5. Turn your bottle underwater into the current and away from you. In slow-moving stream reaches, push the bottle underneath the surface and away from you in the upstream direction.
6. Leave a 1-inch air space (except for DO and BOD samples). Do not fill the bottle completely (so that the sample can be shaken just before analysis). Recap the bottle carefully, remembering not to touch the inside.
7. Fill in the bottle number and/or site number on the appropriate field data sheet. This is important because it tells the lab specialist which bottle goes with which site.
8. If the samples are to be analyzed in the lab, place them in the cooler for transport to the lab.

Sample Preservation and Storage

Samples can change very rapidly. However, no single preservation method will serve for all samples and constituents. If analysis must be delayed, follow the instructions for sample preservation and storage listed in Standard Methods, or those specified by the laboratory that will eventually process the samples (see Table 11.3). In general, handle the sample in a way that prevents changes from biological activity, physical alterations, or chemical reactions. Cool the sample to reduce biological and chemical reactions. Store in darkness to suspend photosynthesis.

Fill the sample container completely to prevent the loss of dissolved gases. Metal cations such as iron and lead and suspended particles may adsorb onto container surfaces during storage.

Standardization of Methods

References used for sampling and testing must correspond to those listed in the most current Federal regulation. For the majority of tests, to compare the results of either different water quality monitors or the same monitors over the course of time

TABLE 11.3
Recommended Sample Storage and Preservation Techniques

Test factor	Container type	Preservation recommended/ regulatory	Max. storage time
Alkalinity	P, G	Refrigerate	24 hr/14 days
BOD	P, G	Refrigerate	6 hr/48 hr
Conductivity	P, G	Refrigerate	28 days/28 days
Hardness	P, G	Lower pH to <2	6 mos/6 mos
Nitrate	P, G	Analyze ASAP	48 hr/48 hr
Nitrite	P, G	Analyze ASAP	none/48 hr
Odor	G	Analyze ASAP	6 hr/NR
Oxygen, dissolved			
Electrode	G	Immediately analyze	0.5 hr/stat
Winkler	G	"Fix" Immediately	8 hr/8 hr
pH	P, G	Immediately analyze	2 hr/stat
Phosphate	G(A)	Immediately refrigerate	48 hr/NR
Salinity	G, wax seal	Immediately analyze	6 mos/NR
Temperature	P, G	Immediately analyze	stat/stat
Turbidity	P, G	Analyze same day or store in dark up to 24 hr, refrigerate	24 hr/48 hr

(A)—glass rinsed with 1 + 1 HNO_3.

requires some form of standardization of the methods. The American Public Health Association (APHA) recognized this requirement when in 1899 the Association appointed a committee to draw up standard procedures for the analysis of water. The report (published in 1905) constituted the first edition of what is now known as Standard Methods for the Examination of Water and Wastewater or Standard Methods. This book is now in its twenty-third edition and serves as the primary reference for water testing methods, and as the basis for most USEPA-approved methods.

TEST METHODS FOR DRINKING WATER AND WASTEWATER

The material presented in this section is based on personal experience and adaptations from Standard Methods, Federal Register, and the Monitor's Handbook, LaMotte Company, Chestertown, Maryland, 1992. Descriptions of general methods to help you understand how each works in specific test kits follow. Always use the specific instructions included with the equipment and individual test kits. Most water analyses are conducted either by titrimetric analyses or by colorimetric analyses. Both methods are easy to use and provide accurate results.

TITRIMETRIC METHODS

Titrimetric analyses are based on adding a solution of known strength (the titrant, which must have an exact known concentration) to a specific volume of a treated sample in the presence of an indicator. The indicator produces a color change indicating the reaction is complete. Titrants are generally added by a titrator (microburette) or a precise glass pipette.

Colorimetric Methods

Colorimetric standards are prepared as a series of solutions with increasing known concentrations of the constituent to be analyzed. Two basic types of colorimetric tests are commonly used:

1. The pH is a measure of the concentration of hydrogen ions (the acidity of a solution) determined by the reaction of an indicator that varies in color, depending on the hydrogen ion levels in the water.
2. Tests that determine a concentration of an element or compound are based on Beer's Law. Simply, this law states that the higher the concentration of a substance, the darker the color produced in the test reaction, and therefore the more light absorbed. Assuming a constant view path, the absorption increases exponentially with concentration.

Visual Methods

The Octet Comparator uses standards that are mounted in a plastic comparator block. It employs eight permanent translucent color standards and built-in filters to eliminate optical distortion. The sample is compared using either of two viewing windows. Two devices that can be used with the comparator are the B-color Reader, which neutralizes color or turbidity in water samples, and view path, which intensifies faint colors of low concentrations for easy distinction.

Electronic Methods

Although the human eye is capable of differentiating color intensity, interpretation is quite subjective. Electronic colorimeters consist of a light source that passes through a sample and is measured on a photodetector with an analog or digital readout. Besides electronic colorimeters, specific electronic instruments are manufactured for lab and field determination of many water quality factors, including pH, total dissolved solids (TDS)/conductivity, dissolved oxygen, temperature, and turbidity.

Dissolved Oxygen Testing

Wastewater from sewage treatment plants often contains organic materials that are decomposed by microorganisms, which use oxygen in the process. (The amount of oxygen consumed by these organisms in breaking down the waste is known as the biochemical oxygen demand (BOD). We include a discussion of BOD and how to monitor it later). Other sources of oxygen-consuming waste include stormwater runoff from farmland or urban streets, feedlots, and failing septic systems.

TABLE 11.4
Maximum DO Concentrations vs. Temperature Variations

Temperature °C	DO (mg/L)	Temperature °C	DO (mg/L)
0	14.60	23	8.56
1	14.19	24	8.40
2	13.81	25	8.24
3	13.44	26	8.09
4	13.09	27	7.95
5	12.75	28	7.81
6	12.43	29	7.67
7	12.12	30	7.54
8	11.83	31	7.41
9	11.55	32	7.28
10	11.27	33	7.16
11	11.01	34	7.0512
12	10.76	35	6.93
13	10.52	36	6.82
14	10.29	37	6.71
15	10.07	38	6.61
16	9.85	39	6.51
17	9.65	40	6.41
18	9.45	41	6.31
19	9.26	42	6.22
20	9.07	43	6.13
21	8.90	44	6.04
22	8.72	45	5.95

Oxygen is measured in its dissolved form as dissolved oxygen (DO). If more oxygen is consumed than produced, dissolved oxygen levels decline and some sensitive animals may move away, weaken, or die. DO levels fluctuate over a 24-hour period and seasonally. They vary with water temperature and altitude. Cold water holds more oxygen than warm water (see Table 11.4), and water holds less oxygen at higher altitudes. Thermal discharges (such as water used to cool machinery in a manufacturing plant or a power plant) raise the temperature of water and lower its oxygen content. Aquatic animals are most vulnerable to lowered DO levels in the early morning on hot summer days when stream flows are low, water temperatures are high, and aquatic plants have not been producing oxygen since sunset.

Sampling and Equipment Considerations

In contrast to lakes, where DO levels are most likely to vary vertically in the water column, changes in DO in rivers and streams move horizontally along the course of the waterway. This is especially true in smaller, shallow streams.

In larger, deeper rivers, some vertical stratification of dissolved oxygen might occur. The DO levels in and below riffle areas, waterfalls, or dam spillways are typically higher than those in pools and slower-moving stretches. If you wanted to measure the effect of a dam, sampling for DO behind the dam, immediately below the spillway, and upstream of the dam would be important. Because DO levels are critical to fish, a good place to sample is in the pools that fish tend to favor, or in the spawning areas they use.

An hourly time profile of DO levels at a sampling site is a valuable set of data, because it shows the change in DO levels from the low point (just before sunrise) to the high point (sometime near midday). However, this might not be practical for a volunteer monitoring program. Note the time of your DO sampling to help judge when in the daily cycle the data were collected.

Dissolved oxygen is measured either in milligrams per liter (mg/L) or "percent saturation." Milligrams per liter are the amount or oxygen in a liter of water. Percent saturation is the amount of oxygen in a liter of water relative to the total amount of oxygen that the water can hold at that temperature. DO samples are collected using a special BOD bottle: a glass bottle with a "turtleneck" and a ground stopper. You can fill the bottle directly in the stream if the stream is wadeable or boatable, or you can use a sampler dropped from a bridge or boat into water deep enough to submerse it. Samplers can be made or purchased.

Winkler Method (Azide Modification)

The Winkler method (azide modification) involves filling a sample bottle completely with water (no air is left to bias the test). The dissolved oxygen is then "fixed" using a series of reagents that form a titrated acid compound. Titration involves the drop-by-drop addition of a reagent that neutralizes the acid compound, causing a change in the color of the solution. The point at which the color changes is the "endpoint" and is equivalent to the amount of oxygen dissolved in the sample. The sample is usually fixed and titrated in the field at the sample site. Preparing the sample in the field and delivering it to a lab for titration is possible. The azide modification method is best suited for relatively clean waters; otherwise, substances such as color, organics, suspended solids, sulfide, chlorine, and ferrous and ferric iron can interfere with test results. If fresh azide is used, nitrite will not interfere with the test.

In testing, iodine is released in proportion to the amount of DO present in the sample. By using sodium thiosulfate with starch as the indicator, the sample can be titrated to determine the amount of DO present. The chemicals used include:

1. Manganese sulfate solution
2. Alkaline azide-iodide solution
3. Sulfuric acid—concentrated
4. Starch indicator
5. Sodium thiosulfate solution 0.025 N, or phenylarsine solution 0.025 N, or potassium biniodate solution 0.025 N
6. Distilled or deionized water

The equipment used includes:

1. Burette, graduated to 0.1 mL
2. Burette stand
3. 300 mL BOD bottles
4. 500 mL Erlenmeyer flasks
5. 1.0 mL pipets with elongated tips
6. Pipet bulb
7. 250 mL graduated cylinder
8. Laboratory-grade water rinse bottle
9. Magnetic stirrer and stir bars (optional)

Procedure

The procedure for the Winkler method is as follows:

1. Collect sample in a 300 mL BOD bottle.
2. Add 1 mL manganous sulfate solution at the surface of the liquid.
3. Add 1 mL alkaline-iodide-azide solution at the surface of the liquid.
4. Insert stoppers and mix by inverting the bottle.
5. Allow the floc to settle halfway in the bottle, remix, and allow to settle again.
6. Add 1 mL concentrated sulfuric acid at the surface of the liquid.
7. Recap bottle, rinse top with laboratory-grade water, and mix until precipitate is dissolved.
8. The liquid in the bottle should appear clear and have an amber color.
9. Measure 201 mL from the BOD bottle into an Erlenmeyer flask.
10. Titrate with 0.025 N PAO or thiosulfate to a pale-yellow color and note the amount of titrant.
11. Add 1 mL of starch indicator solution.
12. Titrate until blue color first disappears.
13. Record total amount of titrant.

Calculation

To calculate the DO concentration when the modified Winkler titration method is used:

$$\text{DO, mg/L} = \frac{\left(\text{Buret}_{\text{Final}}\text{, mL} - \text{Buret}_{\text{Start}}\text{, mL}\right) \times \text{N} \times 8{,}000}{\text{Sample Volume, mL}} \tag{11.1}$$

Note: Using a 200 mL sample and a 0.025 N (N = Normality of the solution used to titrate the sample) titrant reduces this calculation to

$$\text{DO, mg/L} = \text{mL Titrant Used}$$

Example 11.1

PROBLEM:

The operator titrates a 200 mL DO sample. The burette reading at the start of the titration was 0.0 mL. At the end of the titration, the burette read 7.1 mL. The concentration of the titrating solution was 0.025 N. What is the DO concentration in mg/L?

SOLUTION:

$$\text{DO, mg/L} = \frac{(7.1\,\text{mL} - 0.0\,\text{mL}) \times 0.025 \times 8{,}000}{200\,\text{mL}} = 7.1\,\text{mL}$$

Dissolved oxygen-field kits using the Winkler method are relatively inexpensive, especially compared to a meter and probe. Field kits run between $35 and $200, and each kit comes with enough reagents to run 50–100 DO tests. Replacement reagents are inexpensive, and you can buy them already measured out for each test in plastic pillows. You can also purchase the reagents in larger quantities in bottles and measure them out with a volumetric scoop. The pillows' advantage is that they have a longer shelf life and are much less prone to contamination or spillage. Buying larger quantities in bottles has the advantage of considerably lower cost per test.

The major factor in the expense for the kits is the method of titration used—eyedropper, syringe-type titrator. Eyedropper and syringe-type titration is less precise than digital titration, because a larger drop of titrant is allowed to pass through the dropper opening, and on a micro-scale, the drop size (and thus volume of titrant) can vary from drip to drop. A digital titrator or a burette (a long glass tube with a tapered tip like a pipette) permits much more precision and uniformity for titrant it allows to pass.

If a high degree of accuracy and precision in DO results is required, a digital titrator should be used. A kit that uses an eyedropper-type or syringe-type titrator is suitable for most other purposes. The lower cost of this type of DO field kit might be attractive if several teams of samplers and testers at multiple sites at the same time are relied on.

Meter and Probe

A dissolved oxygen meter is an electronic device that converts signals from a probe placed in the water into units of DO in mg/L. Most meters and probes also measure temperature. The probe is filled with a salt solution and has a selectively permeable membrane that allows DO to pass from the stream water into the salt solution. The DO that has diffused into the salt solution changes the electric potential of the salt solution, and this change is sent by electric cable to the meter, which converts the signal to milligrams per liter on a scale that the user can read.

Methodology

If samples are to be collected for analysis in the laboratory, a special APHA sampler, or the equivalent must be used. This is the case because, if the sample is exposed or mixed with air during collection, test results can change dramatically. Therefore, the sampling device must allow collection of a sample that is not mixed with atmospheric air and allows for at least 3X bottle overflow (see Figure 11.12). Again, because the DO level in a sample can change quickly, only grab samples should be used for dissolved oxygen testing. Samples must be tested immediately (within 15 minutes) after collection.

Note: Samples collected for analysis using the modified Winkler titration method may be preserved for up to 8 hours by adding 0.7 mL of concentrated sulfuric acid

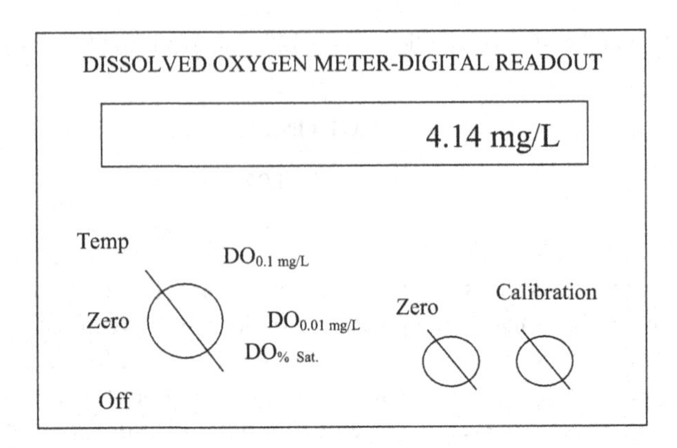

FIGURE 11.13 Dissolved oxygen meter.

or by adding all the chemicals required by the procedure. Samples collected from the aeration tank of the activated sludge process must be preserved using a solution of copper sulfate-sulfamic acid to inhibit biological activity.

The advantage of using the DO oxygen meter method is that the meter can be used to determine DO concentration directly (see Figure 11.13). In the field, a direct reading can be obtained using a probe (see Figure 11.14) or by collection of samples for testing in the laboratory using a laboratory probe (see Figure 11.15).

Note: The field probe can be used for laboratory work by placing a stirrer in the bottom of the sample bottle, but the laboratory probe should never be used in any situation where the entire probe might be submerged.

The probe used in the determination of DO consists of two electrodes, a membrane and a membrane filling solution. Oxygen passes through the membrane into the filling solution and causes a change in the electrical current passing between the two electrodes. The change is measured and displayed as the concentration of DO. In order to be accurate, the probe membrane must be in proper operating condition, and the meter must be calibrated before use. The only chemical used in the DO meter method during normal operation is the electrode filling solution. However, in the Winkler DO method, chemicals are required for meter calibration.

Calibration prior to use is important. Both the meter and the probe must be calibrated to ensure accurate results. The frequency of calibration is dependent on the frequency of use. For example, if the meter is used once a day, then calibration should be performed before use. There are three methods available for calibration: saturated water, saturated air, and the Winkler method. It is important to note that if the Winkler method is not used for routine calibration method, periodic checks using this method are recommended.

Procedure

It is important to keep in mind that the meter and probe supplier's operating procedures should always be followed. Normally, the manufacturer's recommended procedure will include the following generalized steps:

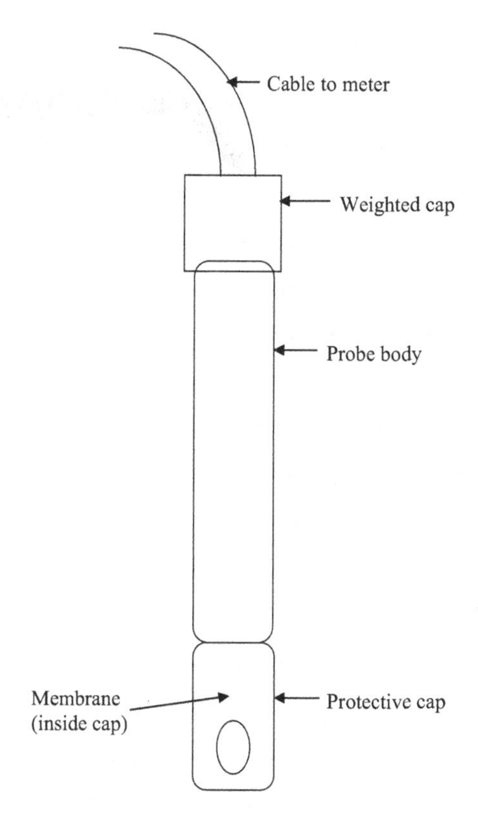

FIGURE 11.14 Dissolved oxygen-field probe.

1. Turn DO meter on and allow 15 minutes for it to warm up.
2. Turn meter switch to zero and adjust as needed.
3. Calibrate meter using the saturated air, saturated water, or Winkler azide procedure for calibration.
4. Collect sample in 300 mL bottle, or place field electrode directly in stream.
5. Place laboratory electrode in BOD bottle without trapping air against membrane, and turn on stirrer.
6. Turn meter switch to temperature setting, and measure temperature.
7. Turn meter switch to DO mode and allow 10 seconds for meter reading to stabilize.
8. Read DO mg/L from meter and record the results.

No calculation is necessary using this method because results are read directly from the meter.

Dissolved oxygen meters are expensive compared to field kits that use the titration method. Meter/probe combinations run between $500 and $1200, including a long cable to connect the probe to the meter. The advantage of a meter/probe is that DO, and temperature can be quickly read at any point where the probe is inserted

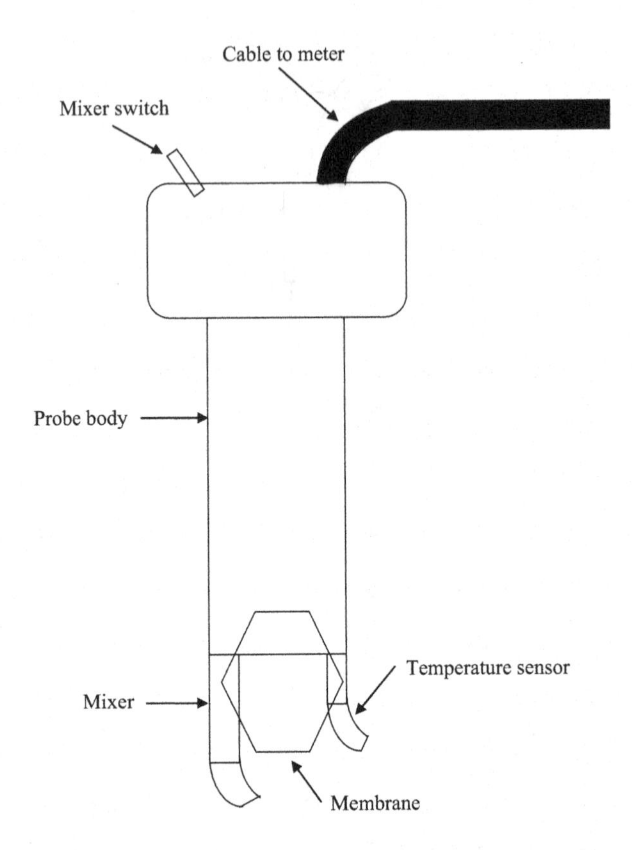

FIGURE 11.15 Dissolved oxygen-lab probe.

into the stream. DO levels can be measured at a certain point on a continuous basis. The results are read directly as milligrams per liter, unlike the titration methods, in which the final titration result might have to be converted by an equation to milligrams per liter. However, DO meters are more fragile than field kits, and repairs to a damaged meter can be costly. The meter/probe must be carefully maintained, and must be calibrated before each sample run, and if many tests are done, between sampling. Because of the expense, a small water/wastewater facility might only have one meter/probe, which means that only one team of samplers can sample DO, and they must test all the sites. With field kits, on the other hand, several teams can sample simultaneously.

Biological Oxygen Demand Testing

Biochemical oxygen demand (BOD) measures the amount of oxygen consumed by microorganisms in decomposing organic matter in stream water. BOD also measures the chemical oxidation of inorganic matter (the extraction of oxygen from water via chemical reaction). A test is used to measure the amount of oxygen consumed by these organisms during a specified period of time (usually five days at 20°C). The rate of oxygen consumption in a stream is affected by a number of variables: temperature, pH, the presence of certain kinds of microorganisms, and the type of organic and inorganic material in the water. BOD directly affects the amount of dissolved oxygen in water bodies. The greater the BOD, the more rapidly oxygen is depleted in the water body, leaving less oxygen available to higher forms of aquatic life. The consequences of high BOD are the same as those for low dissolved oxygen: aquatic organisms become stressed, suffocate, and die. Most river waters used as water supplies have a BOD less than 7 mg/L; therefore, dilution is not necessary.

Sources of BOD include leaves and wood debris; dead plants and animals; animal manure; effluents from pulp and paper mills, wastewater treatment plants, feedlots, and food-processing plants; failing septic systems; and urban stormwater runoff.

Note: To evaluate raw water's potential for use as a drinking water supply, it is usually sampled, analyzed, and tested for biochemical oxygen demand when turbid, polluted water is the only source available.

Sampling Considerations

Biochemical oxygen demand is affected by the same factors that affect dissolved oxygen. Aeration of stream water—by rapids and waterfalls, for example—will accelerate the decomposition of organic and inorganic material. Therefore, BOD levels at a sampling site with slower, deeper waters might be higher for a given column of organic and inorganic material than the levels for a similar site in high aerated waters. Chlorine can also affect BOD measurement by inhibiting or killing the microorganisms that decompose the organic and inorganic matter in a sample. If sampling in chlorinated waters (such as those below the effluent from a sewage treatment plant), neutralizing the chlorine with sodium thiosulfate is necessary (APHA, 1998).

Biochemical oxygen demand measurement requires taking two samples at each site. One is tested immediately for dissolved oxygen, and the second is incubated in the dark at 20°C for five days, and then tested for dissolved oxygen remaining. The difference in oxygen levels between the first test and the second test (in milligrams per liter (mg/L)) is the amount of BOD. This represents the amount of oxygen consumed by microorganisms and used to break down the organic matter present in the sample bottle during the incubation period. Because of the five-day incubation, the tests are conducted in a laboratory.

Sometimes by the end of the five-day incubation period, the dissolved oxygen level is zero. This is especially true for rivers and streams with a lot of organic pollution. Since knowing when the zero point was reached is not possible, determining the BOD level is also impossible. In this case, diluting the original sample by a factor that results in a final dissolved oxygen level of at least 2 mg/L is necessary. Special dilution water should be used for the dilutions (APHA, 1998).

Some experimentation is needed to determine the appropriate dilution factor for a particular sampling site. The result is the difference in dissolved oxygen between the first measurement and the second, after multiplying the second result by the dilution factor. Standard Methods prescribes all phases of procedures and calculations for BOD determination. A BOD test is not required for monitoring water supplies.

BOD Sampling, Analysis, and Testing

The approved biochemical oxygen demand sampling and analysis procedure measures the DO depletion (biological oxidation of organic matter in the sample) over a five-day period under controlled conditions (20°C in the dark). The test is performed using a specified incubation time and temperature. Test results are used to determine plant loadings, plant efficiency, and compliance with NPDES effluent limitations. The duration of the test (five days) makes it difficult to use the data effectively for process control.

The standard BOD test does not differentiate between oxygen used to oxidize organic matter and that used to oxidize organic and ammonia nitrogen to more stable forms. Because many biological treatment plants now control treatment processes to achieve oxidation of the nitrogen compounds, there is a possibility that BOD test results for plant effluent and some process samples may produce BOD test results based on both carbon and nitrogen oxidation. To avoid this situation, a nitrification inhibitor can be added. When this is done, the test results are known as carbonaceous BOD (CBOD). A second uninhibited BOD should also be run whenever CBOD is determined.

When taking a BOD sample, no special sampling container is required. Either a grab or composite sample can be used. BOD_5 samples can be preserved by refrigeration at or below 4°C (not frozen)—composite samples must be refrigerated during collection. Maximum holding time for preserved samples is 48 hours.

Using the incubation of dissolved approved test method, a sample is mixed with dilution water in several different concentrations (dilutions). The dilution water contains nutrients and materials to provide optimum environment. The chemicals used are dissolved oxygen, ferric chloride, magnesium sulfate, calcium chloride, phosphate buffer, and ammonium chloride.

Note: Remember all chemicals can be dangerous if not used properly and in accordance with the recommended procedures. Review appropriate sections of the individual chemical materials safety data sheet (MSDS) to determine proper methods for handling and for safety precautions that should be taken.

Sometimes it is necessary to add (seed) healthy organisms to the sample. The DO of the dilution and the dilution water is determined. If seed material is used, a series of dilutions of seed material must also be prepared. The dilutions and dilution blanks are incubated in the dark for five days at $20°C \pm 1°C$. At the end of five days, the DO of each dilution and the dilution blanks are determined. For the test results to be valid, certain criteria must be met:

1. Dilution water blank DO change must be ≤ 0.2 mg/L.
2. Initial DO must be >7.0 mg/L but ≤ 9.0 mg/L (or saturation at 20°C and test elevation).
3. Sample dilution DO depletion must be ≥ 2.0 mg/L.
4. Sample dilution residual DO must be ≥ 1.0 mg/L.
5. Sample dilution initial DO must be ≥ 7.0 mg/L.
6. Seed correction should be ≥ 0.6 but ≤ 1.0 mg/L.

TABLE 11.5
BOD_5 Test Procedure

(1)	Fill two bottles with BOD dilution water; insert stoppers.
(2)	Place sample in two BOD bottles; fill with dilution water; insert stoppers.
(3)	Test for dissolved oxygen (DO).
(4)	Incubate for 5 days.
(5)	Test for DO.
(6)	Add 1 mL $MnSO_4$ below surface.
(7)	Add 1 mL alkaline Kl below surface.
(8)	Add 1 mL H_2SO_4.
(9)	Transfer 203 mL to flask.
(10)	Titrate with PAO or thiosulfate.

The BOD_5 test procedure consists of ten steps (for unchlorinated water) as shown in Table 11.5.

Note: BOD_5 is calculated individually for all sample dilutions that meet the criteria. Reported result is the average of the BOD_5 of each valid sample dilution.

BOD_5 Calculation

Unlike the direct reading instrument used in the DO analysis, BOD results require calculation. There are several criteria used in selecting which BOD_5 dilutions should be used for calculating test results. Consult a laboratory testing reference manual such as Standard Methods (APHA 1998), for this information. Currently, there are two basic calculations for BOD_5. The first is used for samples that have not been seeded. The second must be used whenever BOD_5 samples are seeded. In this section, we illustrate the calculation procedure for unseeded samples.

$$BOD_5\left(\text{Unseeded}\right) = \frac{\left(DO_{\text{start}},\ \text{mg/L} - DO_{\text{final}},\ \text{mg/L}\right) \times 300\ \text{mL}}{\text{Sample Volume, mL}} \tag{11.2}$$

Example 11.2

PROBLEM:

The BOD_5 test is completed. Bottle 1 of the test had a DO of 7.1 mg/L at the start of the test. After 5 days, bottle 1 had a DO of 2.9 mg/L. Bottle 1 contained 120 mg/L of sample. What is BOD5 (unseeded)?

SOLUTION:

$$BOD_5\left(\text{Unseeded}\right) = \frac{\left(7.1\ \text{mg/L} - 2.9\ \text{mg/L}\right) \times 300\ \text{mL}}{120\ \text{mL}} = 10.5\ \text{mg/L}$$

If the BOD_5 sample has been exposed to conditions that could reduce the number of healthy, active organisms, the sample must be seeded with organisms. Seeding requires use of a correction factor to remove the BOD_5 contribution of the seed material:

$$\text{Seed Correction} = \frac{\text{Seed Material BOD}_5 \times \text{Seed in Dilution, mL}}{300\text{ mL}} \tag{11.3}$$

Temperature Measurement

As mentioned, an ideal water supply should have, at all times, an almost constant temperature or one with minimum variation. Knowing the temperature of the water supply is important because the rates of biological and chemical processes depend on it. Temperature affects the oxygen content of the water (oxygen levels become lower as temperature increases); the rate of photosynthesis by aquatic plants; the metabolic rates of aquatic organisms; and the sensitivity of organisms to toxic wastes, parasites, and diseases. Causes of temperature change include weather, removal of shading streambank vegetation, impoundments (a body of water confined by a barrier, such as a dam), and discharge of cooling water, urban stormwater, and groundwater inflows to the stream.

Sampling and Equipment Considerations

Temperature—for example, in a stream—varies with width and depth, and the temperature of well-sunned portions of a stream can be significantly higher than the shaded portion of the water on a sunny day. In a small stream, the temperature will be relatively constant as long as the stream is uniformly in sun or shade. In a large stream, temperature can vary considerably with width and depth, regardless of shade. If safe to do so, temperature measurements should be collected at varying depths and across the surface of the stream to obtain vertical and horizontal temperature profiles. This can be done at each site at least once to determine the necessity of collecting a profile during each sampling visit. Temperature should be measured at the same place every time.

Temperature is measured in the stream with a thermometer or a meter. Alcohol-filled thermometers are preferred over mercury-filled because they are less hazardous if broken. Armored thermometers for field use can withstand more abuse than unprotected glass thermometers and are worth the additional expense. Meters for other tests (such as pH (acidity) or dissolved oxygen) also measure temperature and can be used instead of a thermometer.

Hardness Measurement

Hardness refers primarily to the amount of calcium and magnesium in the water. Calcium and magnesium enter water mainly by leaching of rocks. Calcium is an important component of aquatic plant cell walls and the shells and bones of many aquatic organisms. Magnesium is an essential nutrient for plants and is a component of the chlorophyll molecule. Hardness test kits express test results in ppm of $CaCO_3$, but these results can be converted directly to calcium or magnesium concentrations:

$$\text{Calcium Hardness as ppm } CaCO_3 \times 0.40 = \text{ppm Ca} \tag{11.4}$$

$$\text{Magnesium Hardness as ppm } CaCO_3 \times 0.24 = \text{ppm Mg} \tag{11.5}$$

Note: Because of less contact with soil minerals and more contact with rain, surface raw water is usually softer than groundwater.

As a rule, when hardness is greater than 150 mg/L, softening treatment may be required for public water systems. Hardness determination via testing is required to ensure efficiency of treatment.

Note: Keep in mind that when measuring calcium hardness, the concentration of calcium is routinely measured separately from total hardness. Its concentration in waters can range from 0 to several thousand mg/L, as $CaCO_3$. Likewise, when measuring magnesium hardness, magnesium is routinely determined by subtracting calcium hardness from total hardness. There is usually less magnesium than calcium in natural water. Lime dosage for water softening operation is partly based on the concentration of magnesium hardness in the water.

In the hardness test, the sample must be carefully measured, and then a buffer is added to the sample to correct pH for the test, and an indicator to signal the titration end point. The indicator reagent is normally blue in a sample of pure water, but if calcium or magnesium ions are present in the sample, the indicator combines with them to form a red-colored complex. The titrant in this test is EDTA (ethylenediaminetetraacetic acid, used with its salts in the titration method), a "chelant" which actually "pulls" the calcium and magnesium ions away from a red-colored complex. The EDTA is added dropwise to the sample until all the calcium and magnesium ions have been "chelated" away from the complex and the indicator returns to its normal blue color. The amount of EDTA required to cause the color change is a direct indication of the amount of calcium and magnesium ions in the sample.

Some hardness kits include an additional indicator that is specific for calcium. This type of kit will provide three readings: total hardness, calcium hardness, and magnesium hardness. For interference, precision, and accuracy, consult the latest edition of Standard Methods.

pH Measurement

pH is defined as the negative log of the hydrogen ion concentration of the solution. This is a measure of the ionized hydrogen in solution. Simply, it is the relative acidity or basicity of the solution. The chemical and physical properties, and the reactivity of almost every component in water are dependent upon pH. It relates to corrosivity, contaminant solubility, and the water's conductance, and has a secondary MCL range set at 6.5–8.5.

Analytical and Equipment Considerations

The pH can be analyzed in the field or in the lab. If analyzed in the lab, it must be measured within two hours of the sample collection, because the pH will change from the carbon dioxide from the air as it dissolves in the water, bringing the pH toward 7. If your program requires a high degree of accuracy and precision in pH results, the pH should be measured with a laboratory-quality pH meter and electrode. Meters of this quality range in cost from around $250 to $1000. Color comparators and pH "pocket pals" are suitable for most other purposes. The cost of either of these is in the $50 range. The lower cost of the alternatives might be attractive if multiple samplers are used to sample several sites at the same time.

pH Meters

A pH meter measures the electric potential (millivolts) across an electrode when immersed in water. This electric potential is a function of the hydrogen ion activity in the sample; therefore, pH meters can display results in either millivolts (mV) or pH units. A pH meter consists of a potentiometer, which measures electric potential where it meets the water sample; a reference electrode, which provides a constant electric potential; and a temperature compensating device, which adjusts the readings according to the temperature of the sample (since pH varies with temperature). The reference and glass electrodes are frequently combined into a single probe called a combination electrode. A wide variety of meters are available, but the most important part of the pH meter is the electrode. Thus, purchasing a good, reliable electrode and following the manufacturer's instructions for proper maintenance is important. Infrequently used or improperly maintained electrodes are subject to corrosion, which makes them highly inaccurate.

pH "Pocket Pals" and Color Comparators

pH "pocket pals" are electronic handheld "pens" that are dipped in the water, providing a digital readout of the pH. They can be calibrated to only one pH buffer. (Lab meters, on the other hand, can be calibrated to two or more buffer solutions and thus are more accurate over a wide range of pH measurements.) Color comparators involve adding a reagent to the sample that colors the sample water. The intensity of the color is proportional to the pH of the sample, then matched against a standard color chart. The color chart equates particular colors to associated pH values, which can be determined by matching the colors from the chart to the color of the sample. For instructions on how to collect and analyze samples, refer to Standard Methods (APHA, 1998).

Turbidity Measurement

Turbidity is a measure of water clarity—how much the material suspended in water decreases the passage of light through the water. Turbidity consists of suspended particles in the water and may be caused by a number of materials, organic and inorganic. These particles are typically in the size range of 0.004 mm (clay) to 1.0 mm (sand). The occurrence of turbid source waters may be permanent or temporary. It can affect the color of the water. Higher turbidity increases water temperatures because suspended particles absorb more heat. This in turn reduces the concentration of dissolved oxygen (DO) because warm water holds less DO than cold. Higher turbidity also reduces the amount of light penetrating the water, which reduces photosynthesis and the production of DO. Suspended materials can clog fish gills, reducing resistance to disease in fish, lowering growth rates, and affecting egg and larval development. As the particles settle, they can blanket the stream bottom (especially in slower waters) and smother fish eggs and benthic macroinvertebrates.

Turbidity also affects treatment plant operations. For example, turbidity hinders disinfection by shielding microbes, some of them pathogens, from the disinfectant. Obviously, this is the most important significance of turbidity monitoring; the test for it is an indication of the effectiveness of filtration of water supplies. It is important to note that turbidity removal is the principal reason for chemical addition, settling, coagulation, settling, and filtration in potable water treatment. Sources of turbidity include:

1. Soil erosion
2. Waste discharge
3. Urban runoff
4. Eroding stream banks
5. Large numbers of bottom feeders (such as carp), which stir up bottom sediments
6. Excessive algal growth

Sampling and Equipment Considerations

Turbidity can be useful as an indicator of the effects of runoff from construction, agricultural practices, logging activity, discharges, and other sources. Turbidity often increases sharply during rainfall, especially in developed watersheds, which typically have relatively high proportions of impervious surfaces. The flow of stormwater runoff from impervious surfaces rapidly increases stream velocity, which increases the erosion rates of streambanks and channels. Turbidity can also rise sharply during dry weather if Earth-disturbing activities occur in or near a stream without erosion control practices in place.

Regular monitoring of turbidity can help detect trends that might indicate increasing erosion in developing watersheds. However, turbidity is closely related to stream flow and velocity and should be correlated with these factors. Comparisons of the change in turbidity over time, therefore, should be made at the same point at the same flow.

Keep in mind that turbidity is not a measurement of the amount of suspended solids present, or the rate of sedimentation of a stream, because it measures only the amount of light that is scattered by suspended particles. Measurement of total solids is a more direct measurement of the amount of material suspended and dissolved in water.

Turbidity is generally measured by using a turbidity meter or tubidimeter. The turbidimeter is a modern nephelometer, which originally was a box containing a light bulb which directed light at a sample. The amount of light scattered at right angles by the turbidity particles was measured, as a measure of the turbidity in the sample, and registered as nephelometric turbidity units (NTU). The tubidimeter uses a photoelectric cell to register the scattered light on an analog or digital scale, and the instrument is calibrated with permanent turbidity standards composed of the colloidal substance, formazin. Meters can measure turbidity over a wide range—from 0 to 1,000 NTUs. A clear mountain stream might have a turbidity of around 1 NTU, whereas a large river like the Mississippi might have a dry-weather turbidity of 10 NTUs. Because these values can jump into hundreds of NTUs during runoff events, the turbidity meter to be used should be reliable over the range in which you will be working. Meters of this quality cost about $800. Many meters in this price range are designed for field or lab use (USEPA, 2000).

An operator may also take samples to a lab for analysis. Another approach, discussed previously, is to measure transparency (an integrated measure of light scattering and absorption) instead of turbidity. Water clarity/transparency can be measured using a Secchi disk (see Figure 11.11) or transparency tube. The Secchi disk can only be used in deep, slow-moving rivers; the transparency tube (a comparatively new development) is gaining acceptance in and around the country but is not in wide use.

Using a Secchi Disk

A Secchi disk is a black-and-white disk that is lowered by hand into the water to the depth at which it vanishes from sight (see Figure 11.11). The distance to vanishing is then recorded—the clearer the water, the greater the distance. Secchi disks are simple to use and inexpensive. For river monitoring they have limited use, however, because in most cases the river bottom will be visible, and the disk will not reach a vanishing point. Deeper, slower-moving rivers are the most appropriate places for Secchi disk measurement; although the current might require that the disk be extra-weighted, so it does not sway and make measurement difficult. Secchi disks cost about $50 but can be homemade.

The line attached to the Secchi disk must be marked in waterproof ink according to units designated by the sampling program. Many programs require samplers to measure to the nearest 1/10 meter. Meter intervals can be tagged (e.g., with duct tape) for ease of use. To measure water clarity with a Secchi disk:

1. Check to make sure that the Secchi disk is securely attached to the measured line.
2. Lean over the side of the boat and lower the Secchi disk into the water, keeping your back to the sun to block glare.
3. Lower the disk until it disappears from view. Lower it one third of a meter and then slowly raise the disk until it just reappears. Move the disk up and down until you find the exact vanishing point.
4. Attach a clothespin to the line at the point where the line enters the water. Record the measurement on your data sheet. Repeating the measurement provides you with a quality control check.

The key to consistent results is to train samplers to follow standard sampling procedures, and if possible, have the same individual take the reading at the same site throughout the season.

Transparency Tube

Pioneered by Australia's Department of Conservation, the transparency tube is a clear, narrow plastic tube marked in units with a dark pattern painted on the bottom. Water is poured into the tube until the pattern painted on the bottom disappears. Water is poured into the tube until the pattern disappears. Some U.S. volunteer monitoring programs (e.g., the Tennessee Valley Authority (TWA) Clean Water Initiative and the Minnesota Pollution Control Agency (MPCA)) are testing the transparency tube in streams and rivers. MPCA uses tubes marked in centimeters and has found tube readings to relate fairly well to lab measurements of turbidity and total suspended solids, although it does not recommend the transparency tube for applications where precise and accurate measurement is required, or in highly colored waters. The TVA and MPCA recommended the following sampling considerations:

1. Collect the sample in a bottle or bucket in mid-stream and at mid-depth if possible. Avoid stagnant water and sample as far from the shoreline as is safe. Avoid collecting sediment from the bottom of the stream.
2. Face upstream as you fill the bottle or bucket.
3. Take readings in open but shaded conditions. Avoid direct sunlight by turning your back to the sun.
4. Carefully stir or swish the water in the bucket or bottle until it is homogeneous, taking care not a produce air bubbles (these scatter light and affect the measurement). Then pour the water slowly in the tube while looking down the tube. Measure the depth of the water column in the tube at the point where the symbol just disappears.

Orthophosphate Measurement

Earlier we discussed the nutrients phosphorus and nitrogen. Both phosphorus and nitrogen are essential nutrients for the plants and animals that make up the aquatic food web. Because phosphorus is the nutrient in short supply in most freshwater systems, even a modest increase in phosphorus can (under the right conditions) set off a whole chain of undesirable events in a stream, including accelerated plant growth, algae blooms, low dissolved oxygen, and the death of certain fish, invertebrates, and other aquatic animals. Phosphorus comes from many sources, both natural and human. These include soil and rocks, wastewater treatment plants, runoff from fertilized lawns and cropland, failing septic systems, runoff from animal manure storage areas, disturbed land areas, drained wetlands, water treatment, and commercial cleaning preparations.

Forms of Phosphorus

Phosphorus has a complicated story. Pure, elemental phosphorus (P) is rare. In nature, phosphorus usually exists as part of a phosphate molecule (PO4). Phosphorus in aquatic systems occurs as organic phosphate and inorganic phosphate. Organic phosphate consists of a phosphate molecule associated with a carbon-based molecule, as in plant or animal tissue. Phosphate that is not associated with organic material is inorganic, the form required by plants. Animals can use either organic or inorganic phosphate. Both organic and inorganic phosphate can either be dissolved in the water or suspended (attached to particles in the water column).

The Phosphorus Cycle

Phosphorus cycles through the environment, changing form as it does so. Aquatic plants take in dissolved inorganic phosphorus, as it becomes part of their tissues. Animals get the organic phosphorus they need by eating either aquatic plants, other animals, or decomposing plant and animal material. In water bodies, as plants and animals excrete wastes or die, the organic phosphorus they contain sinks to the bottom, where bacterial decomposition converts it back to inorganic phosphorus, both dissolved and attached to particles. This inorganic phosphorus gets back into the water column when animals, human activity, interactions, or water currents stir up the bottom. Then plants take it up and the cycle begins again. In a stream system, the phosphorus cycle tends to move phosphorus downstream as the current carries decomposing plant and animal tissue and dissolved phosphorus. It becomes stationary only when it is taken up by plants or is bound to particles that settle to the bottom of ponds.

In the field of water quality chemistry, phosphorus is described by several terms. Some of these terms are chemistry based (referring to chemically based compounds), and others are methods based (they describe what is measured by a particular method). The term "orthophosphate" is a chemistry-based term that refers to the phosphate molecule all by itself. More specifically, orthophosphate is simple phosphate, or reactive phosphate, i.e., Na_3PO_4, sodium phosphate (tribasic), NaH_2PO_4, and sodium phosphate (monobasic). Orthophosphate is the only form of phosphate that can be directly tested for in the laboratory and is the form that bacteria use directly for metabolic processes. "Reactive phosphorus" is a corresponding method-based term that describes what is actually being measured when the test for orthophosphate is being performed. Because the lab procedure isn't quite perfect, mostly orthophosphate is obtained along with a small fraction of some other forms. More complex inorganic phosphate compounds are referred to as condensed phosphates or polyphosphates. The method-based term for these forms is acid hydrolyzable.

Testing Phosphorus

Testing phosphorus is challenging because it involves measuring very low concentrations, down to 0.01 mg/L or even lower. Even such very low concentrations of phosphorus can have a dramatic impact on streams. Less sensitive methods should be used only to identify serious problem areas. While many tests for phosphorus exist, only four are likely to be performed by most samplers. The total orthophosphate test is largely a measure of orthophosphate. Because the sample is not filtered, the procedure measures both dissolved and suspended orthophosphate. The USEPA-approved method for measuring is known as the ascorbic acid method. Briefly, a reagent (either liquid or powder) containing ascorbic acid and ammonium molybdate reacts with orthophosphate in the sample to form a blue compound. The intensity of the blue color is directly proportional to the amount of orthophosphate in the water.

The total phosphate test measures all the forms of phosphorus in the sample (orthophosphate, condensed phosphate, and organic phosphate) by first "digesting" (heating and acidifying) the sample to convert all the other forms to orthophosphate, then measuring the orthophosphate by the ascorbic acid method. Because the sample is not filtered, the procedure measures both dissolved and suspended orthophosphate. The dissolved phosphorus test measures that fraction of the total phosphorus that is in solution in the water (as opposed to being attached to suspended particles). It is determined by first filtering the sample, then analyzing the filtered sample for total phosphorus. Insoluble phosphorus is calculated by subtracting the dissolved phosphorus result from the total phosphorus result.

All these tests have one thing in common—they all depend on measuring orthophosphate. The total orthophosphate test measures the orthophosphate that is already present in the sample. The others measure that which is already present and that which is formed when the other forms of phosphorus are converted to orthophosphate by digestion. Monitoring phosphorus involves two basic steps:

1. Collecting a water sample
2. Analyzing it in the field or lab for one of the types of phosphorus described above

Sampling and Equipment Considerations

Sample containers made of either some form of plastic or of Pyrex® glass are acceptable to the USEPA. Because phosphorus molecules tend to "absorb" (attach) to the inside surface of sample containers, if containers are to be reused, they must be acid-washed to remove absorbed phosphorus. The container must be able to withstand repeated contact with hydrochloric acid. Plastic containers, either high-density polyethylene or polypropylene, might be preferable to glass from a practical standpoint because they are better able to withstand breakage. Some programs use disposable, sterile, plastic Whirl-pak® bags. The size of the container depends on the sample amount needed for the phosphorus analysis method chosen, and the amount needed for other analyses to be performed.

All containers that will hold water samples or come into contact with reagents used in the orthophosphate test must be dedicated. They should not be used for other tests, to eliminate the possibility that reagents containing phosphorus will contaminate the labware. All labware should be acid-washed.

The only form of phosphorus this text recommends for field analysis is total orthophosphate, which uses the ascorbic acid method on an untreated sample. Analysis of any of the other forms requires adding potentially hazardous reagents, heating the sample to boiling, and using too much time and too much equipment to be practical. In addition, analysis for other forms of phosphorus is prone to errors and inaccuracies in field situations. Pretreatment and analysis for these other forms should be handled in a laboratory.

Ascorbic Acid Method for Determining Orthophosphate

In the ascorbic acid method, a combined liquid or prepackaged powder reagent consisting of sulfuric acid, potassium antimonyl tartrate, ammonium molybdate, and ascorbic acid (or comparable compounds) is added to either 50 mL or 25 mL of the water sample. This colors the sample blue in direct proportion to the amount of orthophosphate in the sample. Absorbance or transmittance is then measured after 10 minutes, but before 30 minutes, using a color comparator with a scale in mg/L that increases with the increase in color hue, or an electronic meter that measures the amount of light absorbed or transmitted at a wavelength of 700–880 nm (again, depending on manufacturer's directions).

A color comparator may be useful for identifying heavily polluted sites with high concentrations (greater than 0.1 mg/L). However, matching the color of a treated sample to a comparator can be very subjective, especially at low concentrations, and lead to variable results.

A field spectrophotometer or colorimeter with a 2.5 cm light path and an infrared photocell (set for a wavelength of 700–880 nm) is recommended for accurate determination of low concentrations (0.2–0.02 mg/L). Use of a meter requires that a prepared known standard concentration be analyzed ahead of time to convert the absorbance readings of a stream sample to milligrams per liter, or that the meter reads directly in mg/L.

For information on how to prepare standard concentrations and on how to collect and analyze samples, refer to Standard Methods and USEPA's Methods for Chemical Analysis of Water and Wastes (2nd ed., 1991, Method 365.2).

Nitrates Measurement

Nitrates are a form of nitrogen found in several different forms in terrestrial and aquatic ecosystems. These forms of nitrogen include ammonia (NH_3), nitrates (NO_3), and nitrites (NO_2). Nitrates are essential plant nutrients, but excess amounts can cause significant water quality problems. Together with phosphorus, nitrates in excess amounts can accelerate eutrophication, causing dramatic increases in aquatic plant growth and changes in the types of plants and animals that live in the stream. This, in turn, affects dissolved oxygen, temperature, and other indicators. Excess nitrates can cause hypoxia (low levels of dissolved oxygen) and can become toxic to warm-blooded animals at higher concentrations (10 mg/L or higher) under certain conditions. The natural level of ammonia or nitrate in surface water is typically low (less than 1 mg/L); in the effluent of wastewater treatment plants, it can range up to 30 mg/L. Conventional potable water treatment plants cannot remove nitrate. High concentrations must be prevented by controlling the input at the source. Sources of nitrates include wastewater treatment plants, runoff from fertilized lawns and cropland, failing on-site septic systems, runoff from animal manure storage areas, and industrial discharges that contain corrosion inhibitors.

Sampling and Equipment Considerations

Nitrates from land sources end up in rivers and streams more quickly than other nutrients like phosphorus because they dissolve in water more readily than phosphorus, which has an attraction for soil particles. As a result, nitrates serve as a better indicator of the possibility of a source of sewage or manure pollution during dry weather. Water that is polluted with nitrogen-rich organic matter might show low nitrates. Decomposition of the organic matter lowers the dissolved oxygen level, which in turn slows the rate at which ammonia is oxidized to nitrite (NO_2) and then to nitrate (NO_3). Under such circumstances, monitoring for nitrites or ammonia (considerably more toxic to aquatic life than nitrate) might be also necessary. (See Standard Methods sections 4500-NH3 and 4500-NH2 for appropriate nitrite methods.) Water samples to be tested for nitrate should be collected in glass or polyethylene containers that have been prepared by using Method B (described previously). Two methods are typically used for nitrate testing: the cadmium reduction method and the nitrate electrode. The more commonly used cadmium reduction method produces a color reaction measured either by comparison to a color wheel or by use of a spectrophotometer. A few programs also use a nitrate electrode, which can measure in the range of 0–100 mg/L nitrate. A newer colorimetric immunoassay technique for nitrate screening is also now available.

Cadmium Reduction Method

In the cadmium reduction method, nitrate is reduced to nitrite by passing the sample through a column packed with activated cadmium. The sample is then measured quantitatively for nitrite.

More specifically, the cadmium reduction method is a colorimetric method that involves contact of the nitrate in the sample with cadmium particles, which cause nitrates to be converted to nitrites. The nitrites then react with another reagent to form a red color, in proportional intensity to the original amount of nitrate. The color is measured either by comparison to a color wheel with a scale in mg/L that increases with the increase in color hue or by use of an electronic spectrophotometer that measures the amount of light absorbed by the treated sample at a 543 nm wavelength. The absorbance value converts to the equivalent concentration of nitrate against a standard curve. Methods for making standard solutions and standard curves are presented in Standard Methods.

Before each sampling run, the sampling/monitoring supervisor should create this curve. The curve is developed by making a set of standard concentrations of nitrate, reacting them, and developing the corresponding color, then plotting the absorbance value for each concentration against concentration. A standard curve could also be generated for the color wheel. Use of the color wheel is appropriate only if nitrate concentrations are greater than 1 mg/L. For concentrations below 1 mg/L, use a spectrophotometer. Matching the color of a treated sample at low concentrations to a color wheel (or cubes) can be very subjective and can lead to variable results. Color comparators can, however, be effectively used to identify sites with high nitrates.

This method requires that the samples being treated are clear. If a sample is turbid, filter it through a 0.45 micron filter. Be sure to test to make sure the filter is nitrate-free. If copper, iron, or other metals are present in concentrations above several mg/L, the reaction with the cadmium will slow down and the reaction time must be increased.

The reagents used for this method are often prepackaged for different ranges, depending on the expected concentration of nitrate in the stream. Manufacturers, for example, provide reagents for the following ranges: low (0–0.40 mg/L), medium (0–15 mg/L), and high (0–30 mg/L). Determining the appropriate range for the stream being monitored is important.

Nitrate Electrode Method

A nitrate electrode (used with a meter) is similar in function to a dissolved oxygen meter. It consists of a probe with a sensor that measures nitrate activity in the water; this activity affects the electric potential of a solution in the probe. This change is then transmitted to the meter, which converts the electric signal to a scale that is read in millivolts; then the millivolts are converted to mg/L of nitrate by plotting them against a standard curve. The accuracy of the electrode can be affected by high concentrations of chloride or bicarbonate ions in the sample water. Fluctuating pH levels can also affect the meter reading.

Nitrate electrodes and meters are expensive compared to field kits that employ the cadmium reduction method. (The expense is comparable, however, if a spectrophotometer is used rather than a color wheel.) Meter/probe combinations run between $700 and $1200, including a long cable to connect the probe to the meter. If the program has a pH meter that displays readings in millivolts, it can be used with a nitrate probe and no separate nitrate meter is needed. Results are read directly as mg/L.

Although nitrate electrodes and spectrophotometers can be used in the field, they have certain disadvantages. These devices are more fragile than the color comparators and are therefore more at risk of breaking in the field. They must be carefully maintained and must be calibrated before each sample run, and if many tests are being run, between samplings. This means that samples are best tested in the lab. Note that samples to be tested with a nitrate electrode should be at room temperature, whereas color comparators can be used in the field with samples at any temperature.

SOLIDS MEASUREMENT

Solids in water are defined as any matter that remains as residue upon evaporation and drying at 103°C. They are separated into two classes: suspended solids and dissolved solids.

$$\text{Total Solids} = \underset{\text{(nonfilterable residue)}}{\text{Suspended Solids}} + \underset{\text{(filterable residue)}}{\text{Dissolved Solids}}$$

As shown above, total solids are dissolved solids plus suspended and settleable solids in water. In natural freshwater bodies, dissolved solids consist of calcium, chlorides, nitrate, phosphorus, iron, sulfur, and other ions—particles that will pass through a filter with pores of around 2 microns (0.002 cm) in size. Suspended solids include silt and clay particles, plankton, algae, fine organic debris, and other particulate matter. These are particles that will not pass through a 2-micron filter.

The concentration of total dissolved solids affects the water balance in the cells of aquatic organisms. An organism placed in water with a very low level of solids (distilled water, for example) swells because water tends to move into its cells, which have a higher concentration of solids. An organism placed in water with a high concentration of solids shrinks somewhat, because the water in its cells tends to move out. This in turn affects the organism's ability to maintain proper cell density, making keeping its position in the water column difficult. It might float up or sink down to a depth to which it is not adapted, and it might not survive.

Higher concentrations of suspended solids can serve as carriers of toxics, which readily cling to suspended particles. This is particularly a concern where pesticides are being used on irrigated crops. Where solids are high, pesticide concentrations may increase well beyond those of the original application as the irrigation water travels down irrigation ditches. Higher levels of solids can also clog irrigation devices and might become so high that irrigated plant roots will lose water rather than gain it.

A high concentration of total solids will make drinking water unpalatable and might have an adverse effect on people who are not used to drinking such water. Levels of total solids that are too high or too low can also reduce the efficiency of wastewater treatment plants, as well as the operation of industrial processes that use raw water.

Total solids affect water clarity. Higher solids decrease the passage of light through water, thereby slowing photosynthesis by aquatic plants. Water heats up more rapidly and holds more heat; this, in turn, might adversely affect aquatic life adapted to a lower temperature regime.

Sources of total solids include industrial discharges, sewage, fertilizers, road run-off, and soil erosion. Total solids are measured in mg/L.

Solids Sampling and Equipment Considerations

When conducting solids testing, there are many things that affect the accuracy of the test or result in wide variations in results for a single sample, including:

1. Drying temperature
2. Length of drying time
3. Condition of desiccator and desiccant
4. Nonrepresentative samples' lack of consistency in test procedure
5. Failure to achieve constant weight prior to calculating results

Several precautions that can help to increase the reliability of test results:

1. Use extreme care when measuring samples, weighing materials, and drying or cooling samples.
2. Check and regulate oven and furnace temperatures frequently to maintain the desired range.
3. Use an indicator drying agent in the desiccator that changes color when it is no longer good—change or regenerate the desiccant when necessary.

4. Keep desiccator cover greased with the appropriate type of grease—this will seal the desiccator and prevent moisture from entering the desiccator as the test glassware cools.
5. Check ceramic glassware for cracks and glass fiber filter for possible holes. A hole in a glass filter will cause solids to pass through and give inaccurate results.
6. Follow the manufacturer's recommendation for care and operation of analytical balances.

Total solids are important to measure in areas where discharges from sewage treatment plants, industrial plants, or extensive crop irrigation may occur. In particular, streams and rivers in arid regions where water is scarce, and evaporation is high tend to have higher concentrations of solids and are more readily affected by human introduction of solids from land use activities.

Total solids measurements can be useful as an indicator of the effects of runoff from construction, agricultural practices, logging activities, sewage treatment plant discharges, and other sources. As with turbidity, concentrations often increase sharply during rainfall, especially in developed watersheds. They can also rise sharply during dry weather if Earth-disturbing activities occur in or near the stream without erosion control practices in place. Regular monitoring of total solids can help detect trends that might indicate increasing erosion in developing watersheds. Total solids are closely related to stream flow and velocity and should be correlated with these factors. Any change in total solids over time should be measured at the same site at the same flow.

Total solids are measured by weighing the amount of solids present in a known volume of sample; this is accomplished by weighing a beaker, filling it with a known volume, evaporating the water in an oven and completely drying the residue, then weighing the beaker with the residue. The total solids concentration is equal to the difference between the weight of the beaker with the residue and the weight of the beaker without it. Since the residue is so light in weight, the lab needs a balance that is sensitive to weights in the range of 0.0001 g. Balances of this type are called analytical or Mettler balances, and they are expensive (around $3000). The technique requires that the beakers be kept in a desiccator, a sealed glass container that contains material that absorbs moisture and ensures that the weighing is not biased by water condensing on the beaker. Some desiccants change color to indicate moisture content. Measurement of total solids cannot be done in the field. Samples must be collected using clean glass or plastic bottles, or Whirl-pak® bags and taken to a laboratory where the test can be run.

Total Suspended Solids

The term "solids" means any material suspended or dissolved in water and wastewater. Although normal domestic wastewater contains a very small amount of solids (usually less than 0.1 percent), most treatment processes are designed specifically to remove or convert solids to a form that can be removed or discharged without causing environmental harm.

In sampling for total suspended solids (TSS), samples may be either grab, or composite and can be collected in either glass or plastic containers. TSS samples can be preserved by refrigeration at or below 4°C (not frozen). However, composite samples must be refrigerated during collection. The maximum holding time for preserved samples is seven days.

Test Procedure

To conduct a TSS test procedure, a well-mixed measured sample is poured into a filtration apparatus and, with the aid of a vacuum pump or aspirator, is drawn through a preweighted glass fiber filter. After filtration, the glass filter is dried at 103–105°C, cooled and reweighed. The increase in weight of the filter and solids compared to the filter alone represents the total suspended solids. An example of the specific test procedure used for total suspended solids is given below:

1. Select a sample volume that will yield between 10 and 200 mg of residue with a filtration time of 10 minutes or less.

Note: If filtration time exceeds 10 minutes, increase filter area or decrease volume to reduce filtration time.

Note: For non-homogenous samples or samples with very high solids concentrations (i.e., raw wastewater or mixed liquor), use a larger filter to ensure a representative sample volume can be filtered.

2. Place preweighed glass fiber filter on filtration assembly in a filter flask.
3. Mix sample well and measure the selected volume of sample.
4. Apply suction to filter flask, and wet filter with a small amount of laboratory-grade water to seal it.
5. Pour the selected sample volume into filtration apparatus.
6. Draw sample through filter.
7. Rinse measuring device into filtration apparatus with three successive 10 mL portions of laboratory-grade water. Allow complete drainage between rinsings.
8. Continue suction for 3 minutes after filtration of final rinse is completed.
9. Remove the glass filter from the filtration assembly (membrane filter funnel or clean Gooch crucible). If using the large disks and membrane filter assembly, transfer the glass filter to a support (aluminum pan or evaporating dish) for drying.
10. Place the glass filter with solids and support (pan, dish, or crucible) in a drying oven.
11. Dry filter and solids to constant weight at 103–105°C (at least for 1 hour).
12. Cool to room temperature in a desiccator.
13. Weigh the filter, and support and record constant weight in test record.

TSS Calculations

To determine the total suspended solids concentration in mg/L, we use the following equations:

1. To determine weight of dry solids in grams

$$\text{Dry Solids, g} = \text{Wt. of Dry Solids and Filter, g} - \text{Wt. of Dry Filter, g} \quad (11.6)$$

2. To determine weight of dry solids in milligrams

$$\text{Dry Solids, mg} = \text{Wt. of Solids and Filter, g} - \text{Wt. of Dry Filter, g} \quad (11.7)$$

3. To determine the TSS concentration in mg/L

$$\text{TSS, mg/L} = \frac{\text{Dry Solids, mg} \times 1{,}000 \text{ mL}}{\text{mL sample}} \quad (11.8)$$

Example 11.4

Problem: Using the data provided below, calculate total suspended solids (TSS):

Sample Volume, mL	250 mL
Weight of Dry Solids and Filter, g	2.305 g
Weight of Dry Filter, g	2.297 g

SOLUTION:

Dry Solids, g = 2.305 g − 2.297 g = 0.008 g
Dry Solids, mg = 0.008 g × 1,000 mg/g = 8 mg

$$\text{TSS, mg/L} = \frac{8.0 \times 1{,}000 \text{ mL/L}}{250 \text{ mL}} = 32.0 \text{ mg/L}$$

Volatile Suspended Solids Testing

When the total suspended solids are ignited at 550 ± 50°C, the volatile (organic) suspended solids of the sample are converted to water vapor and carbon dioxide and are released to the atmosphere. The solids that remain after the ignition (ash) are the inorganic or fixed solids. In addition to the equipment and supplies required for the total suspended solids test, you need the following:

1. Muffle furnace (550 ± 50°C)
2. Ceramic dishes
3. Furnace tongs
4. Insulated gloves

Test Procedure

An example of the test procedure used for volatile suspended solids is given below.

1. Place the weighed filter with solids and support from the total suspended solids test in the muffle furnace.
2. Ignite filter, solids, and support at 550 ± 50°C for 15–20 minutes.
3. Remove the ignited solids, filter, and support from the furnace, and partially air cool.
4. Cool to room temperature in a desiccator.
5. Weigh ignited solids, filter, and support on an analytical balance.
6. Record weight of ignited solids, filter, and support.

Total Volatile Suspended Solids Calculations

To calculate total volatile suspended solids (TVSS) requires the following information:

1. Weights of dry solids, filter, and support in grams
2. Weight of ignited solids, filter and support in grams

$$\text{Total Vol. Susp. Solids, mg/L} = \frac{(\text{A} - \text{C}) \times 1{,}000\ \text{mg/g} \times 1{,}000\ \text{mL/L}}{\text{Sample Vol., mL}} \quad (11.10)$$

where
A = Weight of Dried Solids, Filter, and Support
Weight of Ignited Solids, Filter, and Support

Example 11.5

Problem: Using the data provided below calculate the total volatile suspended solids:

Weight of dried solids, filter, and support = 1.6530 g
Weight of ignited solids, filter, and support = 1.6330 g
Sample volume = 100 mL

SOLUTION:

$$\text{TVSS.} = \frac{(1.6530\ \text{g} - 1.6330\ \text{g}) \times 1{,}000\ \text{mg/g} \times 1{,}000\ \text{mL}}{100\ \text{mL}}$$

$$= \frac{0.02 \times 1{,}000{,}000\ \text{mg/L}}{100}$$

$$= 200\ \text{mg/L}$$

Note: Total fixed suspended solids (TFSS) is the difference between the total volatile suspended solids (TVSS) and the total suspended solids (TSS) concentrations.

$$\text{TFSS (mg/L)} = \text{TTS} - \text{TVSS} \tag{17.11}$$

Example: 11.6

Problem: Using the data provided below, calculate the total fixed suspended solids:

Total Fixed Suspended Solids = 202 mg/L
Total Volatile Suspended Solids = 200 mg/L

SOLUTION:

Total Fixed Suspended Solids, mg/L = 202 mg/L – 200 mg/L = 2 mg/L

Conductivity Testing

Conductivity is a measure of the capacity of water to pass an electrical current. Conductivity in water is affected by the presence of inorganic dissolved solids such as chloride, nitrate, sulfate, and phosphate anions (ions that carry a negative charge), or sodium, magnesium, calcium, iron, and aluminum cations (ions that carry a positive charge). Organic compounds like oil, phenol, alcohol, and sugar do not conduct electrical current very well, and therefore have a low conductivity when in water. Conductivity is also affected by temperature: the warmer the water, the higher the conductivity.

Conductivity in streams and rivers is affected primarily by the geology of the area through which the water flows. Streams that run through areas with granite bedrock tend to have lower conductivity because granite is composed of more inert materials that do not ionize (dissolve into ionic components) when washed into the water. On the other hand, streams that run through areas with clay soils tend to have higher conductivity, because of the presence of materials that ionize when washed into the water. Groundwater inflows can have the same effects, depending on the bedrock they flow through.

Discharges to streams can change the conductivity depending on their make-up. A failing sewage system would raise the conductivity because of the presence of chloride, phosphate, and nitrate: an oil spill would lower conductivity.

The basic unit of measurement of conductivity is the mho or siemens. Conductivity is measured in micromhos per centimeter (µmhos/cm) or microsiemens per centimeter (µs/cm). Distilled water has conductivity in the range of 0.5–3 µmhos/cm. The conductivity of rivers in the United States generally ranges from 50 µmhos/cm to 1,500 µmhos/cm. Studies of inland freshwaters indicated that streams supporting good mixed fisheries have a range between 150 µmhos/cm and 500 µmhos/cm. Conductivity outside this range could indicate that the water is not suitable for certain species of fish or macroinvertebrates. Industrial waters can range as high as 10,000 μmhos/cm.

Sampling, Testing, and Equipment Considerations

Conductivity is useful as a general measure of source water quality. Each stream tends to have a relatively constant range of conductivity that, once established, can be used as a baseline for comparison with regular conductivity measurements. Significant changes in conductivity could indicate that a discharge or some other source of pollution has entered a stream. The conductivity test is not routine in potable water treatment, but when performed on source water is a good indicator of contamination. Conductivity readings can also be used to indicate wastewater contamination or saltwater intrusion.

Note: Distilled water used for potable water analyses at public water supply facilities must have a conductivity of no more than 1 μmho/cm.

Conductivity is measured with a probe and a meter. Voltage is applied between two electrodes in a probe immersed in the sample water. The drop in voltage caused by the resistance of the water is used to calculate the conductivity per centimeter. The meter converts the probe measurement to μmhos/cm and displays the result for the user.

Note: Some conductivity meters can also be used to test for total dissolved solids and salinity. The total dissolved solids concentration in mg/L can also be calculated by multiplying the conductivity result by a factor between 0.55 and 0.9, which is empirically determined, see Standard Methods, Methods #2510 (APHA, 1998).

Suitable conductivity meters cost about $350. Meters in this price range should also measure temperature and automatically compensate for temperature in the conductivity reading. Conductivity can be measured in the field or the lab. In most cases, collecting samples in the field and taking them to a lab for testing is probably better. In this way, several teams can collect samples simultaneously. If testing in the field is important, meters designed for field use can be obtained for around the same cost mentioned above. If samples are collected in the field for later measurement, the sample bottle should be a glass or polyethylene bottle that has been washed in phosphate-free detergent and rinsed thoroughly with both tap and distilled water. Factory-prepared Whirl-pak® bags may be used.

Total Alkalinity

Alkalinity is defined as the ability of water to resist a change in pH when acid is added; it relates to the pH buffering capacity of the water. Almost all natural waters have some alkalinity. These alkaline compounds in the water such as bicarbonates (baking soda is one type), carbonates, and hydroxides remove H^+ ions and lower the acidity of the water (which means increased pH). They usually do this by combining with the H^+ ions to make new compounds. Without this acid-neutralizing capacity, any acid added to a stream would cause an immediate change in the pH. Measuring alkalinity is important in determining a stream's ability to neutralize acidic pollution from rainfall or wastewater—one of the best measures of the sensitivity of the stream to acid inputs. Alkalinity in streams is influenced by rocks and soils, salts, certain plant activities, and certain industrial wastewater discharges.

Total alkalinity is determined by measuring the amount of acid (e.g., sulfuric acid) needed to bring the sample to a pH of 4.2. At this pH all the alkaline compounds in the sample are "used up." The result is reported as milligrams per liter of calcium carbonate (mg/L $CaCO_3$).

Testing for alkalinity in potable water treatment is most important with regard to its relation to coagulant addition—that is, it is important that there exists enough natural alkalinity in the water to buffer chemical acid addition so that floc formation will be optimum and the turbidity removal can proceed. In water softening, proper chemical dosage will depend on the type and amount of alkalinity in the water. For corrosion control, the presence of adequate alkalinity in a water supply neutralizes any acid tendencies and prevents it from becoming corrosive.

Analytical and Equipment Considerations

For total alkalinity, a double end point titration using a pH meter (or pH "pocket pal") and a digital titrator or burette is recommended. This can be done in the field or in the lab. If alkalinity must be analyzed in the field, a digital titrator should be used instead of a burette, because burettes are fragile and more difficult to set up. The alkalinity method described below was developed by the Acid Rain Monitoring Project of the University of Massachusetts Water Resources Research Center (River Watch Network, 1992).

Burettes, Titrators, and Digital Titrators for Measuring Alkalinity

The total alkalinity analysis involves titration. In this test, titration is the addition of small, precise quantities of sulfuric acid (the reagent) to the sample, until the sample reaches a certain pH (known as an end point). The amount of acid used corresponds to the total alkalinity of the sample. Alkalinity can be measured using a burette, titrator, or digital titrator (described below).

1. A burette is a long, graduated glass tube with a tapered tip like a pipette and a valve that opens to allow the reagent to drop out of the tube. The amount of reagent used is calculated by subtracting the original volume in the burette from the column left after the end point has been reached. Alkalinity is calculated based on the amount used.
2. Titrators forcefully expel the reagent by using a manual or mechanical plunger. The amount of reagent used is calculated by subtracting the original volume in the titrator from the volume left after the end point has been reached. Alkalinity is then calculated based on the amount used or is read directly from the titrator.
3. Digital titrators have counters that display numbers. A plunger is forced into a cartridge containing the reagent by turning a knob on the titrator. As the knob turns, the counter changes in proportion to the amount of reagent used. Alkalinity is then calculated based on the amount used. Digital titrators cost approximately $100.

Digital titrators and burettes allow for much more precision and uniformity in the amount of titrant that is used.

REFERENCES

APHA, 1971. *Standard Methods for the Examination of Water and Wastewater*, 17th ed. Washington, DC: APHA.

APHA, 1998. *Standard Methods for the Examination of Water and Wastewater*, 20th ed. Washington, DC: American Water Works Association.

AWRI, 2000. *Plankton Sampling*. Allendale, MI: Robert bl Annis Water Resource Institute, Grand Valley state University.

Bahls, L.L., 1993. *Periphyton Bioassessment Methods for Montana Streams*. Helena, MT: Montana Water Quality Bureau, Department of Health and Environmental Science.

Barbour, M.T., Gerritsen, J., Snyder, B.D., & Stibling, J.B., 1997. *Revision to Rapid Bioassessment Protocols for Use in Streams and Rivers, Periphytons, Benthic Macroinvertebrates, and Fish*. Washington, DC: United States Environmental Protection Agency.

Bly, T.D., & Smith, G.F., 1994. *Biomonitoring Our Streams: What's It All About?* Nashville, TN: U.S. Geological Survey.

Botkin, D.B., 1990. *Discordant Harmonies*. New York: Oxford University Press.

Camann, M., 1996. Freshwater aquatic invertebrates: Biomonitoring. Accessed 11/10/07 @ http://www.humboldt.edu.

Carins, J., Jr., & Dickson, K.L., 1971. A simple method for the biological assessment of the effects of waste discharges on aquatic bottom-dwelling organisms. *Journal of Water Pollution Control Federation*, 43: 755–772.

Hill, B.H., 1997. The use of periphyton assemblage data in an index of biotic integrity. *Bulletin of the North American Benthological Society*, 14: 158.

Huff, W.R., 1993. Biological indices define water quality standard. *Water Environment & Technology*, 5: 21–22.

Karr, J.R., 1981. Assessment of biotic integrity using fish communities. *Fisheries*, 66: 21–27.

Karr, J.R., Fausch, K.D., Hagermeier, P.L., Yant, P.R., & Schlosser, I.J., 1986. Assessing biological integrity in running waters: A method and its rationale. *Special Publications* 5. Champaign, Illinois: Illinois Natural History.

Kentucky Department of Environmental Protection, 1993. *Methods for Assessing Biological Integrity of Surface Waters*. Frankfort, Kentucky: Kentucky Department of Environmental Protection.

Kittrell, F.W., 1969. *A Practical Guide to Water Quality Studies of Streams*. Washington, DC: U.S. Department of Interior.

O'Toole, C. (Ed.), 1986. *The Encyclopedia of Insects*. New York: Facts on File, Inc.

Patrick, R., 1973. Use of algae, especially diatoms, in the assessment of water quality. In J. Carins & K.L. Dickson (Eds.), *Biological Methods for the Assessment of Qater Quality. Special Technical Publications* (528). Philadelphia, PA: American Society of Testing and Materials.

River Watch Network, 1992. *Acid Rain Monitoring...total alkalinity & field and Laboratory procedures*. Boson: University of Massachusetts.

Rodgers, J.H., Jr., Dickson, K.L., & Cairns, J. Jr., 1979. A review and analysis of some methods used to measure functional aspects of periphyton. In R.L. Weitzel (Ed.), *Methods and Measurements of Periphyton Communities: A Review. Special Technical Publications* 690. Washington, DC: American Society for Testing and Materials.

Stevenson, R.J., 1996. An introduction to algal ecology in freshwater benthic habitats. In R.J. Stevenson, M. Bothwell, & R.L. Lowe (Eds.), *Algal Ecology: Freshwater Benthic Ecosystems* (3–30). Sand Diego, CA: Academic Press.

Stevenson, R.J., 1998. Diatom indicators of stream and wetland stressors in a risk management framework. *Environmental Monitoring and Assessment*, 51: 107–108.

Stevenson, R.J., & Pan, Y., 1999. Assessing ecological conditions in rivers and streams with diatoms. In E.F. Stoermer & J.P. Smol (Eds.), *The Diatoms: Application to the Environmental and Earth Sciences* (11–40). Cambridge: Cambridge University Press.

Tchobanoglous, G., & Schroeder, E.D., 1985. *Water Quality*. Reading, MA: Addision-Wesley.

USEPA, 1983. *Technical Support Manual: Waterbody Survey and Assessments for Conducting Use Attainability Analyses*. Washington, DC: Environmental Protection Agency.

USEPA, 2000. *Monitoring Water Quality: Intensive Stream Bioassay*. Washington, DC: U.S. Environmental Protection Agency.

Velz, C.J., 1970. *Applied Stream Sanitation*. New York: Wiley Inter-Science.

Warren, M.L., Jr., & Burr, B.M., 1994. Status of freshwater fishes of the US: Overview of an imperiled fauna. *Fisheries*, 19(1): 6–18.

Weitzel, R.L., 1979. Periphyton measurements and applications. In R.L. Weitzel (Ed.), *Methods and Measurements of Periphyton Communities: A Review. Special Technical Publications* 690. Washington, DC: America Society for Testing and Material.

12 Wastewater Bacteria Sampling, Testing, and Analysis

Note: After providing a basic summary of sampling, testing, and analysis procedures for water in the preceding chapter, the focus of this chapter is bacteria in wastewater and water and the basic procedures or protocols used in sampling, testing and analysis. The procedures and protocols presented are based on those presented in F.R. Spellman's Handbook of Water and Wastewater Treatment Plant Operations, 4th ed. (2020). Boca Raton, FL: CRC Press.

FECAL COLIFORM BACTERIA TESTING

Much of the information in this section is from USEPA (1985, 1986). Fecal coliform bacteria are non-disease-causing organisms which are found in the intestinal tract of all warm-blooded animals. Each discharge of body wastes contains large amounts of these organisms. The presence of fecal coliform bacteria in a stream or lake indicates the presence of human or animal wastes. The number of fecal coliform bacteria present is a good indicator of the amount of pollution present in the water. EPA's 2001 Total Coliform Rule 816-F-01-035

1. Is intended to improve public health protection by reducing fecal pathogens to minimal levels through control of total coliform bacteria, including fecal coliforms and *Escherichia coli* (*E. coli*).
2. Establishes a maximum contaminant level (MCL) based on the presence or absence of total coliforms, modifies monitoring requirements including testing for fecal coliforms or *E. coli*, requires use of a sample siting plan, and also requires sanitary surveys for systems collecting fewer than five samples per month.
3. Applies to all public water systems.
4. Has resulted in reduction in risk of illness from disease-causing organisms associated with sewage or animal wastes. Disease symptoms may include diarrhea, cramps, nausea, and possibly jaundice, and associated headaches and fatigue.

Fecal coliforms are used as indicators of possible sewage contamination because they are commonly found in human and animal feces. Although they are not generally harmful themselves, they indicate the possible presence of pathogenic (disease-causing) bacteria, and protozoans that also live in human and animal digestive systems. Their presence in streams suggests that pathogenic microorganisms might also be present, and that swimming in and/or eating shellfish from the waters might present a health risk. Since testing directly for the presence of a large variety of pathogens is difficult, time-consuming, and expensive, water is usually tested for coliforms and fecal streptococci instead. Sources of fecal contamination to surface waters include wastewater treatment plants, on-site septic systems, domestic and wild animal manure, and storm runoff. In addition to the possible health risks associated with the presence of elevated levels of fecal bacteria, they can also cause cloudy water, unpleasant odors, and an increased oxygen demand.

Note: In addition to the most commonly tested fecal bacteria indicators, total coliforms, fecal coliforms, and E. coli, fecal streptococci and enterococci are also commonly used as bacteria indicators. The focus of this presentation is on total coliforms and fecal coliforms.

Fecal coliforms are widespread in nature. All members of the total coliform group can occur in human feces, but some can also be present in animal manure, soil, and submerged wood, and in other places outside the human body. The usefulness of total coliforms as an indicator of fecal contamination depends on the extent to which the bacteria species found are fecal and human in origin. For recreational waters, total coliforms are no longer recommended as an indicator. For drinking water, total coliforms are still the standard test, because their presence indicates contamination of a water supply by an outside source.

Fecal coliforms, a subset of total coliform bacteria, are more fecal-specific in origin. However, even this group contains a genus, Klebsiella, with species that are not necessarily fecal in origin. Klebsiella are commonly associated with textile and pulp and paper mill wastes. If these sources discharge to a local stream, consideration should be given to monitoring more fecal and human-specific bacteria. For recreational waters, this group was the primary bacteria indicator until relatively recently, when USEPA began recommending E. coli and enterococci as better indicators of health risk from water contact. Fecal coliforms are still being used in many states as indicator bacteria.

USEPA's Total Coliform Rule

Under EPA's Total Coliform Rule, sampling requirements are specified as follows.

Routine Sampling Requirements

1. Total coliform samples must be collected at sites which are representative of water quality throughout the distribution system according to a written sample siting plan, subject to state review and revision.
2. Samples must be collected at regular time intervals throughout the month, except groundwater systems serving 4,900 persons or fewer may collect them on the same day.

TABLE 12.1
Public Water System ROUTINE Monitoring Frequencies

Population	Minimum samples/month
25–1,000*	1
1,001–2,500	2
2,501–3,300	3
3,301–4,100	4
4,101–4,900	5
4,901–5,800	6
5,801–6,700	7
6,701–7,600	8
7,601–8,500	9
8,501–12,900	10
12,901–17,200	15
17,201–21,500	20
21,501–25,000	25
25,001–33,000	30
33,001–41,000	40
41,001–50,000	50
50,001–59,000	60
59,001–70,000	70
70,000–83,000	80
83,001–96,000	90
96,001–130,000	100
130,000–220,000	120
220,001–320,000	150
320,001–450,000	180
450,001–600,000	210
600,001–780,000	240
780,001–970,000	270
970,001–1,230,000	330
1,520,001–1,850,000	360
1,850,001–2,270,000	390
2,270,001–3,020,000	420
3,020,001–3,960,000	450
≥3,960,001	480

Includes PWSs, which have at least 15 service connections, but serve <25 people.

3. Monthly sampling requirements are based on population served (see Table 12.1 for the minimum sampling frequency).
4. A reduced monitoring frequency may be available for systems serving 1,000 persons or fewer and using only groundwater if a sanitary survey within the past 5 years shows the system is free of sanitary defects (the frequency may

be no less than 1 sample/quarter for community and 1 sample/year for non-community systems).

5. Each total coliform-positive routine sample must be tested for the presence of fecal coliforms or *E. coli*.

Repeat Sampling Requirements

1. Within 24 hours of learning of a total coliform-positive ROUTINE sample result, at least 3 REPEAT samples must be collected and analyzed for total coliforms.
2. One REPEAT sample must be collected from the same tap as the original sample.
3. One REPEAT sample must be collected within five service connections upstream.
4. One REPEAT sample must be collected within five service connections downstream.
5. Systems that collect 1 ROUTINE sample per month or fewer must collect a 4th REPEAT sample.
6. If any REPEAT sample is total coliform-positive:
    1. The system must analyze that total coliform-positive culture for fecal coliforms or *E. coli*.
    2. The system must collect another set of REPEAT samples, as before, unless the MCL has been violated and the system has notified the state.

ADDITIONAL ROUTINE SAMPLE REQUIREMENTS

A positive routine or repeat total coliform result requires a minimum of five routine samples be collected the following month the system provides water to the public unless waived by the state.

Other Total Coliform Rule Provisions

1. Systems collecting fewer than 5 ROUTINE samples per month must have a sanitary survey every five years (or every ten years if it is a non-community water system using protected and disinfected groundwater).
2. Systems using surface water or groundwater under the direct influence of surface water (GWUDI) and meeting filtration avoidance criteria must collect and have analyzed one coliform sample each day the turbidity of the source water exceeds 1 NTU. This sample must be collected from a tap near the first service connection.

Compliance

Compliance is based on the presence or absence of total coliforms. Moreover, compliance is determined each calendar month the system serves water to the public (or each calendar month that sampling occurs for systems on reduced monitoring). The results of routine and REPEAT samples are used to calculate compliance.

In regard to violations, a monthly MCL violation is triggered if a system collecting fewer than 40 samples per month has greater than 1 ROUTINE/REPEAT sample per month, which is total coliform-positive. In addition, a system collecting at least 40 samples per month has greater than 5.0 percent of the ROUTINE/REPEAT samples in a month total coliform-positive is technically in violation of the Total Coliform Rule. An acute MCL violation is triggered if any public water system has any fecal coliform- or E. coli-positive REPEAT sample or has a fecal coliform- or E. coli-positive ROUTINE sample followed by a total coliform-positive REPEAT sample.

The Total Coliform Rule also has requirements for public notification and reporting. For example, for a monthly MCL violation, the violation must be reported to the state no later than the end of the next business day after the system learns of the violation. The public must be notified within 14 days. For an acute MCL violation, the violation must be reported to the state no later than the end of the next business day after the system learns of the violation. The public must be notified within 72 hours. Systems with ROUTINE or REPEAT samples that are fecal coliform- or E. Coli-positive must notify the state by the end of the day they are notified of the result or by the end of the next business day if the state office is already closed.

Sampling and Equipment Considerations

Bacteria can be difficult to sample and analyze, for many reasons. Natural bacteria levels in streams can vary significantly; bacteria conditions are strongly correlated with rainfall, making the comparison of wet and dry weather bacteria data a problem; many analytical methods have a low level of precision, yet can be quite complex to accomplish; and absolutely sterile conditions are essential to maintain while collecting and handling samples. The primary equipment decision to make when sampling for bacteria is what type and size of sample container you will use. Once you have made that decision, the same straightforward collection procedure is used, regardless of the type of bacteria being monitored.

When monitoring bacteria, it is critical that all containers and surfaces with which the sample will come into contact be sterile. Containers made of either some form of plastic or Pyrex glass are acceptable to the USEPA. However, if the containers are to be reused, they must be sturdy enough to survive sterilization using heat and pressure. The containers can be sterilized by using an autoclave, a machine that sterilizes with pressurized steam. If using an autoclave, the container material must be able to withstand high temperatures and pressure. Plastic containers—either high-density polyethylene or polypropylene—might be preferable to glass from a practical standpoint because they will better withstand breakage. In any case, be sure to check the manufacturer's specifications to see whether the container can withstand 15 minutes in an autoclave at a temperature of 121°C without melting. (Extreme caution is advised when working with an autoclave.) Disposable, sterile, plastic Whirl-pak® bags are used by a number of programs. The size of the container depends on the sample amount needed for the bacteria analysis method you choose, and the amount needed for other analyses. The two basic methods for analyzing water samples for bacteria in common use are the membrane filtration and multiple-tube fermentation methods (described later).

Given the complexity of the analysis procedures and the equipment required, field analysis of bacteria is not recommended. Bacteria can either be analyzed by the volunteer at a well-equipped lab or sent to a state-certified lab for analysis. If you send a bacteria sample to a private lab, make sure that the lab is certified by the state for bacteria analysis. Consider state water quality labs, university and college labs, private labs, wastewater treatment plant labs, and hospitals. You might need to pay these labs for analysis. On the other hand, if you have a modern lab with the proper equipment and properly trained technicians, the fecal coliform testing procedures described in the following section will be helpful. A note of caution: if you decide to analyze your samples in your own lab, be sure to carry out a quality assurance/quality control program.

Fecal Coliform Bacteria Testing Methods

Federal regulations cite two approved methods for the determination of fecal coliform in water: (1) multiple-tube fermentation or most probable number (MPN) procedure and (2) membrane filter (MF) procedure.

Note: Because the MF procedure can yield low or highly variable results for chlorinated wastewater, the USEPA requires verification of results using the MPN procedure to resolve any controversies. However, do not attempt to perform the fecal coliform test using the summary information provided in this handbook. Instead, refer to the appropriate reference cited in the Federal Regulations for a complete discussion of these procedures.

Basic Equipment and Techniques

Whenever microbiological testing of water samples is performed, certain general considerations and techniques will be required. Because these are basically the same for each test procedure, they are reviewed here prior to discussion of the two methods.

1. Reagents and Media—All reagents and media utilized in performing microbiological tests on water samples must meet the standards specified in the reference cited in Federal Regulations.
2. Reagent Grade Water—Deionized water that is tested annually and found to be free of dissolved metals and bactericidal or inhibitory compounds is preferred for use in preparing culture media and test reagents, although distilled water may be used.
3. Chemicals—All chemicals used in fecal coliform monitoring must be ACS reagent grade or equivalent.
4. Media—To ensure uniformity in the test procedures, the use of dehydrated media is recommended. Sterilized, prepared media in sealed test tubes, ampoules, or dehydrated media pads are also acceptable for use in this test.

5. Glassware and Disposable Supplies—All glassware, equipment, and supplies used in microbiological testing should meet the standards specified in the references cited in Federal Regulations.

Sterilization

All glassware used for bacteriological testing must be thoroughly cleaned using a suitable detergent and hot water. The glassware should be rinsed with hot water to remove all traces of residual from the detergent and, finally, should be rinsed with distilled water. Laboratories should use a detergent certified to meet bacteriological standards or, at a minimum, rinse all glassware after washing with two tap water rinses followed by five distilled water rinses. For sterilization of equipment, the hot air sterilizer or autoclave can be used. When using the hot air sterilizer, all equipment should be wrapped in high-quality (Kraft) paper or placed in containers prior to hot air sterilization. All glassware, except those in metal containers, should be sterilized for a minimum of 60 minutes at 170°C. Sterilization of glassware in metal containers should require a minimum of two hours. Hot air sterilization cannot be used for liquids. An autoclave can be used, sample bottles, dilution water, culture media, and glassware may be sterilized by autoclaving at 121°C for 15 minutes.

Sterile Dilution Water Preparation

The dilution water used for making sample serial dilutions is prepared by adding 1.25 mL of stock buffer solution and 5.0 mL of magnesium chloride solution to 1,000 mL of distilled or deionized water. The stock solutions of each chemical should be prepared as outlined in the reference cited by the Federal Regulations. The dilution water is then dispensed in sufficient quantities to produce 9 mL or 99 mL in each dilution bottle following sterilization. If the membrane filter procedure is used, additional 60–100 mL portions of dilution water should be prepared and sterilized to provide rinse water required by the procedure.

Serial Dilution Procedure

At times, the density of the organisms in a sample makes it difficult to accurately determine the actual number of organisms in the sample. When this occurs, the sample size may need to be reduced to one millionth of a milliliter. In order to obtain such small volumes, a technique known as serial dilutions has been developed.

Bacteriological Sampling

To obtain valid test results that can be utilized in the evaluation of process efficiency of water quality, proper technique, equipment, and sample preservation are critical. These factors are especially critical in bacteriological sampling:

1. Sample dechlorination—When samples of chlorinated effluents are to be collected and tested, the sample must be dechlorinated. Prior to sterilization, place enough sodium thiosulfate solution (10 percent) in a clean sample container to produce a concentration of 100 mg/L in the sample (for a 120 mL sample bottle, 0.1 mL is usually sufficient). Sterilize the sample container as previously described
2. Sample procedure:
   A. Keep the sample bottle unopened after sterilization until the sample is to be collected.
   B. Remove the bottle stopper and hood or cap as one unit. Do not touch or contaminate the cap or the neck of the bottle.
   C. Submerge the sample bottle in the water to be sampled.
   D. Fill the sample bottle approximately ¾ full, but not less than 100 mL.
   E. Aseptically replace the stopper or cap on the bottle.
   F. Record the date, time, and location of sampling, as well as the sampler's name and any other descriptive information pertaining to the sample.
3. Sample preservation and storage—Examination of bacteriological water samples should be performed immediately after collection. If testing cannot be started within one hour of sampling, the sample should be iced or refrigerated at 4°C or less. The maximum recommended holding time for fecal coliform samples from wastewater is six hours. The storage temperature and holding time should be recorded as part of the test data.

Multiple-Tube Fermentation Technique

The multiple fermentation tube technique for fecal coliform testing is useful in determining the fecal coliform density in most water, solid, or semisolid samples. Wastewater testing normally requires use of the presumptive and confirming test procedures. It is recognized as the method of choice for any samples that may be controversial (enforcement related). The technique is based on the most probable number of bacteria present in a sample that produces gas in a series of fermentation tubes with various volumes of diluted sample. The MPN is obtained from charts based on statistical studies of known concentrations of bacteria.

The technique utilizes a two-step incubation procedure (see Figure 12.1). The sample dilutions are first incubated in lauryl (sulfonate) tryptose broth for 24–48 hours (presumptive test). Positive samples are then transferred to EC broth and incubated for an additional 24 hours (confirming test). Positive samples from this second incubation are used to statistically determine the MPN from the appropriate reference chart. A single media, 24-hour procedure is also acceptable. In this procedure, sample dilutions are inoculated in A-1 media and are incubated for three hours at 35°C then incubated the remaining 20 hours at 44.5°C. Positive samples from these inoculations are then used to statistically determine the MPN value from the appropriate chart.

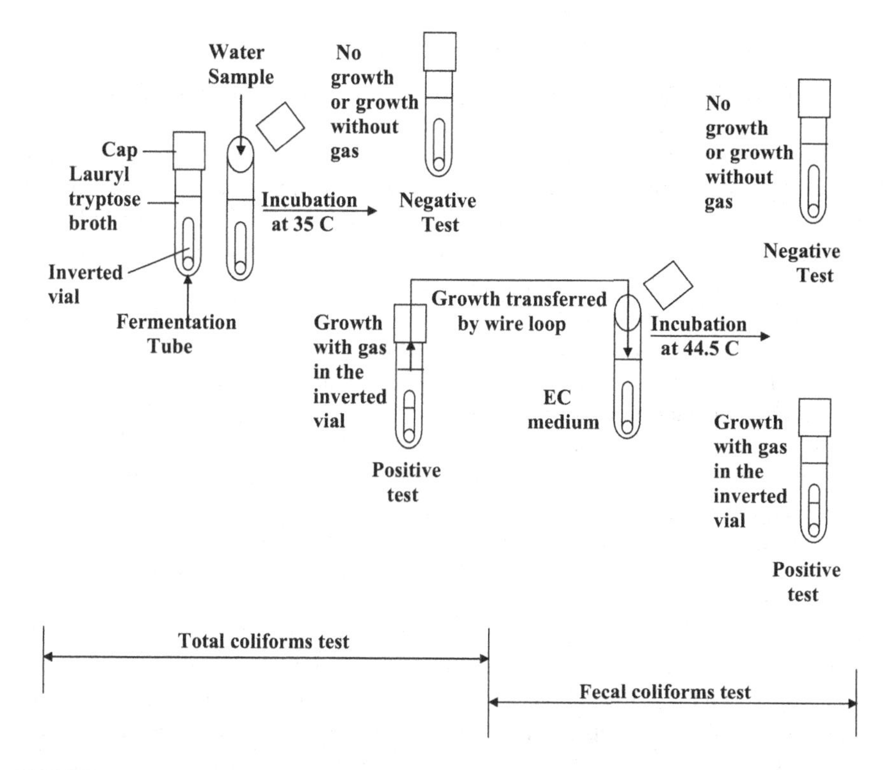

FIGURE 12.1 Multiple-tube fermentation technique.

Fecal Coliform MPN Presumptive Test Procedure

The procedure for the fecal coliform MPN Presumptive test is described below:

1. Prepare dilutions and inoculate five fermentation tubes for each dilution.
2. Cap all tubes, and transfer to incubator.
3. Incubate 24 + 2 hours at 35 ± 0.5°C.
4. Examine tubes for gas.
   A. Gas present = Positive test—transfer
   B. No gas = Continue incubation
5. Incubate total time 48 ± 3 hours at 35 ± 0.5°C
6. Examine tubes for gas.
   A. Gas present = Positive test—transfer
   B. No gas = Negative test

Note: Keep in mind that the fecal coliform MPN confirming procedure of fecal coliform procedure using A-1 broth test is used to determine the MPN/100 mL. The MPN procedure for fecal coliform determinations requires a minimum of three dilutions with five tubes/dilution.

Calculation of Most Probable Number (MPN)/100 mL

Calculation of the MPN test results requires selection of a valid series of three consecutive dilutions. The number of positive tubes in each of the three selected dilution inoculations is used to determine the MPN/100 mL. In selecting the dilution inoculations to be used in the calculation, each dilution is expressed as a ratio of positive tubes per tubes inoculated in the dilution, i.e., three positive/five inoculated (3/5). There are several rules to follow in determining the most valid series of dilutions. In the following examples, four dilutions were used for the test:

1. Using the confirming test data, select the highest dilution showing all positive results (no lower dilution showing less than all positive) and the next two higher dilutions.
2. If a series shows all negative values with the exception of one dilution, select the series that places the only positive dilution in the middle of the selected series.
3. If a series shows a positive result in a dilution higher than the selected series (using rule #1), it should be incorporated into the highest dilution of the selected series.

After selecting the valid series, the MPN/1000 mL is determined by locating the selected series on the MPN reference chart. If the selected dilution series matches the dilution series of the reference chart, the MPN value from the chart is the reported value for the test. If the dilution series used for the test does not match the dilution series of the chart, the test result must be calculated.

$$\text{MPN/10 ml} = \text{MPN}_{\text{chart}} \times \frac{\text{Sample vol. in } 1^{\text{st}} \text{ dilution}_{\text{chart}}}{\text{Sample vol. in } 1^{\text{st}} \text{ dilution}_{\text{sample}}} \tag{12.1}$$

Example 12.1

Problem: Using the results given below, calculate the MPN/100 mL of the example.

Sample in each serial dilution (ml)	Positive Tubes (Inoculated)
10.0	5/5
1.0	5/5
0.1	3/5
0.01	1/5
0.001	1/5

SOLUTION:

1. Select the highest dilution (tube with the lowest amount of sample) with all positive tubes (1.0 mL dilution). Select the next two higher dilutions (0.1 mL and 0.01 mL). In this case, the selected series will be 5-3-1.

TABLE 12.2
MPN Reference Chart

Sample volume, mL				Sample volume, mL			
10	**1.0**	**0.1**	**MPN/100 mL**	**10**	**1.0**	**0.1**	**MPN/100 mL**
0	0	0	0	4	3	0	27
0	0	1	2	4	3	1	33
0	1	0	2	4	4	0	34
0	2	0	4	5	0	0	23
1	0	0	2	5	0	1	31
1	0	1	4	5	0	2	43
1	1	0	4	5	0	0	31
1	1	1	6	5	1	1	46
1	2	0	6	5	1	2	63
2	0	0	5	5	2	0	49
2	0	1	7	5	2	1	70
2	1	0	7	5	2	2	94
2	1	1	9	5	3	0	79
2	2	0	9	5	3	1	110
2	3	0	12	5	3	2	140
3	0	0	8	5	3	3	180
3	0	1	11	5	4	0	130
3	1	0	11	5	4	1	170
3	1	1	14	5	4	2	220
3	2	0	14	5	4	3	280
3	2	1	17	5	4	4	350
4	0	0	13	5	5	0	240
4	0	1	17	5	5	1	350
4	1	0	17	5	5	2	540
4	1	1	21	5	5	3	920
4	1	2	26	5	5	4	1,600
4	2	0	22	5	5	5	≥2,400
4	2	1	26				

2. Include any positive results in dilutions higher than the selected series (0.001 mL dilution 1/5). This changes the selected series to 5-3-2.
3. Using the first three columns of Table 12.2, locate this series (5-3-2).
4. Read the MPN value from the fourth column (140).
5. In Table 12.2, the dilution series begins with 10 mL. For this test, this series begins with 1.0 mL.

$$\text{MPN/100 ml} = 140 \text{ MPN/100 ml} \times \frac{10 \text{ ml}}{1 \text{ ml}}$$

$$= 1,400 \text{ MPN/100 ml}$$

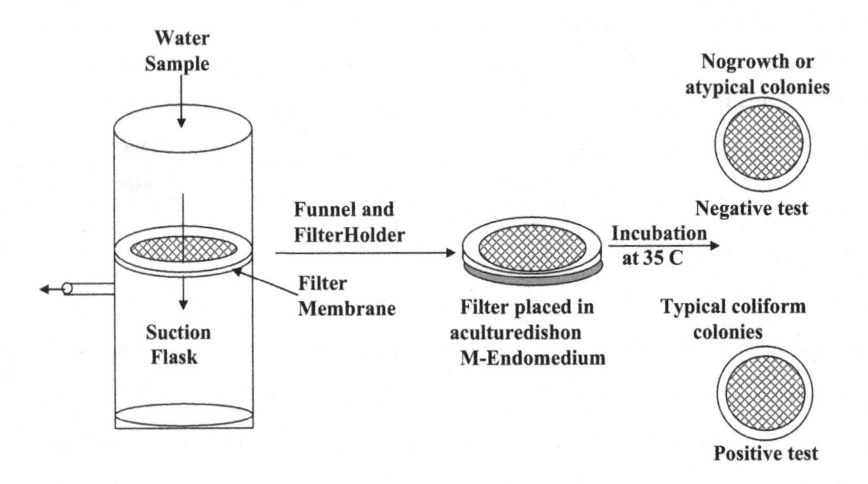

FIGURE 12.2 Membrane filter technique.

Membrane Filtration Technique

The membrane filtration technique can be useful for determining the fecal coliform density in wastewater effluents, except for primary treated wastewater that have not been chlorinated or wastewater containing toxic metals or phenols. Chlorinated secondary or tertiary effluents may be tested using this method, but results are subject to verification by MPN technique. The membrane filter technique utilizes a specially designed filter pad with uniformly sized pores (openings) that are small enough to prevent bacteria from entering the filter (see Figure 12.2). Another unique characteristic of the filter allows liquids, such as the media, placed under the filter to pass upward through the filter to provide nourishment required for bacterial growth.

Note: In the membrane filter method, the number of colonies grown estimates the number of coliforms.

Membrane Filter Procedure

The procedure for the membrane filter method is described below:

1. Sample filtration
   A. Select a filter, and aseptically separate it from the sterile package.
   B. Place the filter on the support plate with the grid side up.
   C. Place the funnel assembly on the support; secure as needed.
   D. Pour 100 mL of sample or serial dilution onto the filter, apply vacuum.

Note: The sample size and/or necessary serial dilution should produce a growth of 20–60 fecal coliform colonies on at least one filter. The selected dilutions must also be capable of showing permit excursions.

   E. Allow all of the liquid to pass through the filter.

F. Rinse the funnel and filter with three portions (20–30 mL) of sterile, buffered dilution water. (Allow each portion to pass through the filter before the next addition).

Note: Filtration units should be sterile at the start of each filtration series and should be sterilized again if the series is interrupted for 30 minutes or more. A rapid interim sterilization can be accomplished by 2 minutes exposure to ultraviolet (UV) light, flowing steam or boiling water.

2. Incubation
   A. Place absorbent pad into culture dish using sterile forceps.
   B. Add 1.8–2.0 mL M-FC media to the absorbent pad.
   C. Discard any media not absorbed by the pad.
   D. Filter sample through sterile filter.
   E. Remove filter from assembly, and place on absorbent pad (grid up).
   F. Cover culture dish.
   G. Seal culture dishes in a weighted plastic bag.
   H. Incubate filters in a water bath for 24 hours at 44.5 ± 0.2°C.

Colony Counting

Upon completion of the incubation period, the surface of the filter will have growths of both fecal coliform and non-fecal coliform bacteria colonies. The fecal coliform will appear blue in color, while non-fecal coliform colonies will appear gray or cream colored. When counting the colonies, the entire surface of the filter should be scanned using a 10x to 15x binocular, wide-field dissecting microscope. The desired range of colonies for the most valid fecal coliform determination is 20–60 colonies per filter. If multiple sample dilutions are used for the test, counts for each filter should be recorded on the laboratory data sheet.

1. Too many colonies—Filters that show a growth over the entire surface of the filter with no individually identifiable colonies should be recorded as "confluent growth." Filters that show a very high number of colonies (greater than 200) should be recorded as TNTC (too numerous to count).
2. Not enough colonies—If no single filter meets the desired minimum colony count (20 colonies), the sum of the individual filter counts, and the respective sample volumes can be used in the formula to calculate the colonies/100 mL.

Note: In each of these cases, adjustments in sample dilution volumes should be made to ensure future tests meet the criteria for obtaining a valid test result.

Calculations

The fecal coliform density can be calculated using the following formula:

$$\text{Colonies/100 ml} = \frac{\text{Colonies counted}}{\text{Sample volume, ml}} \times 100 \text{ ml} \tag{12.2}$$

Example 12.3

Problem: Using the data shown below, calculate the colonies per 100 mL for the influent and effluent samples noted.

Sample location	Influent	Sample	Dilutions	Effluent	Sample	Dilutions
Sample (ml)	1.0	0.1	0.01	10	1.0	0.1
Colonies counted	97	48	16	10	5	3

SOLUTION:

Step 1: Influent sample

Select the influent sample filter that has a colony count in the desired range (20–60). Because one filter meets this criterion, the remaining influent filters that did not meet the criterion are discarded.

$$\text{Colonies/100 ml} = \frac{48 \text{ colonies}}{0.1 \text{mL}} \times 100 \text{ mL}$$

$$= 48{,}000 \text{ colonies/100 mL}$$

Step 2: Effluent sample

Because none of the filters for the effluent sample meets the minimum test requirement, the colonies/100 mL must be determined by totaling the colonies on each filter and the sample volumes used for each filter.

$$\text{Total colonies} = 10 + 5 + 3$$

$$= 18 \text{ colonies}$$

$$\text{Total sample} = 10.0 \text{ mL} + 1.0 \text{ mL} + 0.1 \text{mL}$$

$$= 11.1 \text{mL}$$

$$\text{colonies/100 mL} = \frac{18 \text{ colonies}}{11.1 \text{mL}} \times 100$$

$$= 162 \text{ colonies/100 mL}$$

Note: The USEPA criterion for fecal coliform bacteria in bathing waters is a logarithmic mean of 200 per 100 mL, based on the minimum of five samples taken over a 30-day period, with not more than 10 percent of the total samples exceeding 400 per 100 mL. Because shellfish may be eaten without being cooked, the strictest coliform criterion applies to shellfish cultivation and harvesting. The USEPA criterion states that the mean fecal coliform concentration should not exceed 14 per 100 mL, with not more than 10 percent of the samples exceeding 43 per 100 mL.

Interferences

Large amounts of turbidity, algae, or suspended solids may interfere with this technique blocking the filtration of the sample through the membrane filter. Dilution of these samples to prevent this problem may make the test inappropriate for samples with low fecal coliform densities because the sample volumes after dilution may be too small to give representative results. The presence of large amounts of non-coliform group bacteria in the samples may also prohibit the use of this method.

Note: Many NPDES discharge permits require fecal coliform testing. Results for fecal coliform testing must be reported as a geometric mean (average) of all the test results obtained during a reporting period. A geometric mean, unlike an arithmetic mean or average, dampens the effect of very high or low values that otherwise might cause a non-representative result.

13 Wastewater Protozoa Sampling, Testing, and Analysis

PATHOGENIC PROTOZOA

As mentioned earlier, certain types of protozoans can cause disease. Of particular interest to the drinking water practitioner and wastewater-based epidemiologist are the *Entamoeba histolytica* (amebic dysentery and amebic hepatitis), *Giardia lamblia* (giardiasis), *Cryptosporidium* (Cryptosporidiosis), and the emerging Cyclosporine (Cyclosporasis). Sewage contamination transports eggs, cysts, and oocysts of parasitic protozoa and helminthes (tapeworms, hookworms, etc.) into raw water supplies, leaving water treatment (in particular filtration) and disinfection as the means by which to diminish the danger of contaminated water for the consumer. These pathogenic protozoans can also lead to relaying to the epidemiologist location, distribution, and measurement in designated populations.

To prevent the occurrence of *Giardia* and *Cryptosporidium* spp. in surface water supplies, and to address increasing problems with waterborne diseases, the USEPA implemented its Surface Water Treatment Rule (SWTR) in 1989. The rule requires both filtration and disinfection of all surface water supplies as a means of primarily controlling *Giardia* spp. and enteric viruses. Since implementation of its Surface Water Treatment Rule, the USEPA has also recognized that *Cryptosporidium* spp. is an agent of waterborne disease. In 1996, in its next series of surface water regulations, the USEPA included *Cryptosporidium.* Again, removing or neutralizing pathogenic protozoans should always be the goal in wastewater treatment operations. However, epidemiologists who are able to sample, test, and analyze raw wastewater influent to find, measure, and determine distribution and affected populations, may be able to indicate not only occurrence but also where the contaminated wastewater is being generated—important to the nth degree.

To test the need for and the effectiveness of the USEPA's Surface Water Treatment Rule, LeChevallier et al. (1991) conducted a study on the occurrence and distribution of *Giardia* and *Cryptosporidium* organisms in raw water supplies to 66 surface water filter plants. These plants were located in fourteen states and one Canadian province. A combined immunofluorescence test indicated that cysts and oocysts were widely dispersed in the aquatic environment. *Giardia* spp. was detected in more than 80 percent of the samples. *Cryptosporidium* spp. was found in 85 percent of the sample locations. Considering several variables, *Giardia* or *Cryptosporidium* spp. were detected in 97 percent of the raw water samples. After evaluating their data, the researchers concluded that the Surface Water Treatment Rule might have to be upgraded (subsequently, it has been) to require additional treatment.

GIARDIA

Giardia (gee-ar-dee-ah) lamblia (also known as hiker'/traveler's scourge or disease) is a microscopic parasite that can infect warm-blooded animals and humans. Although *Giardia* was discovered in the nineteenth century, not until 1981 did the World Health Organization (WHO) classify *Giardia* as a pathogen. An outer shell called a cyst that allows it to survive outside the body for long periods protects Giardia. If viable cysts are ingested, *Giardia* can cause the illness known as giardiasis, an intestinal illness that can cause nausea, anorexia, fever, and severe diarrhea. The symptoms last only for several days, and the body can naturally rid itself of the parasite in one to two months. However, for individuals with weakened immune systems, the body often cannot rid itself of the parasite without medical treatment.

In the United States, *Giardia* is the most commonly identified pathogen in waterborne disease outbreaks. Contamination of a water supply by *Giardia* can occur in two ways: (1) by the activity of animals in the watershed area of the water supply or (2) by the introduction of sewage into the water supply. Wild and domestic animals are major contributors in contaminating water supplies. Studies have also shown that, unlike many other pathogens, *Giardia* is not host-specific. In short, *Giardia* cysts excreted by animals can infect and cause illness in humans. Additionally, in several major outbreaks of waterborne diseases, the *Giardia* cyst source was sewage contaminated water supplies.

Treating the water supply, however, can effectively control waterborne *Giardia*. Chlorine and ozone are examples of two disinfectants known to effectively kill *Giardia* cysts. Filtration of the water can also effectively trap and remove the parasite from the water supply. The combination of disinfection and filtration is the most effective water treatment process available today for prevention of *Giardia* contamination.

In drinking water, *Giardia* is regulated under the Surface Water Treatment Rule (SWTR). Although the SWTR does not establish a Maximum Contaminant Level (MCL) for *Giardia*, it does specify treatment requirements to achieve at least 99.9 percent (3-log) removal and/or inactivation of *Giardia*. This regulation requires that all drinking water systems using surface water or groundwater under the influence of surface water must disinfect and filter the water. The Enhanced Surface Water Treatment Rule (ESWTR), which includes *Cryptosporidium* and further regulates *Giardia*, was established in December 1996.

GIARDIASIS

Giardiasis is recognized as one of the most frequently occurring waterborne diseases in the United States. *Giardia lamblia* cysts have been discovered in the United States in places as far apart as Estes Park, Colorado (near the Continental Divide);

Missoula, Montana; Wilkes-Barre, Scranton, and Hazleton, Pennsylvania; and Pittsfield and Lawrence, Massachusetts, just to name a few (CDC, 1995).

Giardiasis is characterized by intestinal symptoms that usually last one week or more and may be accompanied by one or more of the following: diarrhea, abdominal cramps, bloating, flatulence, fatigue, and weight loss. Although vomiting and fever are commonly listed as relatively frequent symptoms, people involved in waterborne outbreaks in the United States have not commonly reported them.

While most *Giardia* infections persist only for one or two months, some people undergo a more chronic phase, which can follow the acute phase, or may become manifest without an antecedent acute illness. Loose stools and increased abdominal gassiness with cramping, flatulence, and burping characterize the chronic phase. Fever is not common, but malaise, fatigue, and depression may ensue. For a small number of people, the persistence of infection is associated with the development of marked malabsorption and weight loss (Weller 1985). Similarly, lactose (milk) intolerance can be a problem for some people. This can develop coincidentally with the infection or be aggravated by it, causing an increase in intestinal symptoms after ingestion of milk products.

Some people may have several of these symptoms without evidence of diarrhea or have only sporadic episodes of diarrhea every three or four days. Still others may not have any symptoms at all. Therefore, the problem may not be whether you are infected with the parasite or not, but how harmoniously you both can live together or how to get rid of the parasite (either spontaneously or by treatment) when the harmony does not exist or is lost.

Note: Three prescription drugs are available in the United States to treat giardiasis: quinacrine, metronidazole, and furazolidone. In a recent review of drug trials in which the efficacies of these drugs were compared, quinacrine produced a cure in 93 percent of patients, metronidazole cured 92 percent, and furazolidone cured about 84 percent of patients (Davidson, 1984).

Giardiasis occurs worldwide. In the United States, *Giardia* is the parasite most commonly identified in stool specimens submitted to state laboratories for parasitologic examination. During a three-year period, approximately 4 percent of one million stool specimens submitted to state laboratories tested positive for *Giardia* (CDC 1979). Other surveys have demonstrated *Giardia* prevalence rates ranging from 1 percent to 20 percent, depending on the location and ages of persons studied. Giardiasis ranks among the top 20 infectious diseases that cause the greatest morbidity in Africa, Asia, and Latin America; it has been estimated that about two million infections occur per year in these regions (Walsh 1981). People who are at highest risk for acquiring *Giardia* infection in the United States may be placed into five major categories:

(1) People in cities whose drinking water originates from streams or rivers, and whose water treatment process does not include filtration, or where filtration is ineffective because of malfunctioning equipment
(2) Hikers/campers/outdoor people
(3) International travelers

(4) Children who attend day-care centers, day-care center staff, and parents and siblings of children infected in day-care centers
(5) Homosexual men

People in categories 1, 2, and 3 have in common the same general source of infection, i.e., they acquire *Giardia* from fecally contaminated drinking water. The city resident usually becomes infected because the municipal water treatment process does not include the filter necessary to physically remove the parasite from the water. The number of people in the United States at risk (i.e., the number who receive municipal drinking water from unfiltered surface water) is estimated to be 20 million. International travelers may also acquire the parasite from improperly treated municipal waters in cities or villages in other parts of the world, particularly in developing countries. In Eurasia, only travelers to Leningrad appear to be at increased risk. In prospective studies, 88 percent of U.S. and 35 percent of Finnish travelers to Leningrad who had negative stool tests for *Giardia* on departure to the Soviet Union developed symptoms of giardiasis and had positive tests for *Giardia* after they returned home (Brodsky et al. 1974). With the exception of visitors to Leningrad, however, *Giardia* has not been implicated as a major cause of traveler's diarrhea—it has been detected in fewer than 2 percent of travelers who develop diarrhea. However, hikers and campers risk infection every time they drink untreated raw water from a stream or river. Persons in categories 4 and 5 become exposed through more direct contact with feces or an infected person by exposure to soiled diapers of an infected child (day-care center-associated cases), or through direct or indirect anal-oral sexual practices in the case of homosexual men.

Although community waterborne outbreaks of giardiasis have received the greatest publicity in the United States during the past decade, about half of the *Giardia* cases discussed with the staff of the Centers for Disease Control over a three-year period had a day-care exposure as the most likely source of infection. Numerous outbreaks of *Giardia* in day-care centers have been reported in recent years. Infection rates for children in day-care center outbreaks range from 21 percent to 44 percent in the United States and from 8 percent to 27 percent in Canada (Black et al. 1981). The highest infection rates are usually observed in children who wear diapers (one to three years of age). In a study of 18 randomly selected day-care centers in Atlanta, 10 percent of diapered children were found infected (CDC Unpublished). Transmission from this age group to older children, day-care staff, and household contacts is also common. About 20 percent of parents caring for an infected child becomes infected.

Local health officials and managers or water utility companies need to realize that sources of *Giardia* infection other than municipal drinking water exist. Armed with this knowledge, they are less likely to make a quick (and sometimes wrong) assumption that a cluster of recently diagnosed cases in a city is related to municipal drinking water. Of course, drinking water must not be ruled out as a source of infection when a larger than expected number of cases is recognized in a community, but the possibility that the cases are associated with a day-care center outbreak, drinking untreated stream water, or international travel should also be entertained.

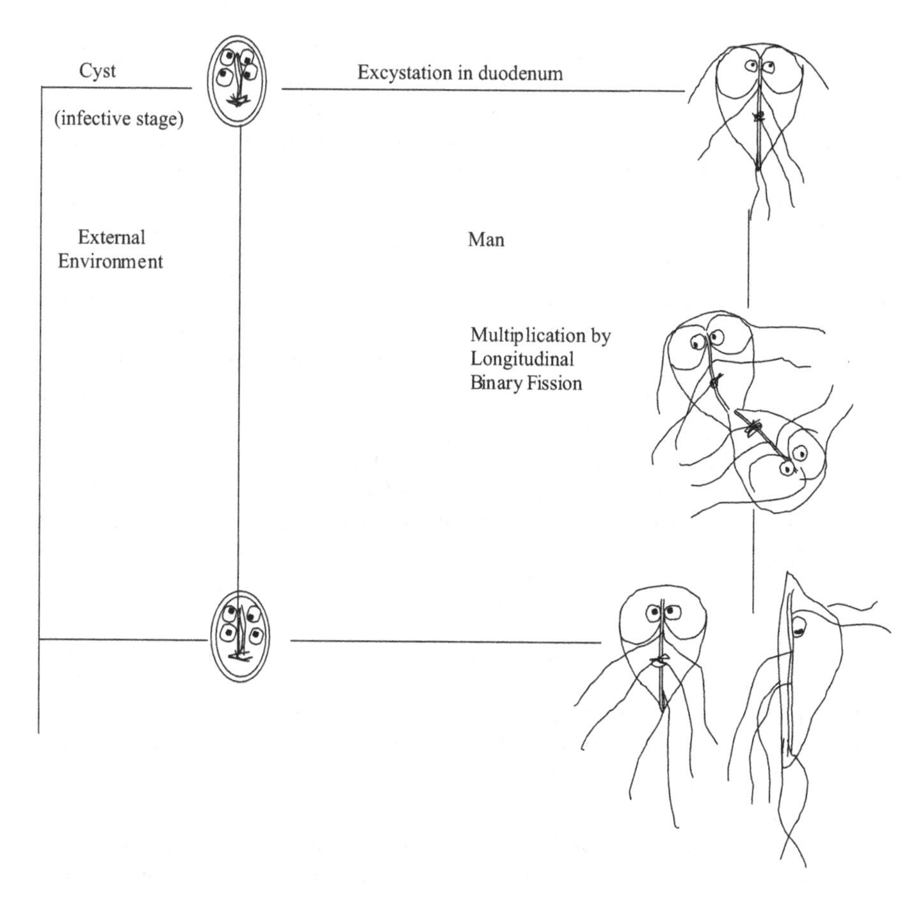

FIGURE 13.1 Life cycle of Giardia lamblia.

To understand the finer aspects of *Giardia* transmission and strategies for control, the drinking water practitioner must become familiar with several aspects of the parasite's biology. Two forms of the parasite exist: a trophozoite and a cyst, both of which are much larger than bacteria (see Figure 13.1). Trophozoites live in the upper small intestine where they attach to the intestinal wall by means of a disc-shaped suction pad on their ventral surface. Trophozoites actively feed and reproduce at this location. At some time during the trophozoites life, it releases its hold on the bowel wall and floats in the fecal stream through the intestine. As it makes this journey, it undergoes a morphologic transformation into an egg like structure called a cyst. The cyst (about 6–9 nm in diameter × 8–12 μm (1/100 mm) in length) has a thick exterior wall that protects the parasite against the harsh elements that it will encounter outside the body. This cyst form of parasite is infectious to other people or animals. Most people become infected either directly (by hand-to-mouth transfer of cysts from the feces of an infected individual) or indirectly (by drinking feces-contaminated water). Less common modes of transmission included ingestion of fecally contaminated food and hand-to-mouth transfer of cysts after touching a fecally contaminated surface. After the cyst is swallowed, the trophozoite is liberated through the action of stomach acid and digestive enzymes and becomes established in the small intestine.

Although infection after ingestion of only one *Giardia* cyst is theoretically possible, the minimum number of cysts shown to infect a human under experimental conditions is ten (Rendtorff 1954). Trophozoites divide by binary fission about every 12 hours. What this means in practical terms is that if a person swallowed only a single cyst, reproduction at this rate would result in more than one million parasites 10 days later, and one billion parasites by day 15.

The exact mechanism by which *Giardia* causes illness is not yet well understood but is not necessarily related to the number of organisms present. Nearly all of the symptoms, however, are related to dysfunction of the gastrointestinal tract. The parasite rarely invades other parts of the body, such as the gall bladder or pancreatic ducts. Intestinal infection does not result in permanent damage.

Note: *Giardia* has an incubation period of 1–8 weeks.

Data reported by the CDC indicate that *Giardia* is the most frequently identified cause of diarrheal outbreaks associated with drinking water in the United States. The remainder of this section is devoted specifically to waterborne transmissions of *Giardia*. *Giardia* cysts have been detected in 16 percent of potable water supplies (lakes, reservoirs, rivers, springs, groundwater) in the United States at an average concentration of 3 cysts per 100L (Rose 1983). Waterborne epidemics of giardiasis are a relatively frequent occurrence. In 1983, for example, *Giardia* was identified as the cause of diarrhea in 68 percent of waterborne outbreaks in which the causal agent was identified. From 1965 to 1982, more than 50 waterborne outbreaks were reported (CDC 1984). In 1984, about 250,000 people in Pennsylvania were advised to boil drinking water for six months because of *Giardia*-contaminated water.

Many of the municipal waterborne outbreaks of *Giardia* have been subjected to intense study to determine their cause. Several general conclusions can be made from data obtained in those studies. Waterborne transmission of *Giardia* in the United States usually occurs in mountainous regions where community drinking water obtained from clear running streams is chlorinated, but not filtered before distribution. Although mountain streams appear to be clean, fecal contamination upstream by human residents or visitors, as well as by *Giardia*-infected animals such as beavers, has been well documented. Water obtained from deep wells is an unlikely source of *Giardia* because of the natural filtration of water as it percolates through the soil to reach underground cisterns. Well-waste sources that pose the greatest risk of fecal contamination are poorly constructed or improperly located ones. A few outbreaks have occurred in towns that included filtration in the water treatment process, where the filtration was not effective in removing *Giardia* cysts because of defects in filter construction, poor maintenance of the filter media, or inadequate pretreatment of the water before filtration. Occasional outbreaks have also occurred because of accidental cross-connections between water and sewage systems.

Important Point: From these data, we conclude that two major ingredients are necessary for waterborne outbreak. *Giardia* cysts must be present in untreated source water, and the water purification process must either fail to kill or to remove *Giardia* cysts from the water.

Although beavers are often blamed for contaminating water with *Giardia* cysts, that they are responsible for introducing the parasite into new areas seem unlikely. Far more likely is that they are also victims: *Giardia* cysts may be carried in untreated human sewage discharged into the water by small-town sewage disposal plants or originate from cabin toilets that drain directly into streams and rivers. Backpackers, campers, and sports enthusiasts may also deposit *Giardia*-contaminated feces in the environment, which are subsequently washed into streams by rain. In support of this concept is a growing amount of data that indicate a higher *Giardia* infection rate in beavers living downstream from U.S. National Forest campgrounds when compared with beavers living in more remote areas that have a near-zero rate of infection.

Although beavers may be unwitting victims of the *Giardia* story, they still play an important part in the contamination scheme, because they can (and probably do) serve as amplifying hosts. An amplifying host is one that is easy to infect, serves as a good habitat for the parasite to reproduce, and in the case of *Giardia*, returns millions of cysts to the water for every one ingested. Beavers are especially important in this regard because they tend to defecate in or very near the water, which ensures that most of the *Giardia* cysts excreted are returned to the water.

The microbial quality of water resources and the management of the microbially laden wastes generated by the burgeoning animal agriculture industry are critical local, regional, and national problems. Animal waste from cattle, hogs, sheep, horses, poultry, and other livestock and commercial animals can contain high concentrations of microorganisms such as *Giardia*, that are pathogenic to humans.

The contribution of other animals to waterborne outbreaks of *Giardia* is less clear. Muskrats (another semiaquatic animal) have been found in several parts of the United States to have high infection rates (30–40 percent) (Frost et al. 1984). Recent studies have shown that muskrats can be infected with *Giardia* cysts from humans and beavers. Occasional *Giardia* infections have been reported in coyotes, deer, elk, cattle, dogs, and cats (but not in horses and sheep) encountered in mountainous regions of the United States. Naturally occurring *Giardia* infections have not been found in most other wild animals (bear, nutria, rabbit, squirrel, badger, marmot, skunk, ferret, porcupine, mink, raccoon, river otter, bobcat, lynx, moose, bighorn sheep) (Frost et al. 1984).

Scientific knowledge about what is required to kill or remove *Giardia* cysts from a contaminated water supply has increased considerably. For example, we know that cysts can survive in cold water (4°C) for at least two months, and they are killed instantaneously by boiling water (100°C) (Frost et al. 1984). We do not know how long the cysts will remain viable at other water temperatures (e.g., at 0°C or in a canteen at 15–20°C), nor do we know how long the parasite will survive on various environment surfaces, e.g., under a pine tree, in the sun, on a diaper-changing table, or in carpets in a day-care center.

The effect of chemical disinfection (chlorination, for example) on the viability of *Giardia* cysts is an even more complex issue. The number of waterborne outbreaks of *Giardia* that have occurred in communities where chlorination was employed as a disinfectant-process demonstrates that the amount of chlorine used routinely for municipal water treatment is not effective against *Giardia* cysts. These observations have been confirmed in the laboratory under experimental conditions (Jarroll et al. 1979). This does not mean that chlorine does not work at all. It does work under certain favorable conditions. Without getting too technical, gaining some appreciation of the problem can be achieved by understanding a few of the variables that influence the efficacy of chlorine as a disinfectant:

- Water pH—At pH values above 7.5, the disinfectant capability of chlorine is greatly reduced.
- Water temperature—The warmer the water, the higher the efficacy. Chlorine does not work in ice-cold water from mountain streams.
- Organic content of the water—Mud, decayed vegetation, or other suspended organic debris in water chemically combines with chlorine, making it unavailable as a disinfectant.
- Chlorine contact time: The longer *Giardia* cysts are exposed to chlorine, the more likely the chemical will kill them.
- Chlorine concentration—The higher the chlorine concentration, the more likely chlorine will kill *Giardia* cysts. Most water treatment facilities try to add enough chlorine to give a free (unbound) chlorine residual at the customer tap of 0.5 mg/L of water.

These five variables are so closely interrelated that improving one can often compensate for another; for example, if chlorine efficacy is expected to be low because water is obtained from an icy stream, the chlorine contact time, chlorine concentration, or both could be increased. In the case of *Giardia*-contaminated water, producing safe drinking water with a chlorine concentration of 1 mg/L and contact time as short as 10 minutes might be possible—if all the other variables were optimal (i.e., pH of 7.0, water temperature of 25°C, and a total organic content of the water close to zero). On the other hand, if all of these variables were unfavorable (i.e., pH of 7.9, water temperature of 5°C, and high organic content), chlorine concentrations in excess of 8 mg/L with several hours of contact time may not be consistently effective. Because water conditions and water treatment plant operations (especially those related to water retention time, and therefore, to chlorine contact time) vary considerably in different parts of the United States, neither the USEPA nor the CDC has been able to identify a chlorine concentration that would be safe yet effective against *Giardia* cysts under all water conditions. Therefore, the use of chlorine as a preventive measure against waterborne giardiasis generally has been used under outbreak conditions when the amount of chlorine and contact time have been tailored to fit specific water conditions and the existing operational design of the water utility.

In an outbreak, for example, the local health department and water utility may issue an advisory to boil water, may increase the chlorine residual at the consumer's tap from 0.5 mg/L to 1 or 2 mg/L, and if the physical layout and operation of the water treatment facility permit, increase the chlorine contact time. These are emergency procedures intended to reduce the risk of transmission until a filtration device can be installed or repaired, or until an alternative source of safe water (a well, for example) can be made operational.

The long-term solution to the problem of municipal waterborne outbreaks of giardiasis involves improvements in and more widespread use of filters in the municipal water treatment process. The sand filters most commonly used in municipal water treatment today cost millions of dollars to install, which makes them unattractive for many small communities. The pore sizes in these filters are not sufficiently small to remove *Giardia* (6–9 μm × 8–12μm). For the sand filter to remove *Giardia* cysts from the water effectively, the water must receive some additional treatment before it reaches the filter. The flow of water through the filter bed must also be carefully regulated.

An ideal prefilter treatment for muddy water would include sedimentation (a holding pond where large, suspended particles are allowed to settle out by the action of gravity) followed by flocculation or coagulation (the addition of chemicals such as alum or ammonium to cause microscopic particles to clump together). The sand filter easily removes the large particles resulting from the flocculation/coagulation process, including Giardia cysts bound to other microparticulates. Chlorine is then added to kill the bacteria and viruses that may escape the filtration process. If the water comes from a relatively clear source, chlorine may be added to the water before it reaches the filter.

The successful operation of a complete waterworks operation is a complex process that requires considerable training. Troubleshooting breakdowns or recognizing the potential problems in the system before they occur often requires the skills of an engineer. Unfortunately, most small water utilities with water treatment facilities that include filtration cannot afford the services of a full-time engineer. Filter operation or maintenance problems in such systems may not be detected until a *Giardia* outbreak is recognized in the community. The bottom line is that although filtration is the best that water treatment technology has to offer for municipal water systems against waterborne giardiasis, it is not infallible. For municipal water filtration facilities to work properly, they must be properly constructed, operated, and maintained.

Whenever possible, persons in the out-of-doors should carry drinking water of known purity with them. When this is not practical, when water from streams, lakes, ponds, and other outdoor sources must be used, time should be taken to properly disinfect the water before drinking it.

CRYPTOSPORIDIUM

Ernest E. Tyzzer first described the protozoan parasite *Cryptosporidium* in 1907. Tyzzer frequently found a parasite in the gastric glands of laboratory mice. Tyzzer identified the parasite as a sporozoan, but of uncertain taxonomic status; he named it *Cryptosporidium muris*. Later, in 1910, after more detailed study, he proposed *Cryptosporidium* as a new genus and *C. muris* as the type of species. Amazingly, except for developmental stages, Tyzzer's original description of the life cycle (see Figure 13.2) was later confirmed by electron microscopy. Later, in 1912, Tyzzer described a new species, *Cryptosporidium parvum* (Tyzzer, 1912).

For almost 50 years, Tyzzer's discovery of the genus *Cryptosporidium* (because it appeared to be of no medical or economic importance) remained (like himself)

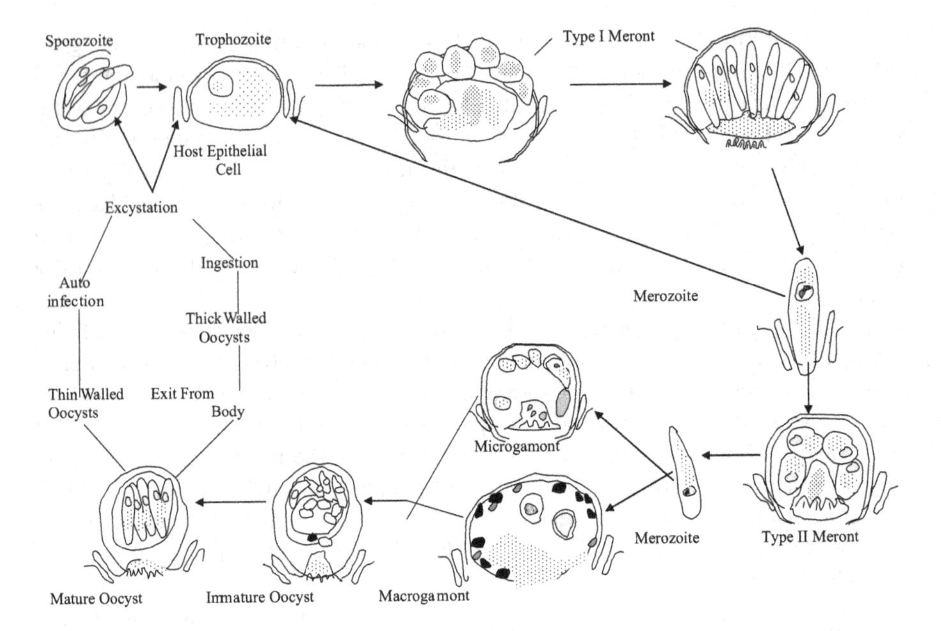

FIGURE 13.2 Life cycle of *Cryptosporidium parvum*.

relatively obscure because it appeared to be of no medical or economic importance. Slight rumblings of the genus' importance were felt in the medical community when Slavin (1955) wrote about a new species, *Cryptosporidium melagridis*, associated with illness and death in turkeys. Interest remained slight even when *Cryptosporidium* was found to be associated with bovine diarrhea (Panciera et al, 1971).

Not until 1982 did worldwide interest focus in on the study of organisms in the genus *Cryptosporidium*. During this period, the medical community and other interested parties were beginning to attempt a full-scale, frantic effort to find out as much as possible about Acquired Immune Deficiency Syndrome (AIDS). The CDC reported that 21 AIDS-infected males from six large cities in the United States had severe protracted diarrhea caused by *Cryptosporidium*. It was in 1993, though, when the "bug—the pernicious parasite *Cryptosporidium* —made [itself and] Milwaukee famous (Mayo Foundation 1996)."

Note: The *Cryptosporidium* outbreak in Milwaukee caused the deaths of 100 people—the largest episode of waterborne disease in the United States in the 70 years since health officials began tracking such outbreaks.

The massive waterborne outbreak in Milwaukee (more than 400,000 persons developed acute and often prolonged diarrhea or other gastrointestinal symptoms) increased an interest in *Cryptosporidium* at an exponential level. The Milwaukee Incident spurred both public interest and the interest of public health agencies, agricultural agencies and groups, environmental agencies and groups, and suppliers of drinking water. This increase in interest level and concern has spurred on new studies of *Cryptosporidium* with emphasis on developing methods for recovery, detection, prevention, and treatment (Fayer et al. 1997).

The USEPA has become particularly interested in this "new" pathogen. For example, in the reexamination of regulations on water treatment and disinfection, the USEPA issued MCLG and CCL for *Cryptosporidium.* The similarity to *Giardia lamblia* and the necessity to provide an efficient conventional water treatment capable of eliminating viruses at the same time forced the USEPA to regulate the surface water supplies in particular. The proposed "Enhanced Surface Water Treatment Rule" (ESWTR) included regulations from watershed protection to specialized operation of treatment plants (certification of operators and state overview) and effective chlorination. Protection against *Cryptosporidium* included control of waterborne pathogens such as *Giardia* and viruses (DeZuane, 1997).

The Basics of *Cryptosporidium*

Cryptosporidium (crip-toe-spor-ID-ee-um) is one of several single-celled protozoan genera in the phylum Apircomplexa (all referred to as coccidian). *Cryptosporidium,* along with other genera in the phylum Apircomplexa, develop in the gastrointestinal tract of vertebrates through all of their life cycle—in short, they live in the intestines of animals and people. This microscopic pathogen causes a disease called Cryptosporidiosis (crip-toe-spor-id-ee-O-sis). The dormant (inactive) form of *Cryptosporidium* called an oocyst (O-o-sist) is excreted in the feces (stool) of infected humans and animals. The tough-walled oocysts survive under a wide range of environmental conditions.

Several species of *Cryptosporidium* were incorrectly named after the host in which they were found; subsequent studies have invalidated many species. Now, eight valid species of *Cryptosporidium* have been named. Upton (1997) reports that *C. muris* infects the gastric glands of laboratory rodents and several other mammalian species, but (even though several texts state otherwise) is not known to infect humans, However, *C. parvum* infects the small intestine of an unusually wide range of mammals, including humans, and is the zoonotic species responsible for human Cryptosporidiosis. In most mammals *C. parvum* is predominately a parasite of neonate (newborn) animals. He points out that even though exceptions occur, older animals generally develop poor infections, even when unexposed previously to the parasite. Humans are the one host that can be seriously infected at any time in their lives, and only previous exposure to the parasite results in either full or partial immunity to challenge infections.

Oocysts are present in most surface bodies of water across the United States, many of which supply public drinking water. Oocysts are more prevalent in surface waters when heavy rains increase runoff of wild and domestic animal wastes from the land, or when sewage treatment plants are overloaded or break down. Only laboratories with specialized capabilities can detect the presence of *Cryptosporidium* oocysts in water. Unfortunately, present sampling and detection methods are unreliable. Recovering oocysts trapped on the material used to filter water samples is difficult. Once a sample is obtained, however, determining whether the oocyst is alive or whether it is the species *C. parvum* that can infect humans is easily accomplished by looking at the sample under a microscope.

The number of oocysts detected in raw (untreated) water varies with location, sampling time, and laboratory methods. Water treatment plants remove most, but not always all, oocysts. Low numbers of oocysts are sufficient to cause Cryptosporidiosis, but the low numbers of oocysts sometimes present in drinking water are not considered cause for alarm in the public.

Protecting water supplies from *Cryptosporidium* demands multiple barriers. Why? Because *Cryptosporidium* oocysts have tough walls that can withstand many environmental stresses and are resistant to the chemical disinfectants such as chlorine that are traditionally used in municipal drinking water systems.

Physical removal of particles, including oocysts, from water by filtration is an important step in the water treatment process. Typically, water pumped from rivers or lakes into a treatment plant is mixed with coagulants, which help settle out particles suspended in the water. If sand filtration is used, even more particles are removed. Finally, the clarified water is disinfected and piped to customers. Filtration is the only conventional method now in use in the U.S. for controlling *Cryptosporidium.*

Ozone is a strong disinfectant that kills protozoa if sufficient doses and contact times are used, but ozone leaves no residual for killing microorganisms in the distribution system, as does chlorine. The high costs of new filtration or ozone treatment plants must be weighed against the benefits of additional treatment. Even well operated water treatment plants cannot ensure that drinking water will be completely free of *Cryptosporidium* oocysts. Water treatment methods alone cannot solve the problem; watershed protection and monitoring of water quality are critical. As mentioned, watershed protection is another barrier to *Cryptosporidium* in drinking water. Land use controls such as septic systems regulations and best management practices to control runoff can help keep human and animal wastes out of water.

Under the Surface Water Treatment Rule of 1989, public water systems must filter surface water sources unless water quality and disinfection requirements are met, and a watershed control program is maintained. This rule, however, did not address *Cryptosporidium.* The USEPA has now set standards for turbidity (cloudiness) and coliform bacteria (which indicate that pathogens are probably present) in drinking water. Frequent monitoring must occur to provide officials with early warning of potential problems to enable them to take steps to protect public health. Unfortunately, no water quality indicators can reliably predict the occurrence of Cryptosporidiosis. More accurate and rapid assays of oocysts will make it possible to notify residents promptly if their water supply is contaminated with *Cryptosporidium* and thus avert outbreaks.

The bottom line: The collaborative efforts of water utilities, government agencies, health care providers, and individuals are needed to prevent outbreaks of Cryptosporidiosis.

Cryptosporidiosis

Juranek (1995) wrote in the journal *Clinical Infectious Diseases:*

> *Cryptosporidium parvum* is an important emerging pathogen in the U.S. and a cause of severe, life-threatening disease in patients with AIDS. No safe and effective form

> of specific treatment for Cryptosporidiosis has been identified to date. The parasite is transmitted by ingestion of oocysts excreted in the feces of infected humans or animals. The infection can therefore be transmitted from person-to-person, through ingestion of contaminated water (drinking water and water used for recreational purposes) or food, from animal to person, or by contact with fecally contaminated environmental surfaces. Outbreaks associated with all of these modes of transmission have been documented. Patients with human immunodeficiency virus infection should be made more aware of the many ways that *Cryptosporidium* species are transmitted, and they should be given guidance on how to reduce their risk of exposure.

Since the Milwaukee outbreak, concern about the safety of drinking water in the United States has increased, and new attention has been focused on determining and reducing the risk of Cryptosporidiosis from community and municipal water supplies. Cryptosporidiosis is spread by putting something in the mouth that has been contaminated with the stool of an infected person or animal. In this way, people swallow the *Cryptosporidium* parasite. As previously mentioned, a person can become infected by drinking contaminated water or eating raw or undercooked food contaminated with *Cryptosporidium* oocysts; direct contact with the droppings of infected animals or stools of infected humans; or hand-to-mouth transfer of oocysts from surfaces that may have become contaminated with microscopic amounts of stool from an infected person or animal.

The symptoms may appear two to ten days after infection by the parasite. Although some persons may not have symptoms, others have watery diarrhea, headache, abdominal cramps, nausea, vomiting, and low-grade fever. These symptoms may lead to weight loss and dehydration. In otherwise healthy persons, these symptoms usually last one to two weeks, at which time the immune system is able to stop the infection. In persons with suppressed immune systems, such as persons who have AIDS or who recently have had an organ or bone marrow transplant, the infection may continue and become life threatening.

Currently, no safe and effective cure for Cryptosporidiosis exists. People with normal immune systems improve without taking antibiotic or antiparasitic medications. The treatment recommend for this diarrheal illness is to drink plenty of fluids and to get extra rest. Physicians may prescribe medication to slow the diarrhea during recovery.

The best way to prevent Cryptosporidiosis is to:

- Avoid water or food that may be contaminated
- Wash hands after using the toilet and before handling food
- If you work in a childcare center where you change diapers, be sure to wash your hands thoroughly with plenty of soap and warm water after every diaper change, even if you wear gloves

During community-wide outbreaks caused by contaminated drinking water, drinking water practitioners should inform the public to boil drinking water for one minute to kill the *Cryptosporidium* parasite.

DID YOU KNOW?

The United States Environmental Protection Agency's Method 1623.1: *Cryptosporidium* and *Giardia* in water is a standard filtration process that filters and retains the oocysts, cysts, and extraneous materials in a water sample. Materials on the filter are eluted (i.e., obtaining one material from another by using a solvent wash) and the eluate is centrifuged to pellet the oocysts and cysts, and the supernatant fluid is aspirated. After being magnetized the oocysts and cysts are separated from the extraneous materials are discarded. The oocysts and cysts are stained on well slides with fluorescently labeled monoclonal antibodies and DAPI. The stained sample is examined using fluorescence and DIC microscopy. The qualitative analysis is performed by scanning each slide well for objects that meet the size, shape, and *fluorescence* characteristics *Cryptosporidium* oocysts or *Giardia* cysts (USEPA, 2012).

REFERENCES

Black, R.E., Dykes, A.C., Anderson, K.E., Wells, J.G., Sinclair, S.P., Gary, G.W., Hatch, M.H., & Ginagaros, E.J., 1981. Handwashing to prevent diarrhea in day-care centers. *American Journal of Epidemiology*, 113: 445–451.

Brodsky, R.E., Spencer, H.C., & Schultz, M.G., 1974. Giardiasis in American travelers to the Soviet Union. *Journal of Infectious Diseases.* 130: 319–323.

CDC, 1979. *Intestinal Parasite Surveillance, Annual Summary 1978.* Atlanta, GA: Centers for Disease Control.

CDC, 1984. *Water-Related Siease Outbreaks Sueveillance, Annual Summary 1983.* Atlanta, GA: Centers for Disease Control and Prevention.

CDC, 1995. *Giardiasis.* D.D. Juranek. Atlanta, GA: Centers for Disease Control.

Davidsion, R.A., 1984. Issues in clinical parasitology: The treatment of giardiasis. *American Journal of Gastroenterology*, 79(4): 156–261.

De Zuane, J., 1997. *Handbook of Drinking Water Quality.* New York: John Wiley & Sons, Inc.

Fayer, R., Speer, C.A., & Dudley, J.P., 1997. *The General Biology Cryptosporidium in Cryptosporidium and Cryptosporidiosis.* Fayer, R. (ed.). Boca Raton, FL: CRC Press.

Frost, F., Plan, B., & Liechty, B., 1984. Giardia prevalence in commercially trapped mammals. *Journal of Environmental Health*, 42: 245–249.

Jarroll, E.L., Jr., Gingham, A.K, & Meyer, E.A., 1979. Giardia cyst destruction; effectiveness of six small-quantity water disinfection methods. *American Journal of Tropical Medicine and Hygiene*, 29: 8–11.

Juranek, D.D., 1995. Cryptosporidium parvum. *Clinical Infectious Diseases*, 21: S37–61.

LeChevallier, M.W., Norton, W.D., & Less, R.G., 1991. Occurrences of *Giardia* and *Cryptosporidium* spp. in surface water supplies. *Applied and Environmental Microbiology*, 57: 2610–2616.

Mayo Clinic Foundation, 1996. *The "Bug" that Made Milwaukee Famous.* Rochester, MN.

Panciera, R.J., Thomassen, R.W., & Garner, R.M., 1971. Cryptosporidial infection in a calf. *Veterinary Pathology*, 8: 479.

Rendtorff, R.C., 1954. The experimental transmission of human intestinal protozoan parasites II *Giardia lamblia* cysts given in capsules. *American Journal of Tropical Medicine and Hygiene*, 59: 209–220.

Rose, J.B., 1983. Survey of potable water supplies for Cryptospridium ad Giardia. *Environmental Science & Technology*, 25: 1393–1399.
Slavin, D., 1955. Cryptosporidium melagridis. *Journal of Comparative Pathology*, 65: 262.
Tyzzer, E.E., 1912. *Cryptosporidium parvum* sp.: a coccidium found in the small intestine of the common mouse. *Archiv Protistenkd*, 26: 394.
Upton, S.J., 1997. *Basic Biology of Cryptosporidium*. Manhattan, Kansas: Kansas State University.
USEPA, 2012. *Method 1623.1: Cryptosporidium and Giardia in Water by Filtration/IMS/FA*. Washington, DC: United States Environmental Protection Agency.
Walsh, J.D., 1981. Survey Infectious Disease. *N. Engl. J. Med.*, 303: 542–544.
Weller, P.F., 1985. Intestinal Protozoa: Giardiasis. *Scientific American Medicine*, 12(4): 554–558.

14 Wastewater Virus Sampling, Testing, and Analysis

WASTEWATER SUGGESTS WIDER SPREAD OF VIRUS

The Associated Press reported September 10, 2020, that an analysis of wastewater in one Virginia County indicated that it had ten times as many people walking around with the coronavirus than test results suggested. Many of the individuals who contributed to the virus in their wastewater had no idea they had the virus.

Various wastewater treatment plants in Virginia have been conducting experimental testing of raw wastewater since early 2020. Along with two treatment plants in Stafford County, Hampton Roads Sanitation District (HRSD) has also been sampling, testing, and analyzing raw wastewater in the Hampton Roads region of southeastern Virginia. In addition to these treatment plants there are more than 100 other wastewater treatment plants across the country that are participating in a no-cost pilot program.

Experience has shown that wastewater can provide an early indicator of a disease spreading before people start seeking health care. It is interesting to note that sampling, testing, and analysis of wastewater in Stafford County has shown that up to 18 percent of the County's population may have had the virus; that means that one out of every five people in the County may have been infected (AP, 2020).

HOW IT WORKS

The Centers for Disease Control and Prevention (CDC) and the U.S. Department of Health and Human Services (HHS), in collaboration with agencies throughout the federal government, are initiating the National Wastewater Surveillance System (NWSS) in response to the COVID-19 pandemic. The data generated by NWSS helps epidemiologists and other public health officials to better understand the extent of COVID-19 infections in communities (CDC, 2020).

At the current time CDC is developing a portal for state, tribal, local, and territorial health departments to submit wastewater testing data into a national database for use in summarizing and interpreting data for public health action. Involvement in a national database will ensure data comparability across jurisdictions (CDC, 2020).

Data from wastewater sampling and testing is not meant to replace existing COVID-19 surveillance systems but is meant to complement them by providing an efficient pooled community sample, data for communities where timely COVID-19 clinical testing is underutilized or unavailable; and data at a sub-county level.

Keep in mind that wastewater, also referred to as sewage, includes water from household/building use (i.e., toilers, showers, sinks) that can contain human fecal waste, as well as water from non-household sources, (i.e., rainwater and industrial use). Wastewater can be tested for RNA from SARS-CoV-2, the virus that causes COVID-19. Note that while SARS-CoV-2 can be shed in the feces of individuals with COVID-19 because of direct exposure to treated or untreated wastewater, there is no evidence to date that anyone working with has become sick with COVID-19; wastewater is a very hostile environment for many pathogenic microorganisms and is a haven for others. History has shown that wastewater testing has been successfully used as a method for early detection of other diseases, such as polio. Another advantage of using feces testing of individuals with symptomatic or asymptomatic infection is that it can capture data on both types of infections. At the present time at least 80 percent of U.S. households are served by municipal sewage collection systems. Epidemiologists can collect quantitative SARS-CoV-2 measurements in untreated sewage that can provide information on changes in total COVID-19 infections in the community contributing to that wastewater treatment plant (that area is known as the "sewershed"). If testing is conducted on a frequent basis, wastewater surveillance can be a leading indicator of changes in the COVID-19 burden in a community.

Keep in mind that wastewater treated in septic-based systems will not allow home-generated wastewater served by such systems to be tested. Moreover, wastewater sampling and testing for COVID-19 can't be accomplished for prisons, universities, or hospitals that treat their own waste. Also, wastewater treatment plants that pre-treat influent are not suitable sites for sampling and testing influent for COVID-19.

It is important to note that wastewater surveillance for COVID-19 is a work in progress. It will take time and much analysis to perfect the testing techniques. For example, one problem is that we have no knowledge of the lower limits of COVID-19 presence in some raw waste-streams. Simply, we do not know what the lower limits of detection are; the old "we do not know what we do not know syndrome." More data on fecal shedding by infected individuals over the course of the disease are needed to better understand the limits of detection. Again, it is important to note in this discussion that wastewater-based epidemiology is full of promise but is still in the early stages of development and refinement.

WHAT GOES IN ALSO COMES OUT

Probably, the most obvious aspect of wastewater-based epidemiology in tracking and assessing the presence and extent of COVID-19 in a particular area resounds, resonates, and echoes in the old saying, "what goes in also comes out" (hopefully) in the food (or other things) we ingest and later discharge into the toilet and is then flushed and conveyed to some sort or type of wastewater treatment operation. The truth be known, knowledgeable epidemiologists can draw some clear insights about the practices of different demographics of the population from samples of raw wastewater. While it is true that some people think that sampling, testing, and analyzing raw wastewater is no picnic in the park, the truth is that data gathered by such activities may save lives, allowing people to enjoy a picnic in the park, or almost anywhere, for that matter. Simply, sampling, testing, and analyzing the contents of raw wastewater involves real-time monitoring; it enables the epidemiologist to predict outbreaks ahead of time. Secondarily, the real-time monitoring will allow tracking the rise of antibiotic resistance.

Wastewater-based surveillance has already been used extensively in detecting the use of illicit drugs in a region or particular community. Moreover, the use of pharmaceuticals can also be revealed.

A LABORIOUS PROCESS

Wastewater-based epidemiology is no walk in the park; it is a difficult and protracted process. For example, epidemiologists are required to gather samples from different treatment plants and then run all the tests or analyses to find out what's inside. However, in this digital age where technology advances almost daily, it should be possible to install remote sensors into the treatment plants tied in with site computer(s) that will collect and provide information in a more convenient manner.

When talking about remote sensors installed in wastewater treatment plants this practice is not new; it has been used for years in taking samples and analyzing them in a computer-based sampling system. Again, the use of biosensors in wastewater treatment operations can be used as biological receptors, like DNA, an antibody, or a protein that generates a signal in the presence of an analytical target, or analyte.

THE 411 ON HOW IT WORKS

Based on personal observation, the present procedure used by wastewater-based epidemiologists is to have water quality technicians visit a wastewater treatment operation on a weekly basis and collect about 4–6 L of untreated wastewater. Each sample taken is a composite of the wastewater flowing through the plant over the past 24 hours (keep in mind that the virus is not thought to be active or infectious or lethal once it is found in the fecal matter—wastewater is a hostile environment for many pathogens).

Note that because the virus is not prominent in the raw wastewater, the samples have to be concentrated with the fecal material and water and filtered out using chemicals and a small electronegative filter. After that, all the microbes in the viruses—not just coronavirus, but also norovirus and other common viruses found in stool—are basically cracked open to look for genetic material—specifically nucleic acid, a key feature of SARS-CoV-2.

This is when PCR comes into play. Preliminary chain reaction (PCR) technology is the same test that's used to analyze nose swabs. Epidemiologists then look at how many segments of nucleic acid are associated with coronavirus and come up with a measurable figure—virus per liter of wastewater—which they analyze over time, looking at which plants the samples came from and determining where the virus is foremost.

At present, the entire process can take up to six hours or more.

Watching How the Coronavirus Moves through the Communities

As stated earlier, wastewater surveillance is not new. It's been used to track polio and combat the opioid crisis; in some locations, epidemiologists have been using the same technology to find bacteria indicative of leaking wastewater interceptor lines that seep into area streams and rivers. Now many people in various locations in the United States are among a growing trend of watching how the coronavirus is moving through communities.

Based on experience, epidemiologists and other concerned professionals have found that it's not enough to rely on testing people, isolating them to understand trends. This is especially the case because of the capacity issues, long testing turn-around times, and risks of false negative results still impacting many communities. Experience has also indicated that wastewater monitoring can complement the other tracking procedures, with epidemiologists and environmental scientists and engineers working in tandem with health care workers and medical and environmental lab technicians.

Experts have found that beyond being an early indicator of the next outbreak, wastewater-based epidemiology can also be useful in directing resources like tests, contact tracers, and eventually treatments to hot spots sooner. This is especially important when detected through health department data because communities have unreported cases that are missed.

One of the problems with wastewater-based surveillance for SARS-CoV-2 is that it is a very young science with few protocols and no set standards. One of the setbacks in attempting to correct these problems is that there are so many ways the samples can be collected, concentrated, and tested; thus, many local health departments haven't adopted wastewater-based epidemiology as a tool. Moreover, it's difficult for scientists to tell how many people are actually infected based on their sample, because they don't know how much of the virus's genetic material people are shedding into their wastewater. Attempts have been made to employ modeling as a technique to make estimates, but uncertainty has been great.

Until more testing is accomplished, and trends established using written, standard protocols and approved standards the goal is to capture enough data through the year to prove that wastewater-based surveillance and epidemiology is able to catch future outbreaks before too many people are infected and the information shows up in clinical databases.

The bottom line: Having a long history of data will allow epidemiologists and other health professionals to determine what the data can help them with and they will know its limitations; therefore, wastewater-based epidemiologists will be able to use data to determine both distribution and measurement of coronavirus in specified populations providing a viable source of early warning.

REFERENCES

AP, 2020. *Sewage Suggests Larger Spread of Virus in Virginia County*. September 8, 2020. Stafford, VA: Associated Publishers.

CDC, 2020. *National Wastewater Surveillance System (NWSS)*. Atlanta, GA: Centers for Disease Control and Prevention.

Index

D

E

L

M

N

O

P

Q

R

S

T

U

V

W

Y

Z

Printed in the United States
by Baker & Taylor Publisher Services